1989

MATHEMATICAL
RECREATIONS AND ESSAYS

46	57	68	70	81	02	13	24	35	99
71	94	37	65	12	40	29	06	88	53
93	26	54	01	38	19	85	77	60	42
15	43	80	27	09	74	66	58	92	31
32	78	16	89	63	55	47	91	04	20
67	05	79	52	44	36	90	83	21	18
84	69	41	33	25	98	72	10	56	07
59	30	22	14	97	61	08	45	73	86
28	11	03	96	50	87	34	62	49	75
00	82	95	48	76	23	51	39	17	64

The integers from 0 to 99 arranged in a way that was believed for 180 years to be impossible. (See page 293.)

MATHEMATICAL RECREATIONS AND ESSAYS

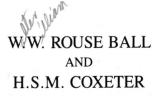

W. W. ROUSE BALL
AND
H.S.M. COXETER

THIRTEENTH EDITION

DOVER PUBLICATIONS, INC.
NEW YORK

Published in Canada by General Publishing Company, Ltd., 30 Lesmill Road, Don Mills, Toronto, Ontario.
Published in the United Kingdom by Constable and Company, Ltd., 10 Orange Street, London WC2H 7EG.

This Dover edition (the thirteenth of the work), first published in 1987, is a corrected republication of the twelfth edition as published by University of Toronto Press, Toronto, Canada, in 1974 (first edition: 1892). Corrections have been made by H.S.M. Coxeter specially for the Dover edition; at Prof. Coxeter's request, the Prefaces to the Eleventh and Twelfth Editions have been omitted. The present edition is published by special arrangement with University of Toronto Press, University of Toronto, Toronto M5S 1A6, Canada.

Manufactured in the United States of America
Dover Publications, Inc., 31 East 2nd Street, Mineola, N.Y. 11501

Library of Congress Cataloging-in-Publication Data

Ball, W. W. Rouse (Walter William Rouse), 1850–1925.
    Mathematical recreations and essays.

    Corrected republication of the work originally published under title: Mathematical recreations & essays.
    Bibliography: p.
    Includes index.
    1. Mathematical recreations. 2. Geometry—Problems, Famous. 3. Cryptography. 4. Ciphers. I. Coxeter, H. S. M. (Harold Scott Macdonald), 1907–   . II. Ball, W. W. Rouse (Walter William Rouse), 1850–1925. Mathematical recreations & essays. III. Title.
QA95.B2   1987   793.7′4   86-29028
ISBN 0-486-25357-0 (pbk.)

PREFACE TO THE DOVER EDITION

Rouse Ball's first edition (with the slightly different title *Mathematical Recreations and Problems*) appeared nearly a hundred years ago. I am grateful to Dover Publications for undertaking to keep the book alive by publishing this new edition. I wish to thank also the many friends who have helped me to bring the material up to date. I have, of course, retained the "new" chapters X and XIV, so kindly contributed by J.J. Seidel (for the twelfth edition) and Abraham Sinkov (for the eleventh).

During the 61 years since Rouse Ball died, mathematical knowledge has increased enormously, but most of the results that interested him still remain valid. Among arithmetical recreations, his lists of Mersenne primes and Fermat composites have been extended with the aid of the computer (see pages 64–69). Without this modern device, we could never have known that the number $217833 \times 10^{7150} + 1$ is prime, or that $2^{2^{27}} + 1$ is divisible by

$$2^{29}430816215 + 1.$$

The computer has also played a role in the geometric pastime of "squaring the square" (pages 115–116).

A remarkable convergence of pure and applied mathematics arose about 1980, when the geometric concept of a "quasi-lattice" was discovered just in time to provide a possible explanation for the physical observation of a "quasicrystal" (page 161).

I shall be glad to receive letters from readers who notice any mistakes or obscurities.

H.S.M. COXETER

University of Toronto
November 1986

PREFACE TO THE TENTH EDITION

This book contains descriptions of various problems of the kind usually termed Mathematical Recreations, and a few Essays on some analogous questions. I have excluded all matter which involves advanced mathematics. I hasten to add that the conclusions are of no practical use, and that most of the results are not new. If therefore the reader proceeds further he is at least forewarned. At the same time I think I may say that many of the questions discussed are interesting, not a few are associated with the names of distinguished mathematicians, while hitherto several of the memoirs quoted have not been easily accessible to English readers. A great deal of new matter has been added since the work was first issued in 1892.

As now presented, the book contains fourteen chapters, of which the subjects are shown in the Table of Contents. Several of the questions mentioned in the first four chapters are of a somewhat trivial character, and had they been treated in any standard English work to which I could have referred the reader, I should have left them out: in the absence of such a work, I thought it better to insert them and trust to the judicious reader to omit them altogether or to skim them as he feels inclined. I may add that in discussing problems where the complete solutions are long or intricate I have been generally content to indicate memoirs or books in which the methods are set out at length, and to give a few illustrative examples. In some cases I have also stated problems which still await solution.

I have inserted detailed references, as far as I know them, to the sources of the various questions and solutions given; also, wherever I have given only the result of a theorem, I have tried to indicate authorities where a proof may be found. In general, unless it is stated otherwise, I have taken the references direct from the original works; but, in spite of considerable time spent in verifying them, I dare not suppose that they are free from all errors or misprints.

W.W. ROUSE BALL

CONTENTS

I   ARITHMETICAL RECREATIONS   3
To find a number selected by someone   5
Prediction of the result of certain operations   8
Problems involving two numbers   11
Problems depending on the scale of notation   12
Other problems with numbers in the denary scale   14
Four fours problem   16
Problems with a series of numbered things   17
Arithmetical restorations   20
Calendar problems   26
Medieval problems in arithmetic   27
    The Josephus problem. Decimation   32
Nim and similar games   36
    Moore's game   38
    Kayles   39
    Wythoff's game   39
Addendum on solutions   40

II   ARITHMETICAL RECREATIONS (*continued*)   41
Arithmetical fallacies   41
Paradoxical problems   44
Probability problems   45
Permutation problems   48
Bachet's weights problem   50
The decimal expression for $1/n$ 53
Decimals and continued fractions   54

Rational right-angled triangles   57
Triangular and pyramidal numbers   59
Divisibility   60
The prime number theorem   62
Mersenne numbers   64
Perfect numbers   66
Fermat numbers   67
Fermat's Last Theorem   69
Galois fields   73

III   GEOMETRICAL RECREATIONS   76
Geometrical fallacies   76
Geometrical paradoxes   84
Continued fractions and lattice points   86
Geometrical dissections   87
Cyclotomy   94
Compass problems   96
The five-disc problem   97
Lebesgue's minimal problem   99
Kakeya's minimal problem   99
Addendum on a solution   102

IV   GEOMETRICAL RECREATIONS (*continued*)   103
Statical games of position   103
   Three-in-a-row. Extension to *p*-in-a-row   103
   Tessellation   105
   Anallagmatic pavements   107
   Polyominoes   109
   Colour-cube problem   113
   Squaring the square   115
Dynamical games of position   116
   Shunting problems   116
   Ferry-boat problems   118
   Geodesic problems   120
   Problems with counters or pawns   121
Paradromic rings   127
Addendum on solutions   129

V   POLYHEDRA   130
Symmetry and symmetries   130
The five Platonic solids   131
  Kepler's mysticism   133
  Pappus, on the distribution of vertices   134
  Compounds   135
The Archimedean solids   136
  Mrs. Stott's construction   139
Equilateral zonohedra   141
The Kepler-Poinsot polyhedra   144
The 59 icosahedra   146
Solid tessellations   147
Ball-piling or close-packing   149
  The sand by the sea-shore   151
Regular sponges   152
Rotating rings of tetrahedra   154
The kaleidoscope   155

VI   CHESS-BOARD RECREATIONS   162
Relative value of pieces   163
The eight queens problem   166
Maximum pieces problem   172
Minimum pieces problem   172
Re-entrant paths on a chess-board   175
  Knight's re-entrant path   175
  King's re-entrant path   186
  Rook's re-entrant path   187
  Bishop's re-entrant path   187
  Routes on a chess-board   187
  Guarini's problem   189
Latin squares   189
  Eulerian squares   190
  Euler's officers problem   192
  Eulerian cubes   192

VII   MAGIC SQUARES   193
Magic squares of an odd order   195

Magic squares of a singly-even order   196
Magic squares of a doubly-even order   199
Bordered squares   200
Number of squares of a given order   201
Symmetrical and pandiagonal squares   202
   Generalization of De la Loubère's rule   204
   Arnoux's method   206
   Margossian's method   207
Magic squares of non-consecutive numbers   210
   Magic squares of primes   211
Doubly-magic and trebly-magic squares   212
Other magic problems   213
   Magic domino squares   213
   Cubic and octahedral dice   214
   Interlocked hexagons   215
Magic cubes   216

VIII   MAP-COLOURING PROBLEMS   222
The four-colour conjecture   222
   The Petersen graph   225
   Reduction to a standard map   227
   Minimum number of districts for possible failure   230
   Equivalent problem in the theory of numbers   231
Unbounded surfaces   232
Dual maps   234
Maps on various surfaces   234
Pits, peaks, and passes   238
Colouring the icosahedron   238

IX   UNICURSAL PROBLEMS   243
Euler's problem   243
Number of ways of describing a unicursal figure   250
Mazes   254
Trees   260
The Hamiltonian game   262
Dragon designs   266

X COMBINATORIAL DESIGNS 271
A projective plane 271
Incidence matrices 272
An Hadamard matrix 273
An error-correcting code 274
A block design 276
Steiner triple systems 278
Finite geometries 281
Kirkman's school-girl problem 287
Latin squares 290
The cube and the simplex 295
Hadamard matrices 296
Picture transmission 297
Equiangular lines in 3-space 299
Lines in higher-dimensional space 303
C-matrices 308
Projective planes 310

XI MISCELLANEOUS PROBLEMS 312
The fifteen puzzle 312
The Tower of Hanoï 316
Chinese rings 318
Problems connected with a pack of cards 322
Shuffling a pack 323
Arrangements by rows and columns 325
Bachet's problem with pairs of cards 326
Gergonne's pile problem 328
The window reader 333
The mouse trap. Treize 336

XII THREE CLASSICAL GEOMETRICAL PROBLEMS 338
The duplication of the cube 339
    Solutions by Hippocrates, Archytas, Plato, Menaechmus,
        Apollonius, and Diocles 341
    Solutions by Vieta, Descartes, Gregory of St. Vincent, and
        Newton 343

The trisection of an angle   344
  Solutions by Pappus, Descartes, Newton, Clairaut, and Chasles   334
The quadrature of the circle   347
  Origin of symbol π 349
  Geometrical methods of approximation to the ·numerical value of π 349
    Results of Egyptians, Babylonians, Jews   350
    Results of Archimedes and other Greek writers   351
    Results of European writers, 1200–1630   352
    Theorems of Wallis and Brouncker   355
    Results of European writers, 1699–1873   356
    Approximations by the theory of probability   359

XIII   CALCULATING PRODIGIES   360
John Wallis, 1616–1703   361
Buxton, circ. 1707–1772   361
Fuller, 1710–1790; Ampère   364
Gauss, Whately   365
Colburn, 1804–1840   365
Bidder, 1806–1878   367
Mondeux, Mangiamele   372
Dase, 1824–1861   372
Safford, 1836–1901   374
Zamebone, Diamandi, Rückle   375
Inaudi, 1867-   375
Types of memory of numbers   377
Bidder's analysis of methods used   378
  Multiplication   379
  Digital method for division and factors   381
  Square roots. Higher roots   382
  Compound interest   384
  Logarithms   385
Alexander Craig Aitken   386

XIV   CRYPTOGRAPHY AND CRYPTANALYSIS   388
Cryptographic systems   389

Transposition systems   391
  Columnar transposition   392
  Digraphs and trigraphs   394
  Comparison of several messages   397
  The grille   401
Substitution systems   402
  Tables of frequency   404
  Polyalphabetic systems   406
  The Vigenère square   407
  The Playfair cipher   410
  Code   412
Determination of cryptographic system   414
A few final remarks   416
Addendum: References for further study   418

INDEX   419

MATHEMATICAL
RECREATIONS AND ESSAYS

"*Les hommes ne sont jamais plus ingénieux que dans l'invention des jeux; l'esprit s'y trouve à son aise. ... Après les jeux qui dépendent uniquement des nombres viennent les jeux où entre la situation. ... Après les jeux où n'entrent que le nombre et la situation viendraient les jeux où entre le mouvement. ... Enfin il serait à souhaiter qu'on eût un cours entier des jeux, traités mathématiquement.*" (Leibniz: letter to De Montmort, 29 July 1715)

CHAPTER I

ARITHMETICAL RECREATIONS

I commence by describing some arithmetical recreations. The interest excited by statements of the relations between numbers of certain forms has been often remarked, and the majority of works on mathematical recreations include several such problems, which, though obvious to anyone acquainted with the elements of algebra, have to many who are ignorant of that subject the same kind of charm that mathematicians find in the more recondite propositions of higher arithmetic. I devote the bulk of this chapter to these elementary problems.

Before entering on the subject, I may add that a large proportion of the elementary questions mentioned here are taken from one of two sources. The first of these is the classical *Problèmes plaisans et délectables*, by Claude Gaspar Bachet, sieur de Méziriac, of which the first edition was published in 1612 and the second in 1624: it is to the edition of 1624 that the references hereafter given apply. Several of Bachet's problems are taken from the writings of Alcuin, Pacioli di Burgo, Tartaglìa, and Cardan, and possibly some of them are of oriental origin, but I have made no attempt to add such references. The other source to which I alluded above is Ozanam's *Récréations mathématiques et physiques*. The greater portion of the original edition, published in two volumes at Paris in 1694, was a compilation from the works of Bachet, Mydorge, and Leurechon: this part is excellent, but the same cannot be said of the additions due to Ozanam. In the *Biographie Universelle* allusion is made to subsequent editions issued in 1720, 1735, 1741, 1778, and 1790; doubtless these references are correct, but the following editions, all of which I have seen, are the only ones of which I have any knowledge. In 1696 an edition was issued at Amsterdam. In 1723 (six years after the death of Ozanam) one was issued in three volumes, with a supplementary fourth volume, containing, among other things, an appendix on puzzles. Fresh editions were issued in 1741, 1750 (the second volume of which bears the date 1749), 1770, and 1790. The edition of 1750 is said to have been corrected by Montucla on condition that his name should not be associated with it; but the edition of 1790 is the earliest one in which reference is made to these corrections, though the editor is referred to only as Monsieur M***. Montucla expunged most of what was actually incorrect in the older editions, and added several historical notes, but unfortunately his scruples prevented him from striking out the accounts of numerous trivial experiments and truisms which overload the work. An English translation of the original edition appeared in 1708, and I believe ran through four editions, the last of them being published in Dublin in 1790. Montucla's revision of 1790 was translated by C. Hutton, and editions of this were issued in 1803, in 1814, and (in one volume) in 1840: my references are to the editions of 1803 and 1840.

I proceed to enumerate some of the typical elementary questions connected with numbers which for nearly three centuries have formed a large part of most compilations of mathematical amusements. They are given here largely for their historical – not for their arithmetical – interest; and perhaps a mathematician may well omit this chapter.

Many of these questions are of the nature of tricks or puzzles, and I follow the usual course and present them in that form. I may note, however, that most of them are not worth proposing, even as tricks, unless either the method employed is disguised or the result arrived at is different from that expected; but, as I am not writing on conjuring, I refrain from alluding to the means of disguising the operations indicated, and give merely a bare enumeration of the steps essential to the success of the method used. To the non-mathematician even to-day some of those results seem astonishing, but the secret is at once revealed as soon as the question is translated by symbols into mathematical language.

TO FIND A NUMBER SELECTED BY SOMEONE

There are innumerable ways of finding a number chosen by someone, provided the result of certain operations on it is known. I confine myself to methods typical of those commonly used. Anyone acquainted with albegra will find no difficulty in framing new rules of an analogous nature.

First method.* (i) Ask the person who has chosen the number to treble it. (ii) Enquire if the product is even or odd: if it is even, request him to take half of it; if it is odd, request him to add unity to it and then to take half of it. (iii) Tell him to multiply the result of the second step by 3. (iv) Ask how many integral times 9 divides into the latter product: suppose the answer to be n. (v) Then the number thought of was $2n$ or $2n + 1$, according as the result of step (i) was even or odd.

The demonstration is obvious. Every even number is of the form $2n$, and the successive operations applied to this give (i) $6n$, which is even; (ii) $\frac{1}{2}6n = 3n$; (iii) $3 \times 3n = 9n$; (iv) $\frac{1}{9}9n = n$; (v) $2n$. Every odd number is of the form $2n + 1$, and the successive operations applied to this give (i) $6n + 3$, which is odd; (ii) $\frac{1}{2}(6n + 3 + 1) = 3n + 2$; (iii) $3(3n + 2) = 9n + 6$; (iv) $\frac{1}{9}(9n + 6) = n + $ a remainder; (v) $2n + 1$. These results lead to the rule given above.

Second method.† Ask the person who has chosen the number to perform in succession the following operations. (i) To multiply the number by 5. (ii) To add 6 to the product. (ii) To multiply the sum by 4. (iv) To add 9 to the product. (v) To multiply the sum by 5. Ask to be told the result of the last operation: if from this product 165 is subtracted, and then the remainder is divided by 100, the quotient will be the number thought of originally.

For let n be the number selected. Then the successive op-

---

*Bachet, *Problèmes*, Lyons, 1624, problem ı, p. 53.
†A similar rule was given by Bachet, problem ıv, p. 74.

erations applied to it give: (i) $5n$; (ii) $5n+6$; (iii) $20n+24$; (iv) $20n+33$; (v) $100n+165$. Hence the rule.

Third method.* Request the person who has thought of the number to perform the following operations. (i) To multiply it by any number you like, say a. (ii) To divide the product by any number, say b. (iii) To multiply the quotient by c. (iv) To divide this result by d. (v) To divide the final result by the number selected originally. (vi) To add to the result of operation (v) the number thought of at first. Ask for the sum so found: then, if ac/bd is subtracted from this sum, the remainder will be the number chosen originally.

For, if n was the number selected, the result of the first four operations is to form nac/bd; operation (v) gives ac/bd; and (vi) gives $n+ac/bd$, which number is mentioned. But ac/bd is known; hence, subtracting it from the number mentioned, n is found. Of course, a, b, c, d may have any numerical values it is liked to assign to them. For example, if $a=12$, $b=4$, $c=7$, $d=3$, it is sufficient to subtract 7 from the final result in order to obtain the number originally selected.

Fourth method. † Ask someone to select a number less than 90. Request him to perform the following operations. (i) To multiply it by 10, and to add any number he pleases, a, which is less than 10. (ii) To divide the result of step (i) by 3, and to mention the remainder, say b. (iii) To multiply the quotient obtained in step (ii) by 10, and to add any number he pleases, c, which is less than 10. (iv) To divide the result of step (iii) by 3, and to mention the remainder, say d, and the third digit (from the right) of the quotient; suppose this digit is e. Then, if the numbers a, b, c, d, e are known, the original number can be at once determined. In fact, if the number is $9x+y$, where $x\leq9$ and $y\leq8$, and if r is the remainder when $a-b+3(c-d)$ is divided by 9, we have $x=e$, $y=9-r$.

The demonstration is not difficult. Suppose the selected

---

*Bachet, problem v, p. 80.

†*Educational Times*, London, May 1, 1895, vol. XLVIII, p. 234. This example is said to have been made up by J. Clerk Maxwell in his boyhood; it is interesting to note how widely it differs from the simple Bachet problems previously mentioned.

number is $9x + y$. Step (i) gives $90x + 10y + a$. Let $y + a = 3n + b$, then the quotient obtained in step (ii) is $30x + 3y + n$. Step (iii) gives $300x + 30y + 10n + c$. Let $n + c = 3m + d$, then the quotient obtained in step (iv) is $100x + 10y + 3n + m$, which I will denote by Q. Now the third digit in Q must be x, because, since $y \leq 8$ and $a \leq 9$, we have $n \leq 5$; and since $n \leq 5$ and $c \leq 9$, we have $m \leq 4$; therefore $10y + 3n + m \leq 99$. Hence the third or hundreds digit in Q is x.

Again, from the relations $y + a = 3n + b$ and $n + c = 3m + d$, we have $9m - y = a - b + 3(c - d)$: hence, if r is the remainder when $a - b + 3(c - d)$ is divided by 9, we have $y = 9 - r$. (This is always true, if we make r positive; but if $a - b + 3(c - d)$ is negative, it is simpler to take y as equal to its numerical value; or we may prevent the occurrence of this case by assigning proper values to a and c.) Thus x and y are both known, and therefore the number selected, namely $9x + y$, is known.

Fifth method.* Ask anyone to select a number less than 60. Request him to perform the following operations. (i) To divide it by 3 and mention the remainder; suppose it to be a. (ii) To divide it by 4, and mention the remainder; suppose it to be b. (iii) To divide it by 5, and mention the remainder; suppose it to be c. Then the number selected is the remainder obtained by dividing $40a + 45b + 36c$ by 60.

This method can be generalized and then will apply to any number chosen. Let a', b', c', ... be a series of numbers prime to one another, and let p be their product. Let n be any number less than p, and let $a, b, c, \ldots$ be the remainders when n is divided by a', b', c', ... respectively. Find a number A which is a multiple of the product $b'c'd' \ldots$ and which exceeds by unity a multiple of a'. Find a number B which is a multiple of $a'c'd' \ldots$ and which exceeds by unity a multiple of b', and similarly find analogous numbers C, D, ... Rules for the calculation of A, B, C, ... are given in the theory of numbers, but, in general, if the numbers a', b', c', ... are small, the corresponding numbers A, B, C, ... can be found by inspection. I proceed to show

---

*Bachet, problem VI, p. 84: Bachet added, on p. 87, a note on the previous history of the problem.

that n is equal to the remainder when $Aa + Bb + Cc + \ldots$ is divided by p.

Let $N = Aa + Bb + Cc + \ldots$, and let $M(x)$ stand for a multiple of x. Now $A = M(a') + 1$, therefore $Aa = M(a') + a$. Hence, if the first term in N (that is, Aa) is divided by a', the remainder is a. Again, B is a multiple of $a'c'd' \ldots$. Therefore Bb is exactly divisible by a'. Similarly Cc, Dd, $\ldots$ are each exactly divisible by a'. Thus every term in N, except the first, is exactly divisible by a'. Hence, if N is divided by a', the remainder is a. Also if n is divided by a', the remainder is a. Therefore

$$N - n = M(a').$$

Similarly
$$N - n = M(b'),$$
$$N - n = M(c'),$$

$$\ldots \ldots \ldots$$

But a', b', c', $\ldots$ are prime to one another. Therefore

$$N - n = M(a'b'c' \ldots) = M(p),$$

that is,
$$N = M(p) + n.$$

Now n is less than p; hence if N is divided by p, the remainder is n.

The rule given by Bachet corresponds to the case of $a' = 3$, $b' = 4$, $c' = 5$, $p = 60$, $A = 40$, $B = 45$, $C = 36$. If the number chosen is less than 420, we may take $a' = 3$, $b' = 4$, $c' = 5$, $d' = 7$, $p = 420$, $A = 280$, $B = 105$, $C = 336$, $D = 120$.

TO FIND THE RESULT OF A SERIES OF OPERATIONS PERFORMED ON ANY NUMBER *(unknown to the operator)* WITHOUT ASKING ANY QUESTIONS

All rules for solving such problems ultimately depend on so arranging the operations that the number disappears from the final result. Four examples will suffice.

First example.* Request someone to think of a number. Suppose it to be n. Ask him (i) to multiply it by any number you please, say a; (ii) then to add, say, b; (iii) then to divide

*Bachet, problem VIII, p. 102.

the sum by, say, c. (iv) Next, tell him to take a/c of the number originally chosen; and (v) to subtract this from the result of the third operation. The result of the first three operations is $(na+b)/c$, and the result of operation (iv) is na/c: the difference between these is b/c and therefore is known to you. For example, if $a=6$, $b=12$, $c=4$, then $a/c=1\frac{1}{2}$, and the final result is 3.

Second example.*   Ask A to take any number of counters that he pleases: suppose that he takes n counters. (i) Ask someone else, say B, to take p times as many, where p is any number you like to choose. (ii) Request A to give q of his counters to B, where q is any number you like to select. (iii) Next, ask B to transfer to A a number of counters equal to p times as many counters as A has in his possession. Then there will remain in B's hands $q(p+1)$ counters; this number is known to you; and the trick can be finished either by mentioning it or in any other way you like.

The reason is as follows. The result of operation (ii) is that B has $pn+q$ counters, and A has $n-q$ counters. The result of (iii) is that B transfers $p(n-q)$ counters to A; hence he has left in his possession $(pn+q)-p(n-q)$ counters, that is, he has $q(p+1)$.

For example, if originally A took any number of counters, then (if you chose p equal to 2), first you would ask B to take twice as many counters as A had done; next (if you chose q equal to 3) you would ask A to give 3 counters to B; and then you would ask B to give to A a number of counters equal to twice the number then in A's possession; after this was done you would know that B had 3(2 + 1), that is, 9 left.

This trick (as also some of the following problems) may be performed equally well with one person, in which case A may stand for his right hand and B for his left hand.

Third example.   Ask someone to perform in succession the following operations. (i) Take any number of three digits, in which the difference between the first and last digits exceeds unity. (ii) Form a new number by reversing the order of the

---

*Bachet, problem XIII, p. 123: Bachet presented the above trick in a form somewhat more general but less effective in practice.

digits. (iii) Find the difference of these two numbers. (iv) Form another number by reversing the order of the digits in this difference. (v) Add together the results of (iii) and (iv). Then the sum obtained as the result of this last operation will be 1089.

An illustration and the explanation of the rule are given below.

(i)	237	$100a + 10b + c$	
(ii)	732	$100c + 10b + a$	
(iii)	495	$100(a-c-1) + 90+(10+c-a)$	
(iv)	594	$100(10+c-a)+ 90+(a-c-1)$	
(v)	1089	900	$+180+9$

The result depends only on the radix of the scale of notation in which the number is expressed. If this radix is r, the result is $(r-1)(r+1)^2$; thus if $r=10$, the result is 9×11^2, that is, 1089. Similar problems can be made with numbers exceeding 999.

Fourth example. The following trick, involving negative numbers, was proposed by Norman Anning. Ask someone to perform in succession the following operations. (i) Write down any number greater than 1 (not necessarily an integer). (ii) Form a new number by reciprocation (the way one derives $\frac{1}{2}$ from 2, or vice versa). (iii) Form a new number by subtraction from 1. (iv) Reciprocate this number. (v) Again subtract from 1. (vi) Again reciprocate. (vii) Add the number originally thought of. Then the result will be 1.

For instance, take the number $\frac{3}{2}$. We have (i) $\frac{3}{2}$, (ii) $\frac{2}{3}$, (iii) $\frac{1}{3}$, (iv) 3, (v) -2, (vi) $-\frac{1}{2}$, (vii) 1.

The following analysis explains the rule, and shows that the final result is independent of the number written down initially:

(i) a, (ii) $1/a$, (iii) $(a-1)/a$, (iv) $a/(a-1)$,

(v) $1/(1-a)$, (vi) $1-a$, (vii) 1;

or, more tidily, in terms of* $\theta = \sec^{-1}\sqrt{a}$,

(i) $\sec^2\theta$, (ii) $\cos^2\theta$, (iii) $\sin^2\theta$, (iv) $\operatorname{cosec}^2\theta$,

(v) $-\cot^2\theta$, (vi) $-\tan^2\theta$, (vii) 1.

Instead of step (vii), it is perhaps more natural to ask "Subtract from 1 again." Then the result will be the number originally thought of.

Before describing further tricks, it may be worthwhile to remark that a fifth edition of Bachet's *Problèmes*† is now available, "revue, simplifiée et augmentée par A. Labosne." This includes an elegant portrait of Bachet and a short biography. We might be tempted to imagine that he wrote the book for the edification of his seven children; but in fact it was first published long before his marriage.

PROBLEMS INVOLVING TWO NUMBERS

I proceed next to give a couple of examples of a class of problems which involve two numbers.

First example.‡ Suppose that there are two numbers, one even and the other odd, and that a person A is asked to select one of them, and that another person B takes the other. It is desired to know whether A selected the even or the odd number. Ask A to multiply his number by 2, or any even number, and B to multiply his by 3, or any odd number. Request them to add the two products together and tell you the sum. If it is even, then originally A selected the odd number, but if it is odd, then originally A selected the even number. The reason is obvious.

Second example.§ Ask someone to think of two positive numbers (possibly equal and not necessarily integers) and to perform in succession the following operations. (i) Form a third number by adding 1 to the second and dividing by the

*For the meaning of this angle θ, see H.S.M. Coxeter, *Non-Euclidean Geometry*, Toronto, 1968, p. 105.

†C.-G. Bachet, *Problèmes plaisants et délectables*. A. Blanchard, Paris, 1959.

‡Bachet, problem IX, p. 107.

§R.C. Lyness, *Mathematical Gazette*, 1942, vol. XXVI, p. 42 (Note 1581); 1945, vol. XXIX, p. 231 (Note 1847). For the related game of "Frieze patterns" see H.S.M. Coxeter, *Acta Arithmetica*, 1971, vol. XVIII, pp. 297–310.

first. (ii) Form a fourth number by adding 1 to the third and dividing by the second.... (v) Form a seventh number by adding 1 to the sixth and dividing by the fifth. Then the sixth and seventh numbers are the same as the first and second: the sequence is periodic, with 5 terms in the period. (This conclusion is easily verified by taking the first and second numbers to be a and b.)

PROBLEMS DEPENDING ON THE SCALE OF NOTATION

Many of the rules for finding two or more numbers depend on the fact that in arithmetic an integral number is denoted by a succession of digits, where each digit represents the product of that digit and a power of ten, and the number is equal to the sum of these products. For example, 2017 signifies (2×10^3) $+(0 \times 10^2)+(1 \times 10)+7$, that is, the 2 represents 2 thousands (i.e. the product of 2 and 10^3), the 0 represents 0 hundreds (i.e. the product of 0 and 10^2), the 1 represents 1 ten (i.e. the product of 1 and 10), and the 7 represents 7 units. Thus every digit has a local value. The application to tricks connected with numbers will be understood readily from three illustrative examples.

First example.* A common conjuring trick is to ask a boy among the audience to throw two dice, or to select at random from a box a domino on each half of which is a number. The boy is then told to recollect the two numbers thus obtained, to choose either of them, to multiply it by 5, to add 7 to the result, to double this result, and lastly to add to this the other number. From the number thus obtained, the conjurer subtracts 14, and obtains a number of two digits which are the two numbers chosen originally.

For suppose that the boy selected the numbers a and b. Each

---

*Some similar questions were given by Bachet in problem XII, p. 117; by Oughtred or Leake in the *Mathematicall Recreations*, commonly attributed to the former, London, 1653, problem XXXIV; and by Ozanam, part I, chapter x. Probably the *Mathematicall Recreations* were compiled by Leake, but as the work is usually catalogued under the name of W. Oughtred, I shall so describe it: it is founded on the similar work by J. Leurechon, otherwise known as H. van Etten, published in 1626.

of these is less than ten, dice or dominoes ensuring this. The successive operations give: (i) $5a$; (ii) $5a+7$; (iii) $10a+14$; (iv) $10a+14+b$. Hence, if 14 is subtracted from the final result, there will be left a number of two digits, and these digits are the numbers selected originally. An analogous trick might be performed in other scales of notation if it was thought necessary to disguise the process further.

Second example. * Similarly, if three numbers, say a, b, c, are chosen, then, if each of them is less than ten, they can be found by the following rule. (i) Take one of the numbers, say a, and multiply it by 2. (ii) Add 3 to the product. (iii) Multiply this by 5, and add 7 to the product. (iv) To this sum add the second number, b. (v) Multiply the result by 2. (vi) Add 3 to the product. (vii) Multiply by 5, and, to the product, add the third number, c. The result is $100a+10b+c+235$. Hence, if the final result is known, it is sufficient to subtract 235 from it, and the remainder will be a number of three digits. These digits are the numbers chosen originally.

Third example.† The following rule for finding a person's age is of the same kind. Tell him to think of a number (preferably not exceeding 10); (i) square it; (ii) subtract 1; (iii) multiply the result by the original number; (iv) multiply this by 3; (v) add his age; (vi) give the sum of the digits of the result. You then have to guess his age within nine years, knowing that it has this same digit-sum.

The algebraic proof of the rule is obvious. Let a be the age, and b the number thought of. The successive operations give (i) b^2; (ii) b^2-1; (iii) $b(b^2-1)$; (iv) $3b(b^2-1)$; (v) $a+3b(b^2-1)$; (vi) the digit-sum of a, because $3b(b^2-1)$ is always a multiple of 9.·

Other examples.‡ Another such problem, but of more difficulty, is the determination of all numbers which are integral multiples of their reversals. For instance, among num-

* Bachet gave some similar questions in problem XII, p. 117.
† Due to Royal V. Heath.
‡ *L'Intermédiaire des Mathématiciens*, Paris, vol. XV, 1908, pp. 228, 278; vol. XVI, 1909, p. 34; vol. XIX, 1912, p. 128. M. Kraïtchik, *La Mathématique des Jeux*, Brussels, 1930, pp. 55, 59.

bers of four digits, $8712 = 4 \times 2178$ and $9801 = 9 \times 1089$ possess this property.

Again, we might ask for two numbers whose product is the reversal of the product of their reversals; for example,

$$312 \times 221 = 68952, \quad 213 \times 122 = 25986.$$

The number 698896 is remarkable as being a perfect square, equal to its own reversal, and having an even number of digits.

Just four numbers have the property of being equal to the sum of the cubes of their digits:* $153 = 1^3 + 5^3 + 3^3$, $370 = 3^3 + 7^3 + 0^3$, $371 = 3^3 + 7^3 + 1^3$, $407 = 4^3 + 0^3 + 7^3$.

The properties of the recurring decimal for $\frac{1}{7}$ are notorious. Troitsky has shown† that 142857 and 285714 are the only numbers less than a million which yield multiples of themselves when we remove the first digit and put it after the last.

OTHER PROBLEMS WITH NUMBERS IN THE DENARY SCALE

I may mention here two or three other problems which seem to be unknown to most compilers of books of puzzles.

First problem. The first of them is as follows. Take any number of three digits, the first and third digits being different: reverse the order of the digits: subtract the number so formed from the original number: then, if the last digit of the difference is mentioned, all the digits in the difference are known.

For suppose the number is $100a + 10b + c$, then the number obtained by reversing the digits is $100c + 10b + a$. The difference of these numbers is equal to $(100a + c) - (100c + a)$, that is, to $99(a - c)$. But $a - c$ is not greater than 9, and therefore the remainder can only be 99, 198, 297, 396, 495, 594, 693, 792, or 891; in each case the middle digit being 9 and the digit before it (if any) being equal to the difference between 9 and the last digit. Hence, if the last digit is known, so is the whole of the remainder.

Second problem. The second problem is somewhat similar, and is as follows. (i) Take any number; (ii) reverse the digits;

*Sphinx (Brussels), 1937, pp. 72, 87.
†L'Echiquier, 1930, p. 663.

(iii) find the difference between the number formed in (ii) and the given number; (iv) multiply this difference by any number you like to name; (v) cross out any digit except a nought; (vi) read the remainder. Then the sum of the digits in the remainder subtracted from the next highest multiple of nine will give the figure struck out. This is clear since the result of operation (iv) is a multiple of nine, and the sum of the digits of every multiple of nine is itself a multiple of nine. This and the previous problem are typical of numerous analogous questions.

Third problem. If n digits are required in the pagination of a book, how many pages are contained therein – for instance, $n = 3001$?

The analysis is obvious. The first 999 pages require the use of $9 + 180 + 2700$ digits. But 112 additional digits are employed, and these suffice to identify 28 more pages. Therefore the total number of pages is $999 + 28$, that is, 1027.

Fourth problem. The numbers from 1 upwards are written consecutively. What is the nth digit, for instance, in the case when $n = 500000$?

The numbers from 1 to 99999 inclusive require 488889 digits. Hence we want the 11111th digit in the series of six-digit numbers starting with 100000. We have $11111 = 6 \times 1851 + 5$. Hence we want the 5th digit in 101851, and this is 5.

Empirical problems. There are also numerous empirical problems, such as the following. With the ten digits, 9, 8, 7, 6, 5, 4, 3, 2, 1, 0, express numbers whose sum is unity: each digit being used only once, and the use of the usual notations for fractions being allowed. With the same ten digits express numbers whose sum is 100. With the nine digits 9, 8, 7, 6, 5, 4, 3, 2, 1, express numbers whose sum is 100. To the making of such questions there is no limit, but their solution involves little or no mathematical skill.

Four digits problem. I suggest the following problem as being more interesting. With the digits 1, 2, 3, 4, express the consecutive numbers from 1 upwards as far as possible: each of the four digits being used once, and only once, in the expression of each number. Allowing the notation of the denary scale (including decimals), as also algebraic sums, products, and positive integral powers, we can get to 88. If the use of the symbols for square roots and factorials (repeated if desired a finite number of times) is also permitted, we can get to 264; if negative integral indices are also permitted, to 276; and if fractional indices are permitted, to 312. Many similar questions may be proposed, such as using four out of the digits 1, 2, 3, 4, 5. With the five digits 1, 2, 3, 4, 5, each being used once and only once, I have got to 3832 and 4282, according as negative and fractional indices are excluded or allowed.

Four fours problem.   Another traditional recreation is, with the ordinary arithmetic and algebraic notation, to express the consecutive numbers from 1 upwards as far as possible in terms of four "4's." Everything turns on what we mean by ordinary notation. If (α) this is taken to admit only the use of the denary scale (e.g. numbers like 44), decimals, brackets, and the symbols for addition, subtraction, multiplication, and division, we can thus express every number up to 22 inclusive.* If (β) also we grant the use of the symbol for square root (repeated if desired a finite number of times), we can get to 30; but note that though by its use a number like 2 can be expressed by one "4," we cannot for that reason say that ·2 is so expressible. If (γ), further, we permit the use of symbols for factorials, we can express every number to 112.† Finally, if (δ) we sanction the employment of integral indices expressible by a "4" or "4's" and allow the symbol for a square root to be used an infinite number of times, we can get to 156; but if (ε) we concede the employment of integral indices and the use of sub-factorials,‡ we can get to 877. These interesting problems are typical of a class of similar questions. Thus, under conditions γ and using no indices, with four "1's" we can get to 34, with four "2's" to 36, with four "3's" to 46, with four "5's" to 36,

---

*E.g., $22 = (4+4)/(\cdot\dot{4}) + 4$.
†E.g., $99 = 4 \times 4! + \sqrt{4/(\cdot\dot{4})}$.
‡Sub-factorial n is equal to $n!(1 - 1/1! + 1/2! - 1/3! + \ldots \pm 1/n!)$. On the use of this for the four "4's" problem see the *Mathematical Gazette*, May 1912.

with four "6's" to 30, with four "7's" to 25, with four "8's" to 36, and with four "9's" to 130; e.g., as T. Haji has remarked, $67 = \sqrt{9!/(9 \times 9) + 9}$.

PROBLEMS WITH A SERIES OF THINGS WHICH ARE NUMBERED

Any collections of things numbered consecutively lend themselves to easy illustrations of questions depending on elementary properties of numbers. As examples I proceed to enumerate a few familiar tricks. The first two of these are commonly shown by the use of a watch, and the last four may be exemplified by the use of a pack of playing cards.

First example.* The first of these examples is connected with the hours marked on the face of a watch. In this puzzle someone is asked to think of some hour, say m, and then to touch a number that marks another hour, say n. Then if, beginning with the number touched, he taps each successive hour marked on the face of the watch, going in the opposite direction to that in which the hands of the watch move, and reckoning to himself the taps as m, $m+1$, etc., the $(n+12)$th tap will be on the hour he thought of. For example, if he thinks of v and touches IX, then, if he taps successively IX, VIII, VII, VI, ..., going backwards and reckoning them respectively as 5, 6, 7, 8, ..., the tap which he reckons as 21 will be on the v.

The reason of the rule is obvious, for he arrives finally at the $(n+12-m)$th hour from which he started. Now, since he goes in the opposite direction to that in which the hands of the watch move, he has to go over $n-m$ hours to reach the hour m; also it will make no difference if, in addition, he goes over 12 hours, since the only effect of this is to take him once completely round the circle. Now $n+12-m$ is always positive, since n is positive and m is not greater than 12, and therefore if we make him pass over $n+12-m$ hours we can give the rule in a form which is equally valid whether m is greater or less than n.

Second example. The following is another well-known

*Bachet, problem xx, p. 155; Oughtred or Leake, *Mathematicall Recreations*, London, 1653, p. 28.

watch-dial problem. If the hours on the face are tapped successively, beginning at VII and proceeding backwards round the dial to VI, V, etc., and if the person who selected the number counts the taps, beginning to count from the number of the hour selected (thus, if he selected X, he would reckon the first tap as the 11th), then the 20th tap as reckoned by him will be on the hour chosen.

For suppose he selected the nth hour. Then the 8th tap is on XII and is reckoned by him as the $(n+8)$th; and the tap which he reckons as $(n+p)$th is on the hour $20-p$. Hence, putting $p = 20-n$, the tap which he reckons as 20th is on the hour n. Of course, the hours indicated by the first seven taps are immaterial: obviously also we can modify the presentation by beginning on the hour VIII and making 21 consecutive taps, or on the hour IX and making 22 consecutive taps, and so on.

Third example. The following is another simple example. Suppose that a pack of n cards is given to someone who is asked to select one out of the first m cards and to remember (but not to mention) what is its number from the top of the pack; suppose it is actually the xth card in the pack. Then take the pack, reverse the order of the top m cards (which can be easily effected by shuffling), and transfer y cards, where $y < n - m$, from the bottom to the top of the pack. The effect of this is that the card originally chosen is now the $(y + m - x + 1)$th from the top. Return to the spectator the pack so rearranged, and ask that the top card be counted as the $(x+1)$th, the next as the $(x+2)$th, and so on, in which case the card originally chosen will be the $(y+m+1)$th. Now y and m can be chosen as we please, and may be varied every time the trick is performed; thus anyone unskilled in arithmetic will not readily detect the method used.

Fourth example.* Place a card on the table, and on it place as many other cards from the pack as with the number of pips on the card will make a total of twelve. For example, if the card placed first on the table is the five of clubs, then seven additional cards must be placed on it. The court cards may

---

*A particular case of this problem was given by Bachet, problem XVII, p. 138.

have any values assigned to them, but usually they are reckoned as tens. This is done again with another card, and thus another pile is formed. The operation may be repeated either only three or four times or as often as the pack will permit of such piles being formed. If finally there are p such piles, and if the number of cards left over is r, then the sum of the number of pips on the bottom cards of all the piles will be $13(p-4)+r$.

For, if x is the number of pips on the bottom card of a pile, the number of cards in that pile will be $13-x$. A similar argument holds for each pile. Also there are 52 cards in the pack; and this must be equal to the sum of the cards in the p piles and the r cards left over. Thus we have

$$(13-x_1)+(13-x_2)+ \ldots +(13-x_p)+r=52,$$

$$13p-(x_1+x_2+ \ldots +x_p)+r=52,$$

$$\begin{aligned} x_1+x_2+ \ldots +x_p &= 13p-52+r \\ &= 13(p-4)+r. \end{aligned}$$

More generally, if a pack of n cards is taken, and if in each pile the sum of the pips on the bottom card and the number of cards put on it is equal to m, then the sum of the pips on the bottom cards of the piles will be $(m+1)p+r-n$. In an écarté pack $n=32$, and it is convenient to take $m=15$.

Fifth example. It may be noticed that cutting a pack of cards never alters the relative position of the cards provided that, if necessary, we regard the top card as following immediately after the bottom card in the pack. This is used in the following trick.* Take a pack, and deal the cards face upwards on the table, calling them one, two, three, etc., as you put them down, and noting in your own mind the card first dealt. Ask someone to select a card and recollect its number. Turn the pack over, and let it be cut (not shuffled) as often as you like. Enquire what was the number of the card chosen. Then, if you deal, and as soon as you come to the original first card, begin (silently) to count, reckoning this as one, the selected card will appear at the number mentioned. Of course,

*Bachet, problem XIX, p. 152.

if all the cards are dealt before reaching this number, you must turn the cards over and go on counting continuously.

Sixth example. Here is another simple question of this class. Remove the court cards from a pack. Arrange the remaining 40 cards, faces upwards, in suits, in four lines thus. In the first line, the 1, 2, . . ., 10 of suit A; in the second line, the 10, 1, 2, . . ., 9 of suit B; in the third line, the 9, 10, 1, . . ., 8 of suit C; in the last line, the 8, 9, 10, 1, . . ., 7 of suit D. Next take up, face upwards, the first card of line 1, put below it the first card of line 2, below that the first card of line 3, and below that the first card of line 4. Turn this pile face downwards. Next take up the four cards in the second column in the same way, turn them face downwards, and put them below the first pile. Continue this process until all the cards are taken up. Ask someone to mention any card. Suppose the number of pips on it is n. Then if the suit is A, it will be the $4n$th card in the pack; if the suit is B, it will be the $(4n+3)$th card; if the suit is C, it will be the $(4n+6)$th card; and if the suit is D, it will be the $(4n+9)$th card. Hence by counting the cards, cyclically if necessary, the card desired can be picked out. It is easy to alter the form of presentation, and a full pack can be used if desired. The explanation is obvious.

ARITHMETICAL RESTORATIONS

I take next a class of problems dealing with the reconstruction of arithmetical sums from which various digits have been erased. Some of these questions are easy, some difficult. This kind of exercise has attracted a good deal of attention in recent years. I give examples of the three kinds of restoration.

Class A. The solutions of one group of these restoration questions depend on the well-known propositions that every number

$$a + 10b + 10^2c + 10^3d + \ldots$$

is equal to any of certain expressions such as

$$M(9) + a + b + c + d + \ldots$$
$$M(11) + a - b + c - d + \ldots$$

$$M \ (33) \ +(a+10b)+(c+10d)+(e+10f)+\ldots$$
$$M \ (101)+(a+10b)-(c+10d)+(e+10f)-\ldots$$
$$M \ (m)+(a+10b+10^2c)+(d+10e+10^2f)+\ldots$$
$$M \ (n) \ +(a+10b+10^2c)-(d+10e+10^2f)+\ldots$$

where, in the penultimate line, $m=27$, or 37, or 111, and in the last line, $n=7$, or 11, or 13, or 77, or 91, or 143.

Examples, depending on such propositions, are not uncommon. Here are four easy instances of this class of questions.

(i) The product of 417 and . 1... is 9 ... 057. Find the missing digits, each of which is represented by a dot. If the undetermined digits in the multiplier are denoted in order by a, b, c, d, and we take the steps of the multiplication in their reverse order, we obtain successively $d=1$, $c=2$, $b=9$. Also the product has 7 digits, therefore $a=2$. Hence the product is 9,141,057.

(ii) The seven-digit number 70 . . 34 . is exactly divisible by 792. Find the missing digits, each of which is represented by a dot. Since 792 is $8 \times 9 \times 11$, we can easily show that the number is 7,054,344.

(iii) The five-digit number 4 . 18 . is divisible by 101. Find the missing digits.*

Denote the two missing digits, from right to left, by x and y. Applying the theorem for 101, noting that each of the unknowns cannot exceed 9, and for convenience putting $y=10-z$, this equation gives $z=1$, $x=7$, $y=9$. Hence the number is 49,187.

(iv) The four-digit number . 8 . . is divisible by 1287. Find the missing digits.†

Denote these digits, from right to left, x, y, z. We have $1287=9 \times 11 \times 13$. Applying the suitable propositions, and noting that each of the unknowns cannot exceed 9, we get $x=1$, $y=6$, $z=3$. Hence the number is 3861.

(v) As a slightly harder example of this type, suppose we know that 6 . 80 . 8 .. 51 is exactly divisible by 73 and 137.

*P. Delens, *Problèmes d'Arithmétique amusante*, Paris, 1914, p. 55.
†*Ibid.*, p. 57.

Find the missing digits.* The data suffice to determine the number, which is 6,780,187,951.

Class B.   Another and more difficult class of restoration problems is illustrated by the following examples. Their solutions involve analytical skill which cannot be reduced to rules.

(i) I begin with an easy instance, said to be of Hindu origin, in which the problem is to restore the missing digits in the annexed division sum where a certain six-digit number when divided by a three-digit number gives a three-digit result. †

```
        ...) ...... ( ...
             .0..
             ----
             ....
             .50.
             ----
              ...
              .4.
              ---
```

The solution involves no difficulty. The answer is that the divisor is 215, and the quotient 573; the solution is unique.

(ii) As a more difficult specimen I give the following problem, proposed in 1921 by Prof. Schuh of Delft. A certain seven-digit integer when divided by a certain six-digit integer gives a result whose integral part is a two-digit number and whose fractional part is a ten-digit expression of which the last nine digits form a repeating decimal, as indicated in the following work, where a bar has been put above the repeating digits. It is required to restore the working. ‡ This problem is remarkable from the fact that not a single digit is given explicitly.

* *Ibid.*, p. 60.
† *American Mathematical Monthly*, 1921, vol. xxviii, p. 37.
‡ *Ibid.*, 1922, vol. xxix, p. 211.

```
......)....... ( .. · ..........
        ......
        _____
       .......
        ......
        _____
        .......
        .......
        _____
         .......
          ......
          _____
         .......
          ......
          _____
           .......
            ......
            _____
             .......
            .......
            _____
              .......
               ......
               _____
                 ......
                 ......
                 _____
                  ......
```

The answer is that the divisor is 667334 and the dividend is 7752341.

Here are three additional examples of arithmetical restorations.* The solutions are lengthy and involve much empirical work.

*All are due to W.E.H. Berwick. The "7" problem appeared in the *School World*, July and October 1906, vol. VIII, pp. 280, 320; the "4" problem appeared in the *Mathematical Gazette*, 1920, vol. x, pp. 43, 359–360; the "5" problem in the same volume, p. 361, and in vol. XI, p. 8.

(iii) The first of these Berwick questions is as follows. In the following division sum all the digits, except the seven "7's" shown, have been erased: each missing digit may be 1, 2, 3, 4, 5, 6, 7, 8, 9, or (except in the first digit of a line) 0. Observe that every step in the working consists of two lines each of which contains an equal number of digits. The problem is to restore the whole working of the sum. The solution is unique and gives a divisor of 125473 and a quotient of 58781.

```
....7.)..7.......(..7..
        ......
        ------
       .....7.
        .......
        -------
          .7....
          .7....
          ------
          .......
          ....7..
          -------
            ......
            ......
            ------
```

(iv) The second problem is similar and requires the restoration of the digits in the following division sum, where the position of four "4's" is given,

```
...)......4(.4..
     ...
     ---
     ..4.
     ....
     ----
       ....
       .4.
       ---
       ....
       ....
       ----
```

To this problem there are four solutions, the divisors being 846, 848, 943, 949; and the respective quotients 1419, 1418, 1418, 1416.

If we propound the problem (using five "4's") thus:

```
...)......4(.4..
    ...
    ___
    ..4.
    ...4
    ___
    ....
    .4.
    ___
    ....
    ....
    ___
```

there is only one solution, and some will think this is a better form in which to enunciate it.

(v) In the last of these Berwick examples, it is required to restore the working of the following division sum where all the digits, except five "5's" have been erased.

```
....).55..5.(.5.
     ..5..
     ___
     .....
     .....
     ___
     ....
     ....
     ___
```

To this problem there is only one solution, the divisor being 3926 and the quotient 652.

Class C.   A third class of digit problems depends on finding the values of certain symbols which represent specified numbers. Two examples will suffice.

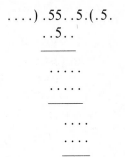

132,683

(i) Here is a very simple illustrative specimen. The result of multiplying *bc* by *bc* is *abc*, where the letters stand for certain numbers. What are the numbers? A brief examinations shows that *bc* stands for 25, and therefore *a* stands for 6.

(ii) Here is another example. The object is to find the digits represented by letters in the following sum :*

$$ab)cdeeb(bfb$$
$$ceb$$
$$\overline{}$$
$$gge$$
$$gch$$
$$\overline{}$$
$$ceb$$
$$ceb$$
$$\overline{}$$

A solution may be obtained thus : Since the product of *b* by *b* is a number which ends in *b*, *b* must be 1, 5, or 6. Since the product of *ab* by *b* is a number of three digits, *b* cannot be 1. The result of the subtraction of *h* from *e* is *e*, hence $h=0$, and therefore if $b=5$ we have *f* even, and if $b=6$ we have $f=5$. Also the result of the subtraction of *c* from *g* is *c*, hence $g=2c$, and therefore *c* cannot be greater than 4 : from which it follows that *b* cannot be 6. A few trials now show that the question arose from the division of 19,775 by 35.

It is possible to frame digit restoration examples of a mixed character involving the difficulties of all the examples given above, and to increase the difficulty by expressing them in a non-denary scale of notation. But such elaborations do not add to the interest of the questions.

CALENDAR PROBLEMS

The formulae given by Gauss and Zeller, which I quoted in former editions of this work, serve to solve all questions likely to occur about dates, days of the week, Easter, etc. Here

*Strand Magazine, September–October 1921.

are two easy but elegant questions on the Gregorian Calendar of a somewhat different nature.

The first is as follows: it is due, I believe, to E. Fourrey. In the century and a half between 1725 and 1875 the French fought and won a certain battle on 22 April of one year, and 4382 days later, also on 22 April, they gained another victory. The sum of the digits of the years is 40. Find the dates of the battles.

To solve it we notice that $4382 = 12 \times 365 + 2$. Hence the date of the second battle was 12 years after that of the first battle; but only two leap years had intervened, and therefore the year 1800 must be within the limiting dates. Thus 1788 and 1800, 1789 and 1801, ..., 1800 and 1812, are the only possible years. Of the years thus suggested, 1796 and 1808 alone give 40 as the sum of their digits. Hence the battles were fought on the 22 April 1796 (Mondovi under Napoleon), and 22 April 1808 (Eckmuhl under Davoust).

The other of these questions is to show that the first or last day of every alternate century must be a Monday. This follows from knowing any one assigned date, and the fact that the Gregorian cycle is completed in 400 years ($= 20871$ weeks). The same principle is involved in B.H. Brown's assertion* that the thirteenth day of the month is more likely to be a Friday than to be any other day of the week.

MEDIEVAL PROBLEMS IN ARITHMETIC

Before leaving the subject of these elementary questions, I may mention a few problems which for centuries have appeared in nearly every collection of mathematical recreations, and may therefore claim what is almost a prescriptive right to a place here.

First example.   The following is a sample of one class of these puzzles. A man goes to a tub of water with two jars, of which one holds exactly 3 pints and the other 5 pints. How can he bring back exactly 4 pints of water? The solution presents no difficulty.

---

*American Mathematical Monthly, 1933, vol. XL, p. 607.

Second example.* Here is another problem of the same kind. Three men robbed a gentleman of a vase, containing 24 ounces of balsam. Whilst running away they met a glass-seller, of whom they purchased three vessels. On reaching a place of safety they wished to divide the booty, but found that their vessels could hold 5, 11, and 13 ounces respectively. How could they divide the balsam into equal portions? Problems like this can be worked out only by trial.

Third example.† The next of these is a not uncommon game, played by two people, say A and B. A begins by mentioning some number not greater than (say) six, B may add to that any number not greater than six, A may add to that again any number not greater than six, and so on. He wins who is the first to reach (say) 50. Obviously, if A calls 43, then whatever B adds to that, A can win next time. Similarly, if A calls 36, B cannot prevent A's calling 43 the next time. In this way it is clear that the key numbers are those forming the arithmetical progression 43, 36, 29, 22, 15, 8, 1; and whoever plays first ought to win.

Similarly, if no number greater than m may be added at any one time, and n is the number to be called by the victor, then the key numbers will be those forming the arithmetical progression whose common difference is $m+1$ and whose smallest term is the remainder obtained by dividing n by $m+1$.

The same game may be played in another form by placing p coins, matches, or other objects on a table, and directing each player in turn to take away not more than m of them. Whoever takes away the last coin wins. Obviously the key numbers are multiples of $m+1$, and the first player who is able to leave an exact multiple of $m+1$ coins can win. Perhaps a better form of the game is to make that player lose who takes away the last coin, in which case each of the key numbers exceeds by unity a multiple of $m+1$.

---

*Some similar problems were given by Bachet, Appendix, problem III, p. 206; problem IX, p. 233; by Oughtred or Leake in the *Mathematicall Recreations*, p. 22; and by Ozanam, 1803 edition, vol. I, p. 174; 1840 edition, p. 79. Earlier instances occur in Tartaglia's writings. See also Kraïtchik, p. 11; H.S.M. Coxeter and S.L. Greitzer, *Geometry Revisited*, New York, 1967, pp. 89–93.

†Bachet, problem XXII, p. 170.

Another variety* consists in placing p counters in the form of a circle, and allowing each player in succession to take away not more than m of them which are in unbroken sequence: m being less than p and greater than unity. In this case the second of the two players can always win.

These games are simple, but if we impose on the original problem the restriction that each player may not add the same number more than (say) three times, the analysis becomes by no means easy. I have never seen this extension described in print, and I will enunciate it at length. Suppose that each player is given eighteen cards, three of them marked 6, three marked 5, three marked 4, three marked 3, three marked 2, and three marked 1. They play alternately; A begins by playing one of his cards; then B plays one of his, and so on. He wins who first plays a card which makes the sum of the points or numbers on all the cards played exactly equal to 50, but he loses if he plays a card which makes this sum exceed 50. The game can be played by noting the numbers on a piece of paper, and it is not necessary to use cards.

Thus suppose they play as follows. A takes a 4, and scores 4; B takes a 3, and scores 7; A takes a 1, and scores 8; B takes a 6, and scores 14; A takes a 3, and scores 17; B takes a 4, and scores 21; A takes a 4, and scores 25; B takes a 5, and scores 30; A takes a 4, and scores 34; B takes a 4, and scores 38; A takes a 5, and scores 43. B can now win, for he may safely play 3, since A has not another 4 wherewith to follow it; and if A plays less than 4, B will win the next time. Again, suppose they play thus. A, 6; B, 3; A, 1; B, 6; A, 3; B, 4; A, 2; B, 5; A, 1; B, 5; A, 2; B, 5; A, 2; B, 3. A is now forced to play 1, and B wins by playing 1.

A slightly different form of the game has also been suggested. In this there are put on the table an agreed number of cards – say, for example, the four aces, twos, threes, fours, fives, and sixes of a pack of cards – twenty-four cards in all. Each player in turn takes a card. The score at any time is the sum of the pips on all the cards taken, whether by A or B. He wins who first selects a card which makes the score equal, say, to 50, and a player who is forced to go beyond 50 loses.

Thus, suppose they play as follows. *A* takes a 6, and scores 6; *B* takes a 2, and scores 8; *A* takes a 5, and scores 13; *B* takes a 2, and scores 15; *A* takes a 5, and scores 20; *B* takes a 2, and scores 22; *A* takes a 5, and scores 27; *B* takes a 2, and scores 29; *A* takes a 5, and scores 34; *B* takes a 6, and scores 40; *A* takes a 1, and scores 41; *B* takes a 4, and scores 45; *A* takes a 3, and scores 48; *B* now must take 1, and thus score 49; then *A* takes a 1, and wins.

In these variations the object of each player is to get to one of the key numbers, provided there are sufficient available remaining numbers to let him retain the possession of each subsequent key number. The number of cards used, the points on them, and the number to be reached can be changed at will; and the higher the number to be reached, the more difficult it is to forecast the result and to say whether or not it is an advantage to begin.

Fourth example.   The following medieval problem is somewhat more elaborate. Suppose that three people, *P*, *Q*, *R*, select three things, which we may denote by *a*, *e*, *i* respectively, and that it is desired to find by whom each object was selected.*

Place 24 counters on a table. Ask *P* to take one counter, *Q* to take two counters, and *R* to take three counters. Next, ask the person who selected *a* to take as many counters as he has already, whoever selected *e* to take twice as many counters as he has already, and whoever selected *i* to take four times as many counters as he has already. Note how many counters remain on the table. There are only six ways of distributing the three things among *P*, *Q*, and *R*; and the number of counters remaining on the table is different for each way. The remainders may be 1, 2, 3, 4, 5, 6, or 7. Bachet summed up the results in the mnemonic line *Par fer* (1) *César* (2) *jadis* (3) *devint* (5) *si grand* (6) *prince* (7). Corresponding to any remainder is a word or words containing two syllables: for instance, to the remain-

*Bachet, problem xxv, p. 187.

der 5 corresponds the word *devint*. The vowel in the first syllable indicates the thing selected by P, the vowel in the second syllable indicates the thing selected by Q, and of course R selected the remaining thing.

Extension. M. Bourlet, in the course of a very kindly notice* of the second edition of this work, gave a much neater solution of the above question, and has extended the problem to the case of n people, P_0, P_1, P_2, ..., P_{n-1}, each of whom selects one object, out of a collection of n objects, such as dominoes or cards. It is required to know which domino or card was selected by each person.

Let us suppose the dominoes to be denoted or marked by the numbers 0, 1, ..., $n-1$, instead of by vowels. Give one counter to P_1, two counters to P_2, and generally k counters to P_k. Note the number of counters left on the table. Next ask the person who had chosen the domino 0 to take as many counters as he had already, and generally whoever had chosen the domino h to take n^h times as many dominoes as he had already: thus if P_k had chosen the domino numbered h, he would take $n^h k$ counters. The total number of counters taken is $\Sigma n^h k$. Divide this by n, then the remainder will show who selected the domino 0; divide the quotient by n, and the remainder will show who selected the domino 1; divide this quotient by n, and the remainder will show who selected the domino 2; and so on. In other words, if the number of counters taken is expressed in the scale of notation whose radix is n, then the $(h+1)$th digit from the right will be k if it was P_k who selected the domino h.

Thus in Bachet's problem with 3 people and 3 dominoes, we should first give one counter to Q, and two counters to R, while P would have no counters; then we should ask the person who had selected the domino marked 0 or a to take as many counters as he had already, whoever had selected the domino marked 1 or e to take three times as many counters as he had already, and whoever had selected the domino marked 2 or i to take nine times as many counters as he had already.

*Bulletin des Sciences mathématiques, Paris, 1893, vol. XVII, pp. 105–107.

By noticing the original number of counters, and observing that 3 of these had been given to Q and R, we should know the total number taken by P, Q, and R. If this number were divided by 3, the remainder would indicate who had taken the domino a (being 0 for P, 1 for Q, 2 for R); if the quotient were divided by 3 the remainder would indicate who had taken the domino e; and the final quotient would indicate who had taken the domino i.

Exploration problems. Another common question is concerned with the maximum distance into a desert which could be reached from a frontier settlement by an explorer capable of carrying provisions that would last him for a days. He is allowed to stock up n times at the frontier settlement and to make depôts. The answer is that the longest possible journey will occupy

$$\left(\frac{1}{2} + \frac{1}{4} + \frac{1}{6} + \ldots + \frac{1}{2n} \right) a$$

days if the explorer returns home, and

$$\left(1 + \frac{1}{3} + \frac{1}{5} + \ldots + \frac{1}{2n-1} \right) a$$

days otherwise.*

The Josephus problem. Another of these antique problems consists in placing men round a circle so that if every mth man is killed, the remainder shall be certain specified individuals. Such problems can be easily solved empirically.

Hegesippus† says that Josephus saved his life by such a device. According to his account, after the Romans had captured Jotapat, Josephus and forty other Jews took refuge in a cave. Josephus, much to his disgust, found that all except himself and one other man were resolved to kill themselves, so as not to fall into the hands of their conquerors. Fearing to show his opposition too openly he consented, but declared' that the operation must be carried out in an orderly way, and

---

*D. Gale, *American Mathematical Monthly*, 1970, vol. LXXVII, pp. 493–501.
†*De Bello Judaico*, bk. III, chaps. 16–18.

suggested that they should arrange themselves round a circle and that every third person should be killed until but one man was left, who must then commit suicide. It is alleged that he placed himself and the other man in the 31st and 16th place respectively.

The medieval question was usually presented in the following form. A ship, carrying as passengers 15 Turks and 15 Christians, encountered a storm, and, in order to save the ship and crew, one-half of the passengers had to be thrown into the sea. Accordingly the passengers were placed in a circle, and every ninth man, reckoning from a certain point, was cast overboard. It is desired to find an arrangement by which all the Christians should be saved.* In this case we must arrange the men thus: *C C C C T T T T T C C T C C C T C T T C C T T T C T T C C T*, where *C* stands for a Christian and *T* for a Turk. The order can be recollected by the positions of the vowels in the following line: *From numbers' aid and art, never will fame depart*, where *a* stands for 1, *e* for 2, *i* for 3, *o* for 4, and *u* for 5. Hence the order is *o* Christians, *u* Turks, etc.

If every tenth man were cast overboard, a similar mnemonic line is *Rex paphi cum gente bona dat signa serena*. An oriental setting of this decimation problem runs somewhat as follows. Once upon a time, there lived a rich farmer who had 30 children, 15 by his first wife who was dead, and 15 by his second wife. The latter woman was eager that her eldest son should inherit the property. Accordingly one day she said to him, "Dear Husband, you are getting old. We ought to settle who shall be your heir. Let us arrange our 30 children in a circle, and counting from one of them, remove every tenth child until there remains but one, who shall succeed to your estate." The proposal seemed reasonable. As the process of selection went on, the farmer grew more and more astonished as he noticed that the first 14 to disappear were children by his first wife, and he observed that the next to go would be the last remaining member of that family. So he suggested that they

*Bachet, problem xxiii, p. 174. The same problem had been previously enunciated by Tartaglia.

should see what would happen if they began to count backwards from this lad. She, forced to make an immediate decision, and reflecting that the odds were now 15 to 1 in favour of her family, readily assented. Who became the heir?

In the general case n men are arranged in a circle which is closed up as individuals are picked out. Beginning anywhere, we continually go round, picking out each mth man until only r are left. Let one of these be the man who originally occupied the pth place. Then had we begun with $n+1$ men, he would have originally occupied the $(p+m)$th place when $p+m$ is not greater than $n+1$, and the $(p+m-n-1)$th place when $p+m$ is greater than $n+1$. Thus, provided there are to be r men left, their original positions are each shifted forwards along the circle m places for each addition of a single man to the original group.*

Now suppose that with n men the last survivor $(r=1)$ occupied originally the pth place, and that with $n+x$ men the last survivor occupied the yth place. Then, if we confine ourselves to the lowest value of x which makes y less than m, we have $y=(p+mx)-(n+x)$.

Based on this theorem we can, for any specified value of n, calculate rapidly the position occupied by the last survivor of the company. In effect, Tait found the values of n for which a man occupying a given position p, which is less than m, would be the last survivor, and then, by repeated applications of the proposition, obtained the position of the survivor for intermediate values of n.

For instance, take the Josephus problem in which $m=3$. Then we know that the final survivor of 41 men occupied originally the 31st place. Suppose that when there had been $(41+x)$ men, the survivor occupied originally the yth place. Then, if we consider only the lowest value of x which makes y less than m, we have $y=(31+3x)-(41+x)=2x-10$. Now, we have to take a value of x which makes y positive and less than m, that is, in this case equal to 1 or 2. This is $x=6$, which

*P.G. Tait, *Collected Scientific Papers*, Cambridge, vol. II, 1900, pp. 432–435.

makes $y=2$. Hence, had there been 47 men, the man last chosen would have originally occupied the second place. Similarly had there been $47+x$ men, the man would have occupied originally the yth place, where, subject to the same conditions as before, we have $y=(2+3x)-(47+x)=2x-45$. If $x=23$, $y=1$. Hence, with 70 men the man last chosen would have occupied originally the first place. Continuing the process, it is easily found that if n does not exceed 2000000 the last man to be taken occupies the first place when $n=4, 6, 9, 31, 70, 105, 355, 799, 1798, 2697, 9103, 20482, 30723, 69127, 155536, 233304, 349956, 524934$, or 787401; and the second place when $n=2, 3, 14, 21, 47, 158, 237, 533, 1199, 4046, 6069, 13655, 46085, 103691, 1181102$, or 1771653. From these results, by repeated applications of the proposition, we find, for any intermediate values of n, the position originally occupied by the man last taken. Thus with 1000 men, the 604th place; with 100000 men, the 92620th place; and with 1000000 men, the 637798th place are those which would be selected by a prudent mathematician in a company subjected to trimation.

Similarly if a set of 100 men were subjected to decimation, the last to be taken would be the man originally in the 26th place. Hence, with 227 men the last to be taken would be the man originally in the first place.

Modifications of the original problem have been suggested. For instance,* let 5 Christians and 5 Turks be arranged round a circle thus, *T C T C C T C T C T.* Suppose that if beginning at the ath man, every hth man is selected, all the Turks will be picked out for punishment; but if beginning at the bth man, every kth man is selected, all the Christians will be picked out for punishment. The problem is to find a, b, h, and k. A solution is $a=1$, $h=11$, $b=9$, $k=29$.

I suggest as a similar problem, to find an arrangement of c Turks and c Christians arranged in a circle, so that if beginning at a particular man, say the first, every hth man is selected, all the Turks will be picked out, but if, beginning at the same man, every kth man is selected, all the Christians will be picked out.

*H.E. Dudeney, *Tit-Bits*, London, Oct. 14 and 28, 1905.

This makes an interesting question because it is conceivable that the operator who picked out the victims might get confused and take k instead of h, or vice versa, and so consign all his friends to execution instead of those whom he had intended to pick out. The problem is, for any given value of c, to find an arrangement of the men and the corresponding suitable values of h and k. Obviously if $c = 2$, then for an arrangement like $T\ C\ C\ T$ a solution is $h = 4$, $k = 3$. If $c = 3$, then for an arrangement like $T\ C\ T\ C\ C\ T$ a solution is $h = 7$, $k = 8$. If $c = 4$, then for an arrangement like $T\ C\ T\ T\ C\ T\ C\ C$ a solution is $h = 9$, $k = 5$. And generally, as first pointed out by Mr. Swinden, with $2c$ men, c of them, occupying initially the consecutive places numbered c, $c + 1$, ..., $2c - 1$, will be picked out if h is the L.C.M. of $c + 1$, $c + 2$, ..., $2c - 1$; and the other c will be picked out if $k = h + 1$, though it may well be that there is a simpler solution for another initial arrangement. It may be impossible to arrange the men so that n specified individuals shall be picked out in a defined order.

NIM AND SIMILAR GAMES*

Several games are played by two people, A and B, with one or more heaps of counters, each player in turn being required to take away some or all of the counters according to specified rules. The player taking the last counter wins. One example is the third medieval problem, discussed above, in which there is just one heap, and a move consists in taking any number of counters between 1 and m inclusive. Many such games can be analysed by means of the Sprague-Grundy number † $G(C)$. For the empty position O, containing no counters, $G(O) = 0$. For any other combination $C = (x,\ y,\ \ldots)$ of heaps of x, y, ... counters respectively, suppose that there are permitted moves from C to other combinations D, E, ... Then $G(C)$ is the smallest non-negative integer different from $G(D)$, $G(E)$,

---

*The present version of this section has been kindly supplied by C.A.B. Smith.

†R. Sprague, *Tôhoku Journal of Mathematics*, 1936, vol. XLI, p. 438; P.M. Grundy, *Eureka*, 1939, vol. II, p. 6.

.... This defines $G(C)$ inductively for all combinations C permitted by the rules. Thus, in the third medieval problem we find that $G(x) =$ the remainder on dividing x by $m+1$.

If $G(C) > 0$, the next player, say A, can make sure of winning by moving to a "safe" combination S with $G(S) = 0$. For, by the definition of $G(S)$, either S is the empty position, in which case A has already won, or the opponent B must move to a new "unsafe" position U with $G(U) > 0$, and the argument repeats. The game must end in a finite number of moves, with A winning.

In the game of Nim* there can be any number of heaps, and each player in turn chooses just one heap and removes from it as many counters as he pleases (but at least one). $G(x, y, \ldots)$ is then the "nim-sum" of $x, y, \ldots$, where the operation of nim-addition is defined as follows. Write $x, y, \ldots$ in the scale of 2. Then add without carrying. Finally, replace the digits in this sum by their remainders on dividing by 2, interpreting the result in the scale of 2. Thus to find the nim-sum $3 +_{nim} 7 +_{nim} 9$ we write

denary 3 = binary	11
„ 7 = „	111
„ 9 = „	1001

sum without carry	1123
nim-sum	1101 = denary 13 = $G(3, 7, 9)$

The proof that this in fact satisfies the conditions for the Sprague-Grundy number amounts effectively to showing that if $G(C) > h \geq 0$, then there exists a move from C to some combination D with $G(D) = h$. For instance, in the above example take $h =$ denary 11 = binary 1011. This differs from $G(C) = 1101$ in the middle two digits. By changing the second heap from 7 = 0111 to 1 = 0001 the required change in the nim-sum is effected, so that $D = (3, 1, 9)$ has $G(D) = 11$. Such a change is always possible. The safe combinations are those with zero nim-sum; examples are (x, x) with two heaps, and $(1, 2, 3)$,

*C.L. Bouton, *Annals of Mathematics*, 1902, series 2, vol. III, pp. 35–39.

(1, 4, 5), (1, 6, 7), (2, 4, 6), (2, 5, 7), and (3, 4, 7) with three heaps. In these triples any one number is the nim-sum of the other two; $1 +_{nim} 2 = 3$, $1 +_{nim} 3 = 2$, etc. Hence we deduce, e.g., that $(1 +_{nim} 3) +_{nim} (7 +_{nim} 5) = 2 +_{nim} 2 = 0$, so that (1, 3, 7, 5) is a safe combination in Nim.

In "restricted Nim" a player moves by taking up to m counters from one heap. $G(x, y, \ldots)$ is then the nim-sum of $x', y', \ldots$, the respective remainders on dividing $x, y, \ldots$ by $m + 1$. More generally, if Nim is modified by the restriction that the number of counters taken away must be one of a certain "permitted set" of positive integers, then $G(x, y, \ldots)$ is the nim-sum of $G(x), G(y), \ldots$, where $G(x)$ is the Sprague-Grundy number for a single pile of x counters. The safe combinations still have $G(x, y, \ldots) = 0$. Thus, if the number taken must be a perfect square, we find that $G(x) = 0, 1, 0, 1, 2$ according as x leaves remainder 0, 1, 2, 3, or 4 on division by 5; and from these we can deduce the safe combinations in a game with more than one heap.

Moore's game.*   The rules for E.H. Moore's Nim_k are the same as for ordinary Nim (which is Nim_1), except that a player takes at will from any number of heaps not exceeding k. A combination $(x, y, \ldots)$ is safe if, when the numbers $x, y, \ldots$ are written in the scale of 2 and added without carrying, all the digits of the sum are multiples of $k + 1$. If the rules of Nim_k are modified so that the player taking the last counter loses, the safe combinations are the same, except that when all the piles contain only single counters the number of piles must be one more than a multiple of $k + 1$.

If Nim_k is further modified by restricting the number of counters taken from each heap to one of a "permitted set," a general analysis is difficult. But if there are only $k + 1$ heaps in all, with numbers $(x_1, x_2, \ldots, x_{k+1})$ of counters, and the last player wins, a safe position obeys the rule $G(x_1) = G(x_2) = \ldots = G(x_{k+1})$. If there are fewer than $k + 1$ heaps, all the $G(x_r)$ must be zero.

*_Annals of Mathematics_, 1910, series 2, vol. XI, pp. 90–94.

Kayles.* This is a game played with rows of counters. A move consists in removing any one counter, or any pair of adjacent counters. This may split a row into two smaller rows. The player taking the last counters wins. The rule that $G(x, y, \ldots)$ is the nim-sum of $G(x)$, $G(y)$, ... still holds. The values of $G(x)$ for $x = 0, 1, 2, \ldots$ are respectively 0, 1, 2, 3, 1, 4, 3, 2, 1, 4, 2, 6, 4, 1, ... R. K. Guy has shown that from $x = 71$ onwards this series is periodic, with period 12. In "duplicate Kayles" one may take any one pair or triplet of adjacent counters. The $G(x)$ series then becomes 0, 0, 1, 1, 2, 2, 3, 3, 1, 1, ..., the same as the series for simple Kayles except that each value is repeated. If the rule is to take either a pair of adjacent counters or any single counter standing on its own, the $G(x)$ series for $x = 0, 1, 2, \ldots$ is 0, 1, 1, 0, 2, 1, 3, 0, 1, 1, 3, 2, 2, 3, ... ; from $x = 33$ onwards this is periodic with period 34.

Wythoff's game.   W.A. Wythoff† invented a game with two heaps of counters. A player can take counters at will from one heap or else an equal number from both. The player taking the last counter wins. The safe combinations are (1, 2), (3, 5), (4, 7), (6, 10), (8, 13), (9, 15), ... The rth safe combination is $(x, x+r)$, where x is the integral part of $\tau r = \frac{1}{2}(\sqrt{5}+1)r$ and hence $x+r$ is the integral part of $\tau^2 r$. Every positive integer occurs once in these safe combinations, so that between any two adjacent positive integers there must lie either one multiple of τ or one multiple of τ^2. T.H. O'Beirne has observed that if the player taking the last counter loses, the safe combinations are the same except that (1, 2) must be replaced by (0, 1) and (2, 2).‡

The number $\tau = \frac{1}{2}(\sqrt{5}+1) = 1\cdot6180339887\ldots$ has fascinated professionals and amateurs ever since the Pythagoreans began to study the pentagon and considered the ratio of a diagonal to a side. We shall meet it again on pages 56, 57, 132, and 161. One

---

*S. Loyd, *Cyclopedia of Tricks and Puzzles*, New York, 1914, p. 232; R.K. Guy and C.A.B. Smith, *Proceedings of the Cambridge Philosophical Society*, 1956, vol. LII, p. 514.

†*Nieuw Archief voor Wiskunde*, 1907, p. 199. See also H.S.M. Coxeter, *Scripta Mathematica*, 1953, vol. XIX, pp. 135–143.

‡T.H. O'Beirne, *Puzzles and Paradoxes*, Oxford, 1965, pp. 109, 134–138.

of its more sophisticated properties belongs to the theory of rational approximations to a given irrational number ζ. A theorem of Adolf Hurwitz* states that, if $0 < c \leq \sqrt{5}$, there are infinitely many rational numbers h/k such that

$$|\zeta - h/k| < 1/ck^2 ;$$

but if $c > \sqrt{5}$, there are some irrational numbers, one of which is τ, such that this approximation holds only for finitely many rational numbers h/k.

ADDENDUM

Note. Page 15. Solutions of the ten-digit problems are $35/70 + 148/296 = 1$ or $\cdot01234 + \cdot98765 = 1$, and $50 + 49 + 1/2 + 38/76 = 100$. A solution of the nine-digit problem is $1\cdot234 + 98\cdot765 = 100$ or $97 + 8/12 + 4/6 + 5/3 = 100$; but a neater solution (due to Perelman†) is $1 + 2 + 3 + 4 + 5 + 6 + 7 + 8 \times 9 = 100$, where the digits occur in their natural order.

Note. Page 28. There are several solutions of the division of 24 ounces under the conditions specified. The neatest is due to E.A. Harber:

	24 oz.	13 oz.	11 oz.	5 oz.
The vessels can contain	24 oz.	13 oz.	11 oz.	5 oz.
Their contents originally are	24	0	0	0
First, make their contents	8	0	11	5
Second, make their contents	8	5	11	0
Third, make their contents	8	13	3	0
Fourth, make their contents	8	8	3	5

*K. Chandrasekharan, *An Introduction to Analytic Number Theory*, Berlin, 1968, p. 23.

†For several hundred other solutions, see *Sphinx*, 1935, pp. 95, 111, 112, 124, 125.

CHAPTER II

ARITHMETICAL RECREATIONS (*continued*)

I devote this chapter to the description of some arithmetical fallacies, a few additional problems, and notes on one or two problems in higher arithmetic.

ARITHMETICAL FALLACIES

I begin by mentioning some instances of demonstrations* leading to arithmetical results which are obviously impossible. I include algebraical proofs as well as arithmetical ones. Some of the fallacies are so patent that in preparing the first and second editions I did not think such questions worth printing, but, as some correspondents expressed a contrary opinion, I give them for what they are worth.

First fallacy. One of the oldest of these, and not a very interesting specimen, is as follows. Suppose that $a = b$, then

$$ab = a^2. \quad \therefore ab - b^2 = a^2 - b^2. \quad \therefore b(a-b) = (a+b)(a-b).$$
$$\therefore b = a + b. \quad \therefore b = 2b. \quad \therefore 1 = 2.$$

Second fallacy. Another example, the idea of which is due to John Bernoulli, may be stated as follows. We have $(-1)^2 = 1$.

*Of the fallacies given in the text, the first and second are well known; the third is not new, but the earliest work in which I recollect seeing it is my *Algebra*, Cambridge, 1890, p. 430; the fourth is given in G. Chrystal's *Algebra*, Edinburgh, 1889, vol. II, p. 159; the sixth is due to G.T. Walker, and, I believe, has not appeared elsewhere than in this book; the seventh is due to D'Alembert; and the eighth to F. Galton. It may be worth recording (i) that a mechanical demonstration that $1 = 2$ was given by R. Chartres in *Knowledge*, July 1891; and (ii) that J.L.F. Bertrand pointed out that a demonstration that $1 = -1$ can be obtained from the proposition in the integral calculus that, if the limits are constant, the order of integration is indifferent; hence the integral to x (from $x = 0$ to $x = 1$) of the integral to y (from $y = 0$ to $y = 1$) of a function ϕ should be equal to the integral to y (from $y = 0$ to $y = 1$) of the integral to x (from $x = 0$ to $x = 1$) of ϕ, but if $\phi = (x^2 - y^2)/(x^2 + y^2)^2$, this gives $\frac{1}{4}\pi = -\frac{1}{4}\pi$.

Take logarithms, $\therefore 2 \log(-1) = \log 1 = 0$. $\therefore \log(-1) = 0$. $\therefore -1 = e^0$. $\therefore -1 = 1$.

The same argument may be expressed thus. Let x be a quantity which satisfies the equation $e^x = -1$. Square both sides,

$$\therefore e^{2x} = 1. \quad \therefore 2x = 0. \quad \therefore x = 0. \quad \therefore e^x = e^0.$$

But $e^x = -1$ and $e^0 = 1$, $\therefore -1 = 1$.

The error in each of the foregoing examples is obvious, but the fallacies in the next examples are concealed somewhat better.

Third fallacy. As yet another instance, we know that

$$\log(1 + x) = x - \tfrac{1}{2}x^2 + \tfrac{1}{3}x^3 - \dots.$$

If $x = 1$, the resulting series is convergent; hence we have

$$\log 2 = 1 - \tfrac{1}{2} + \tfrac{1}{3} - \tfrac{1}{4} + \tfrac{1}{5} - \tfrac{1}{6} + \tfrac{1}{7} - \tfrac{1}{8} + \tfrac{1}{9} - \dots.$$
$$\therefore 2 \log 2 = 2 - 1 + \tfrac{2}{3} - \tfrac{1}{2} + \tfrac{2}{5} - \tfrac{1}{3} + \tfrac{2}{7} - \tfrac{1}{4} + \tfrac{2}{9} - \dots.$$

Taking those terms together which have a common denominator, we obtain

$$2 \log 2 = 1 + \tfrac{1}{3} - \tfrac{1}{2} + \tfrac{1}{5} + \tfrac{1}{7} - \tfrac{1}{4} + \tfrac{1}{9} + \dots$$
$$= 1 - \tfrac{1}{2} + \tfrac{1}{3} - \tfrac{1}{4} + \tfrac{1}{5} - \dots$$
$$= \log 2.$$

Hence $2 = 1$.

Fourth fallacy. This fallacy is very similar to that last given. We have

$$\log 2 = 1 - \tfrac{1}{2} + \tfrac{1}{3} - \tfrac{1}{4} + \tfrac{1}{5} - \tfrac{1}{6} + \dots$$
$$= (1 + \tfrac{1}{3} + \tfrac{1}{5} + \dots) - (\tfrac{1}{2} + \tfrac{1}{4} + \tfrac{1}{6} + \dots)$$
$$= \{(1 + \tfrac{1}{3} + \tfrac{1}{5} + \dots) + (\tfrac{1}{2} + \tfrac{1}{4} + \tfrac{1}{6} + \dots)\} - 2(\tfrac{1}{2} + \tfrac{1}{4} + \tfrac{1}{6} + \dots)$$
$$= \{1 + \tfrac{1}{2} + \tfrac{1}{3} + \dots\} - (1 + \tfrac{1}{2} + \tfrac{1}{3} + \dots)$$
$$= 0.$$

Fifth fallacy. We have

$$\sqrt{a} \times \sqrt{b} = \sqrt{ab}.$$

Hence $$\sqrt{-1} \times \sqrt{-1} = \sqrt{(-1)(-1)};$$

therefore, $(\sqrt{-1})^2 = \sqrt{1}$, that is, $-1 = 1$.

Sixth fallacy. The following demonstration depends on the fact that an algebraical identity is true whatever be the symbols used in it, and it will appeal only to those who are familiar with this fact. We have, as an identity,

$$\sqrt{x-y} = i\sqrt{y-x}, \tag{i}$$

where i stands either for $+\sqrt{-1}$ or for $-\sqrt{-1}$. Now an identity in x and y is necessarily true whatever numbers x and y may represent. First put $x = a$ and $y = b$,

$$\therefore \sqrt{a-b} = i\sqrt{b-a}. \tag{ii}$$

Next put $x = b$ and $y = a$,

$$\therefore \sqrt{b-a} = i\sqrt{a-b}. \tag{iii}$$

Also since (i) is an identity, it follows that in (ii) and (iii) the symbol i must be the same, that is, it represents $+\sqrt{-1}$ or $-\sqrt{-1}$ in both cases. Hence, from (ii) and (iii), we have

$$\sqrt{a-b}\sqrt{b-a} = i^2\sqrt{b-a}\sqrt{a-b},$$
$$\therefore 1 = i^2,$$

that is, $$1 = -1.$$

Seventh fallacy. The following fallacy is due to D'Alembert.* We know that if the product of two numbers is equal to the product of two other numbers, the numbers will be in proportion, and from the definition of a proportion it follows that if the first term is greater than the second, then the third term will be greater than the fourth: thus, if $ad = bc$, then $a : b = c : d$, and if in this proportion $a > b$, then $c > d$. Now if we put $a = d = 1$ and $b = c = -1$ we have four numbers which satisfy the relation $ad = bc$ and such that $a > b$; hence, by the proposition, $c > d$, that is, $-1 > 1$, which is absurd.

*Opuscules Mathématiques, Paris, 1761, vol. I, p. 201.

Eighth fallacy. The mathematical theory of probability leads to various paradoxes: of these I will give a few specimens. Suppose* three coins to be thrown up and the fact whether each comes down head or tail to be noticed. The probability that all three coins come down head is clearly $(1/2)^3$, that is, is $1/8$; similarly the probability that all three come down tail is $1/8$; hence the probability that all the coins come down alike (i.e. either all of them heads or all of them tails) is $1/4$. But, of three coins thus thrown up, at least two must come down alike: now the probability that the third coin comes down head is $1/2$ and the probability that it comes down tail is $1/2$, thus the probability that it comes down the same as the other two coins is $1/2$: hence the probability that all the coins come down alike is $1/2$. I leave to my readers to say whether either of these conflicting conclusions is right, and, if so, which is correct.

The paradox of the second ace. Suppose that a player at bridge or whist asserts that an ace is included among the thirteen cards dealt to him, and let p be the probability that he has another ace among the other cards in his hand. Suppose, however, that he asserts that the ace of hearts is included in the thirteen cards dealt to him; then the probability, q, that he has another ace among the other cards in his hand is greater than was the probability p in the first case. For,† the number of hands containing an ace is $\binom{52}{13}-\binom{48}{13}$, whereas the number of hands containing two or more aces is $\binom{52}{13}-\binom{48}{13}-4\binom{48}{12}$; therefore $p = 1 - 4\binom{48}{12}/\{\binom{52}{13}-\binom{48}{13}\} = \frac{5359}{14498}$. But the number of hands containing the ace of hearts is $\binom{51}{12}$, whereas the number of hands containing this and another ace is $\binom{51}{12}-\binom{48}{12}$; therefore $q = 1 - \binom{48}{12}/\binom{51}{12} = \frac{11686}{20825}$. Thus $p < \frac{1}{2} < q$, which at first sight appears to be absurd.

The St. Petersburg paradox. ‡ The following example is a

*See *Nature*, Feb. 15, March 1, 1894, vol. XLIX, pp. 365–366, 413.

† This argument is due to D.B. De Lury.

‡ Cf. E. Kamke, *Einführung in die Wahrscheinlichkeitstheorie*, Leipzig, 1932, pp. 82–89.

famous one. A penny is tossed until it comes down head. If this happens at the first throw, the bank pays the player £1. Otherwise, the player throws again. If head appears at the second throw, the bank pays £2; if at the third throw, £4; and so on, doubling every time. Thus, if the coin does not come down head till the nth throw, the player then receives £2^{n-1}. What should the player pay the bank for the privilege of playing this game?

There is a probability $1/2$ that the player will receive the £1, a probability $1/4$ that he will receive the £2, and so on. Hence the total number of pounds that he may reasonably expect to receive is

$$\tfrac{1}{2}.1 + \tfrac{1}{4}.2 + \ldots + (\tfrac{1}{2})^n 2^{n-1} + \ldots = \tfrac{1}{2} + \tfrac{1}{2} + \tfrac{1}{2} + \ldots,$$

i.e. it is infinite.

Among various ways of modifying the problem to make the answer finite, one of the most satisfactory was given by Gabriel Cramer (about 1730) in a letter to Nicolas Bernoulli.* Cramer assumed that the bank's wealth is limited, say to £2^{24}. There is then a probability $1/2^n$ that the player will receive £2^{n-1} at the nth throw only as long as $n < 25$; thereafter he will receive merely £2^{24}. Since

$$\sum_{1}^{24} \frac{2^{n-1}}{2^n} + \sum_{25}^{\infty} \frac{2^{24}}{2^n} = 12 + 1 = 13,$$

the player's expectation is £13, a reasonable amount.

OTHER QUESTIONS ON PROBABILITY

Here is a result (due to Harold Davenport) which many people find surprising. If you know more than 23 people's birthdays, it is more likely than not that two of them are the same (as to day and month). Consider the probability that n people's birthdays are all different, i.e. that in a random selection of n days out of 365 there shall be no day counted more than once.

*Isaac Todhunter, *A History of the Mathematical Theory of Probability*, London, 1865, p. 221 (art. 391).

The total number of possible selections is 365^n, and the number of selections in which no day is counted more than once is $365.364 \ldots (365-n+1)$. The probability is therefore* $365.364 \ldots (365-n+1)/365^n$. The occurrence is as likely as not if this expression is equal to $1/2$, i.e. if

$$\left(1-\frac{1}{365}\right)\left(1-\frac{2}{365}\right)\cdots\left(1-\frac{n-1}{365}\right)=\frac{1}{2}.$$

By taking logarithms, we obtain approximately

$$\frac{1}{365}+\frac{2}{365}+\cdots+\frac{n-1}{365}=\log_e 2,$$

or $n(n-1)=506$, whence $n=23$.

Derangements. The following problem† is somewhat similar. Suppose you have written a letter to each of n different friends, and addressed the n corresponding envelopes. In how many ways can you make the regrettable mistake of putting *every* letter into a wrong envelope? Let X_n denote the number of ways. Suppose the first letter is placed in the ath envelope, and the bth letter in the first envelope. If $a=b$, there remain $n-2$ letters to be placed in wrong envelopes, which can be done in X_{n-2} ways. Since $a(=b)$ may take any value from 2 to n, this possibility covers $(n-1)X_{n-2}$ cases. In every other case, $a\neq b$. For the moment, let us fix b (and let a run over all values from 2 to n, except b). In the $n-1$ envelopes other than the first, we have to place the $n-1$ letters other than the bth, but the first letter must not be placed in the bth envelope. The number of ways in which this can be done is just X_{n-1}, since the situation is equivalent to that of the original problem with $n-1$ letters, if we imagine the bth envelope to be the proper

---

*As a check on the working, note that this correctly gives 0 when $n=366$. (For simplicity, I have ignored the possibility of a birthday falling on Feb. 29; that does not affect the result, save in this extreme case.)

†Cf. P.R. de Montmort, *Essai d'analyse sur les jeux de hasard*, Paris, 1713, p. 132; J.L. Coolidge, *An Introduction to Mathematical Probability*, Oxford, 1925, p. 24; C.V. Durell and Alan Robson, *Advanced Algebra*, London, 1937, p. 459. See also A.C. Aitken, *Determinants and Matrices*, Edinburgh, 1956, p. 135.

one for the first letter. Letting b vary, this accounts for $(n-1)X_{n-1}$ cases. Hence $X_n=(n-1)(X_{n-1}+X_{n-2})$. This relation* gives the successive numbers readily. Obviously $X_1=0$ and $X_2=1$; therefore $X_3=2$, $X_4=9$, $X_5=44$, $X_6=265$, and so on. The explicit formula is

$$X_n=n!\left(1-\frac{1}{1!}+\frac{1}{2!}-\frac{1}{3!}+\ldots\pm\frac{1}{n!}\right)$$

("sub-factorial n").

Since n given letters can be placed in n given envelopes in $n!$ ways altogether, the probability of this unhappy occurrence is

$$\frac{X_n}{n!}=1-\frac{1}{1!}+\frac{1}{2!}-\frac{1}{3!}+\ldots\pm\frac{1}{n!}.$$

Now this is the beginning of the series for $e^{-1}(=0{\cdot}367879\ldots)$, so we may say that the probability is approximately $1/e$. The error is less than $1/(n+1)!$, which, when $n=6$, is about $0{\cdot}0002$.

If we compare two packs of cards (one of them having been well shuffled), card by card, what is the probability that we shall get right through the packs without finding a single coincidence? This is merely another form of the same problem; the answer is $1/e$ (with an error of less than 10^{-69}, for packs of 52 cards). Many people are prepared to bet that no coincidence will occur, so an unscrupulous gambler might profit by knowing that $e>2$.

Miscellaneous problems.   To the above examples I may add the following standard questions, or recreations.

The first of these questions is as follows. Two clerks, A and B, are engaged, A at a salary commencing at the rate of (say) £100 a year with a rise of £20 every year, B at a salary commencing at the same rate of £100 a year with a rise of £5 every half-year, in each case payments being made half-yearly; which has the larger income? The answer is B; for in the first year A receives £100, but B receives £50 and £55 as his two half-yearly payments and thus receives in all £105. In the second year A receives £120, but B receives £60 and £65 as his two half-yearly payments and thus receives in all £125.

*A still simpler relation is $X_n=nX_{n-1}+(-1)^n$ $(X_1=0)$.

In fact B will always receive £5 a year more than A.

Another simple arithmetical problem is as follows. A hymn-board in a church has four grooved rows on which the numbers of four hymns chosen for the service are placed. The hymn-book in use contains 700 hymns. What is the smallest number of plates, each carrying one digit, which must be kept stock so that the numbers of any four different hymns selected can be displayed; and how will the result be affected if an inverted 6 can be used for a 9? The answers are 86 and 81. What are the answers if a digit is painted on each side of each plate?

As another question take the following. A man bets $1/n$th of his money on an even chance (say tossing heads or tails with a penny): he repeats this again and again, each time betting $1/n$th of all the money then in his possession. If, finally, the number of times he has won is equal to the number of times he has lost, has he gained or lost by the transaction? He has, in fact, lost.

Here is another simple question to which I have frequently received incorrect answers. One tumbler is half-full of wine, another is half-full of water: from the first tumbler a teaspoonful of wine is taken out and poured into the tumbler containing the water: a teaspoonful of the mixture in the second tumbler is then transferred to the first tumbler. As the result of this double transaction, is the quantity of wine removed from the first tumbler greater or less than the quantity of water removed from the second tumbler? In any experience the majority of people will say it is greater, but this is not the case.

PERMUTATION PROBLEMS

Many of the problems in permutations and combinations are of considerable interest. As a simple illustration of the very large number of ways in which combinations of even a few things can be arranged, I may note that 12 differently coloured rods of equal length can form a skeleton cube in as many as 19,958400 ways;* while there are no less than $(52!)/(13!)^4$,

*Mathematical Tripos, Cambridge, Part I, 1894.

that is, 53644,737765,488792,839237,440000 possible different distributions of hands at bridge with a pack of fifty-two cards.

Voting problems. Here are two simple examples. (i) If there are two candidates, and a votes are cast for one, b for the other $(a>b)$, the probability that the number of votes for the former remains greater than that for the latter throughout the counting is $(a-b)/(a+b)$. (ii) If there are p electors each having r votes of which not more than s may be given to one candidate, and n men are to be elected, then the least number of supporters who can secure the election of a candidate must exceed $pr/(ns+r)$.

The Knights of the Round Table. A far more difficult permutation problem consists in finding as many arrangements as possible of n people in a ring so that no one has the same two neighbours more than once. It is a well-known proposition that n persons can be arranged in a ring in $(n-1)!/2$ different ways. The number of these arrangements in which all the persons have different pairs of neighbours on each occasion cannot exceed $(n-1)(n-2)/2$, since this gives the number of ways in which any assigned person may sit between every possible pair selected from the rest. But in fact it is always possible to determine $(n-1)(n-2)/2$ arrangements in which no one has the same two neighbours on any two occasions.

Solutions for various values of n have been given. Here, for instance $(n=8)$, are 21 arrangements* of eight persons. Each arrangement may be placed round a circle and no one has the same two neighbours on any two occasions.

```
1.2.3.4.5.6.7.8;   1.2.5.6.8.7.4.3;   1.2.7.8.4.3.5.6;
1.3.5.2.7.4.8.6;   1.3.7.4.6.8.2.5;   1.3.8.6.2.5.7.4;
1.4.2.6.3.8.5.7;   1.4.3.8,7.5.6.2;   1.4.5.7.6.2.3.8;
1.5.6.4.3.7.8.2;   1.5.7.3.8.2.6.4;   1.5.8.2.4.6.3.7;
1.6.2.7.5.3.8.4;   1.6.3.5.8.4.2.7;   1.6.4.8.2.7.3.5;
1.7.4.2.5.8.6.3;   1.7.6.3.2.4.5.8;   1.7.8.5.6.3.4.2;
1.8.2.3.7.6.4.5;   1.8.4.5.3.2.7.6;   1.8.6.7.4.5.2.3;
```

*Communicated to me by Mr. E.G.B. Bergholt, May 1906; see *The Secretary* and *The Queen*, August 1906. Mr. Dudeney had given the problem for the case when $n=6$ in 1905, and informs me that the problem has been solved by Mr. E.D. Bewley when n is even, and that he has a general method applicable when n is odd. Various memoirs on the subject have appeared in the mathematical journals.

The methods of determining these arrangements are lengthy, and far from easy.

The ménage problem.* Another difficult permutation problem is concerned with the number x of possible arrangements of n married couples, seated alternately man and woman, round a table, the n wives being in assigned positions, and the n husbands so placed that a man does not sit next to his wife.

The solution involves the theory of discordant permutations,† and is far from easy. I content myself with noting the results when n does not exceed 10. When $n=3$, $x=1$; when $n=4$, $x=2$; when $n=5$, $x=13$; when $n=6$, $x=80$; when $n=7$, $x=579$; when $n=8$, $x=4738$; when $n=9$, $x=43387$; and when $n=10$, $x=439792$.

BACHET'S WEIGHTS PROBLEM ‡

It will be noticed that a considerable number of the easier problems given in the last chapter either are due to Bachet or were collected by him in his classical *Problèmes*. Among the more difficult problems proposed by him was the determination of the least number of weights which would serve to weigh any integral number of pounds from 1 lb. to 40 lb. inclusive. Bachet gave two solutions: namely, (i) the series of weights of 1, 2, 4, 8, 16, and 32 lb.; (ii) the series of weights of 1, 3, 9, and 27 lb.

If the weights may be placed in only one of the scale-pans, the first series gives a solution, as had been pointed out in 1556 by Tartaglia.§

Bachet, however, assumed that any weight might be placed in either of the scale-pans. In this case the second series gives the least possible number of weights required. His reasoning

---

**Théorie des Nombres*, by E. Lucas, Paris, 1891, pp. 215, 491–495.

†See P.A. MacMahon, *Combinatory Analysis*, vol. I, Cambridge, 1915, pp. 253–256; P. Halmos and H.E. Vaughan, *American Journal of Mathematics*, 1950, vol. LXXII, pp. 214–215; D.J. Newman, *American Mathematical Monthly*, 1958, vol. LXV, p. 611.

‡Bachet, Appendix, problem v, p. 215.

§*Trattato de' numeri e misure*, Venice, 1556, vol. II, bk. I, chap. XVI, art. 32.

is as follows. To weigh 1 lb. we must have a 1-lb. weight. To weigh 2 lb. we must have in addition either a 2-lb. weight or a 3-lb. weight; but, whereas with a 2-lb. weight we can weigh 1 lb., 2 lb., and 3 lb., with a 3-lb. weight we can weigh 1 lb., $(3-1)$ lb., 3 lb., and $(3+1)$ lb. Another weight of 9 lb. will enable us to weigh all weights from 1 lb. to 13 lb.; and we get thus a greater range than is obtainable with any weight less than 9 lb. Similarly weights of 1, 3, 9, and 27 lb. suffice for all weights up to 40 lb., and weights of $1, 3, 3^2, \ldots, 3^{n-1}$ lb. enable us to weigh any integral number of pounds from 1 lb. to $(1+3+3^2+\ldots+3^{n-1})$ lb., that is, to $\frac{1}{2}(3^n-1)$ lb.

To determine the arrangement of the weights to weigh any given mass we have only to express the number of pounds in it as a number in the ternary scale of notation, except that in finding the successive digits we must make every remainder either 0, 1 or -1: to effect this a remainder 2 must be written as $3-1$ – that is, the quotient must be increased by unity, in which case the remainder is -1. This is explained in most text-books on algebra.

Bachet's argument does not prove that his result is unique or that it gives the least possible number of weights required. These omissions have been supplied by Major MacMahon, who has discussed the far more difficult problem (of which Bachet's is a particular case) of the determination of all possible sets of weights, not necessarily unequal, which enable us to weigh any integral number of pounds from 1 to n inclusive, (i) when the weights may be placed in only one scale-pan, and (ii) when any weight may be placed in either scale-pan. He has investigated also the modifications of the results which are necessary when we impose either or both of the further conditions (a) that no other weighings are to be possible, and (b) that each weighing is to be possible in only one way, that is, is to be unique.*

The method for case (i) consists in resolving $1 + x + x^2 + \ldots$

*See his article in the *Quarterly Journal of Mathematics*, 1886, vol. XXI, pp. 367–373. An account of the method is given in *Nature*, Dec, 4, 1890, vol. XLII, pp. 113–114.

$+x^n$ into factors, each factor being of the form $1+x^a+x^{2a}$ $+\ldots+x^{ma}$; the number of solutions depends on the composite character of $n+1$. The method for case (ii) consists in resolving the expression $x^{-n}+x^{-n+1}+\ldots+x^{-1}+1+x+\ldots$ $+x^{n-1}+x^n$ into factors, each factor being of the form x^{-ma} $+\ldots+x^{-a}+1+x^a+\ldots+x^{ma}$; the number of solutions depends on the composite character of $2n+1$.

Bachet's problem falls under case (ii), $n=40$. MacMahon's analysis shows that there are eight such ways of factorizing $x^{-40}+x^{-39}+\ldots+1+\ldots+x^{39}+x^{40}$. First, there is the expression itself in which $a=1$, $m=40$. Second, the expression is equal to $(1-x^{81})/x^{40}(1-x)$, which can be resolved into the product of $(1-x^3)/x(1-x)$ and $(1-x^{81})/x^{39}(1-x^3)$; hence it can be resolved into two factors of the form given above, in one of which $a=1$, $m=1$, and in the other $a=3$, $m=13$. Third, similarly, it can be resolved into two such factors, in one of which $a=1$, $m=4$, and in the other $a=9$, $m=4$. Fourth, it can be resolved into three such factors, in one of which $a=1$, $m=1$, in another $a=3$, $m=1$, and in the other $a=9$, $m=4$. Fifth, it can be resolved into two such factors, in one of which $a=1$, $m=13$, and in the other $a=27$, $m=1$. Sixth, it can be resolved into three such factors, in one of which $a=1$, $m=1$, in another $a=3$, $m=4$, and in the other $a=27$, $m=1$. Seventh, it can be resolved into three such factors, in one of which $a=1$, $m=4$, in another $a=9$, $m=1$, and in the other $a=27$, $m=1$. Eighth, it can be resolved into four such factors, in one of which $a=1$, $m=1$, in another $a=3$, $m=1$, in another $a=9$, $m=1$, and in the other $a=27$, $m=1$.

These results show that there are eight possible sets of weights with which any integral number of pounds from 1 to 40 can be weighed subject to the conditions (ii), (a), and (b). If we denote p weights each equal to w by w^p, these eight solutions are 1^{40}; $1, 3^{13}$; $1^4, 9^4$; $1, 3, 9^4$; $1^{13}, 27$; $1, 3^4, 27$; $1^4, 9, 27$; $1, 3, 9, 27$. The last of these is Bachet's solution: not only is it that in which the least number of weights are employed, but it is also the only one in which all the weights are unequal.

THE DECIMAL EXPRESSION FOR $1/n$

J.C.P. Miller, when he was at school (about 1920), invented a remarkable technique for expressing certain fractions $1/n$ as recurring decimals. To avoid unpleasant complications, we assume that n is not divisible by either 2 or 5. The first few digits have to be calculated by division in the ordinary way. Then we add 1 and divide by 2 so as to obtain the beginning of an expression for m/n, where $m = \frac{1}{2}(n+1)$. If m is even, we divide by 2 again; if m is odd, we add 1 and then divide by 2. We continue thus until we recognize, in some multiple of $1/n$, a sequence of several digits appearing in an earlier position than the same sequence did in $1/n$ itself. An almost embarrassingly simple instance occurs when $n = 19$:

$$\frac{20}{19} = 1 \cdot 0\dot{5}2631578947368421\dot{1},$$

$$\frac{10}{19} = \cdot \dot{5}26315789473684210\dot{0}.$$

In this case we merely have to transfer each successive digit from the second line to the first, and then continue to divide by 2.

When $n = 17$ or 47 we find it necessary to divide by 2 three times before recognizing a sequence of three digits (printed in dark type):

$$\frac{18}{17} = 1 \cdot 05\mathbf{882}\ldots, \qquad \frac{48}{47} = 1 \cdot 02\mathbf{127}\ldots,$$

$$\frac{26}{17} = 1 \cdot 52941\ldots, \qquad \frac{24}{47} = \cdot 51063\ldots,$$

$$\frac{30}{17} = 1 \cdot 76470\ldots, \qquad \frac{12}{47} = \cdot 25531\ldots,$$

$$\frac{15}{17} = \cdot \mathbf{882}35\ldots, \qquad \frac{6}{47} = \cdot \mathbf{12765}\ldots.$$

In the latter case, instead of halving three times we can more rapidly divide the first line by 8 and then transfer the digits (one by one, or two by two) from the fourth line to the first, without troubling to complete the second and third lines.

When $n = 81$, we halve $82/81 = 1 \cdot 01\mathbf{234}$ ... six times (adding 1 whenever the numerator would otherwise be odd) and then we recognize the digits 234. Instead of continuing to divide by 2 six times, we can conveniently divide by 8 twice, thus developing only the first, fourth, and seventh lines, as follows:

$$\frac{568}{81} = 7 \cdot \mathbf{0123}456790\ldots,$$

$$\frac{152}{81} = 1 \cdot 8765432098\ldots,$$

$$\frac{19}{81} = \quad \cdot 2345679\mathbf{012}\ldots.$$

We conclude that $1/81 = \cdot \dot{0}1234567\dot{9}$.

If we recognize a sequence of several digits appearing in a *later* position, we can modify the procedure by transferring digits from the first row to a subsequent row and then multiplying the latter by 2 (or dividing it by 5) to continue the previous row. For instance, when $n = 49$ we find

$$\frac{50}{49} = 1 \cdot \dot{0}204081632653061224489795918367346938775 5\dot{1},$$

$$\frac{25}{49} = \quad \cdot \dot{5}1020408163265306122448979591836734693877 \dot{5}.$$

DECIMALS AND CONTINUED FRACTIONS

Every positive number can be expressed in just one way as a decimal. If the number is rational (i.e., expressible as a vulgar fraction), its decimal expression will either terminate or (after a certain stage) recur. Otherwise, as in the case of $\sqrt{2} = 1 \cdot 41421356\ldots$, or of $\pi = 3 \cdot 14159265\ldots$, it will neither terminate nor recur. If another scale of notation is used, the sequence of digits will, of course, be quite different. The "decimal" for a rational number may terminate in one scale and recur in another; thus, in the scale of seven, $\frac{1}{7} = 0 \cdot 1$, instead of the familiar $0 \cdot 14285\dot{7}$. Given an integer, expressed in the ordinary (denary) scale, we can express it in the scale of p by dividing by p again and again, noting the successive remainders. These remainders, when read backwards, are the digits of the required new expression. The rule for a fraction is different. There we *multiply* by p again and again, operating only on the fractional part at each stage. The successive integral parts are the digits of the required new expression (in their proper order). For instance, in the binary scale,

$$\sqrt{2}(=\sqrt{10}) = 1 \cdot 0110101000001100111100\ldots,$$
$$\pi = 11 \cdot 001001000011111101101\ldots.$$

The actual computation proceeds thus:

1·41421356	3·14159265
0·82842712	0·28318530
1·65685424	0·56637060
1·3137085	1·1327412
0·6274170	0·2654824
1·2548340	0·5309648
0·509668	1·061930
etc.	etc.

The sequence of five zeros makes $1 \cdot 0110101$ a particularly good approximation for $\sqrt{2}$. In fact, working in the binary scale, we easily find that $(1 \cdot 0110101)^2 = 1 \cdot 11111111111001$. As a good approximation for π, we note that $11 \cdot 00\bar{1} = 3\frac{1}{7}$.

Somewhat analogously, every positive number can be expressed in just one way as a *regular continued fraction*

$$a_0 + \cfrac{1}{a_1 + \cfrac{1}{a_2 + \cfrac{1}{a_3 + \ldots}}} = a_0 + 1/a_1 + 1/a_2 + 1/a_3 + \ldots$$

(by convention, each solidus / covers everything that follows), where the a's are positive integers, save that a_0 may be zero. Thus*

---

*Although the "partial quotients" $(a_1, a_2, \ldots)$ for $\sqrt{2}$ and e obey simple laws, no law has yet been found for π. The largest known partial quotient is $a_{156381} = 179136$; see the review of R. W. Gosper's *Table of the Simple Continued Fraction for π . . . in Mathematics of Computation,* 1977, vol. XXXI, p. 1044.

$$\sqrt{2} = 1 + 1/2 + 1/2 + 1/2 + 1/\ldots,$$

$$e = 2 + 1/1 + 1/2 + 1/1 + 1/1 + 1/4 + 1/$$
$$1 + 1/1 + 1/6 + 1/1 + 1/1 + 1/8 + 1/\ldots,$$

$$\pi = 3 + 1/7 + 1/15 + 1/1 + 1/292 + 1/1 + 1/$$
$$1 + 1/1 + 1/2 + 1/1 + 1/3 + 1/1 + 1/\ldots.$$

This kind of expression excels the decimal in three respects: (i) the continued fraction terminates whenever the number is rational, and recurs whenever it is quadratically irrational; (ii) it does not depend on any particular scale of notation (except trivially, in writing the a's); (iii) it leads to rational approximations which are "best possible" in a sense that will be illustrated geometrically in the next chapter (page 86).

The rational approximations

$$\frac{b_1}{c_1} = \frac{a_0}{1}, \quad \frac{b_2}{c_2} = \frac{1 + a_0 a_1}{a_1}, \quad \frac{b_3}{c_3} = \frac{a_0 + a_0 a_1 a_2 + a_2}{1 + a_1 a_2}, \quad \ldots,$$

obtained by stopping the continued fraction at each successive stage, are called *convergents*, and have many remarkable properties. They are given successively by the formulae

$$b_{n+1} = b_{n-1} + a_n b_n, \quad c_{n+1} = c_{n-1} + a_n c_n.$$

For instance, the first half-dozen convergents to $\sqrt{2}$, e, and π are $\frac{1}{1}, \frac{3}{2}, \frac{7}{5}, \frac{17}{12}, \frac{41}{29}, \frac{99}{70}$; $\frac{2}{1}, \frac{3}{1}, \frac{8}{3}, \frac{11}{4}, \frac{19}{7}, \frac{87}{32}$; $\frac{3}{1}, \frac{22}{7}, \frac{333}{106}, \frac{355}{113}, \frac{103993}{33102}, \frac{104348}{33215}$.

The simplest of all irrational continued fractions is

$$\tau = 1 + 1/1 + 1/1 + 1/\ldots,$$

which satisfies the equation $\tau = 1 + 1/\tau$, whence, being obviously positive, $\tau = \frac{1}{2}(\sqrt{5} + 1)$. (Cf. Wythoff's Game, on page 39.) Its convergents are $\frac{1}{1}, \frac{2}{1}, \frac{3}{2}, \frac{5}{3}, \frac{8}{5}, \frac{13}{8}, \ldots$, both numerators and denominators being formed from the sequence of *Fibonacci numbers*

$$1, 1, 2, 3, 5, 8, 13, 21, 34, 55, 89, 144, 233, 377, \ldots.$$

Each of these numbers is equal to the sum of the preceding two.

The ratios of alternate Fibonacci numbers are said* to measure the fraction of a turn between successive leaves on the stalk of a plant: $\frac{1}{2}$ for elm and linden, $\frac{1}{3}$ for beech and hazel, $\frac{2}{5}$ for oak and apple, $\frac{3}{8}$ for poplar and rose, $\frac{5}{13}$ for willow and almond, and so on. These are convergents to

$$\tau^{-2} = 1/2 + 1/1 + 1/1 + 1/ \ldots .$$

The number τ is intimately connected with the metrical properties of the pentagon, decagon, dodecahedron, and icosahedron, since it is equal to $2 \cos \frac{1}{5}\pi$. A line-segment is said to be divided according to the *golden section* if one part is τ times the other, or $1/\tau$ of the whole. The symbol τ is appropriate because it is the initial of $\tau o \mu \acute{\eta}$ ("section"). The nth Fibonacci number is

$$\{\tau^n - (-\tau)^{-n}\}/\sqrt{5}.$$

RATIONAL RIGHT-ANGLED TRIANGLES

If the sides of a right-angled triangle are in rational ratios, we may take them to be integers without a common factor. By Pythagoras' Theorem, the sides, x, y, z, of such a "primitive" triangle satisfy the equation $x^2 + y^2 = z^2$, z being the hypotenuse. The general solution (apart from the obvious possibility of interchanging x and y) is†

$$x = b^2 - c^2, \quad y = 2bc, \quad z = b^2 + c^2,$$

where b and c are arbitrary co-prime integers, one even and one odd, with $b > c$. The values $b = 2$, $c = 1$ give the familiar 3, 4, 5 triangle.‡ The numbers x and z are always odd, and y is divisible by 4. Either x or y is divisible by 3, and one of x, y, z by 5. Consequently, xy is divisible by 12, and xyz by 60.

*See, for instance, H.S.M. Coxeter, *Introduction to Geometry*, New York, 1969, pp. 169–172.

†G. Chrystal, *Algebra*, Edinburgh, 1889, vol. II, p. 531.

‡E.T. Bell, *Numerology*, Baltimore, 1933, p. 26ff.

b	c	x	y	z	$x+y+z$ $=2b(b+c)$	b	c	x	y	z	$x+y+z$ $=2b(b+c)$
2	1	3	4	5	12	5	4	9	40	41	90
3	2	5	12	13	30	7	2	45	28	53	126
4	1	15	8	17	40	6	5	11	60	61	132
4	3	7	24	25	56	8	1	63	16	65	144
5	2	21	20	29	70	7	4	33	56	65	154
6	1	35	12	37	84	8	3	55	48	73	176

The first twelve primitive triangles are tabulated above. These are all that have $z < 80$, and all that have $x + y + z < 180$. D.N. Lehmer has proved* that the number of primitive triangles with hypotenuse less than X is approximately $X/2\pi$, and that the number with perimeter less than X is approximately $(X \log 2)/\pi^2$. We observe that $80/2\pi = 12\cdot73 \ldots$, while $(180 \log 2)/\pi^2 = 12\cdot64 \ldots$.

If $b - c = 1$, then $z - y = 1$, as in the first, second, fourth, seventh, and ninth of the above triangles. If c and b are consecutive terms of the sequence 1, 2, 5, 12, 29, 70, $\ldots$, i.e.† if b/c is a convergent to the continued fraction

$$\sqrt{2} + 1 = 2 + 1/2 + 1/2 + 1/ \ldots ,$$

then $|x - y| = 1$, as in the first and fifth triangles.

F. Hoppenot has pointed out that the sum of the squares of $n + 1$ consecutive integers, of which the greatest is $2n(n+1)$, is equal to the sum of the squares of the next n integers; thus $10^2 + 11^2 + 12^2 = 13^2 + 14^2$, $21^2 + 22^2 + 23^2 + 24^2 = 25^2 + 26^2 + 27^2$, and so on. As another analogue of the identity $3^2 + 4^2 = 5^2$, observe that $3^3 + 4^3 + 5^3 = 6^3$. The equation $x^3 + y^3 = z^3$ has no solutions in integers; nor has $x^4 + y^4 = z^4$. Euler‡ has conjectured a corresponding result for the equation $x^4 + v^4 + z^4 = v^4$; if any solution exists, v must exceed 220000.

---

*American Journal of Mathematics, 1900, vol. XXII, p. 38.

†M. Kraïtchik, La Mathématique des Jeux, Brussels, 1930, p. 106.

‡ See L.J. Lander, T.R. Parkin, and J.L. Selfridge, Mathematics of Computation, 1967, vol. XXI, pp. 446–459.

On the other hand, $x^4 + y^4 = z^4 + v^4$ has infinitely many solutions.*

The *triangular numbers*

$$1, 3, 6, 10, 15, 21, 28, 36, 45, 55, 66, \ldots, \tfrac{1}{2}n(n+1), \ldots$$

are the sums of consecutive integers, beginning with 1. Hence, as their name indicates, they are the numbers of dots (or equal circles) that can be arranged in triangular formation: one at the top, two below it, three below them, and so on. The numbers 1, 36, 1225, 41616, 1413721, 48024900, ... are simultaneously triangular and square; the general formula for such a number is $b^2 c^2$, where b/c is any convergent to the continued fraction for $\sqrt{2}$.

The sums of consecutive triangular numbers are the *tetrahedral numbers* (formerly called "pyramidal numbers")

$$1, 4, 10, 20, 35, 56, 84, 120, \ldots, \tfrac{1}{6}n(n+1)(n+2), \ldots.$$

These are the numbers of equal spheres that can be piled in tetrahedral formation. Similarly, the sums of consecutive squares are the *pyramidal numbers*

$$1, 5, 14, 30, 55, 91, 140, 204, \ldots, \tfrac{1}{6}n(n+1)(2n+1), \ldots.$$

The sums of consecutive cubes (from 1 up) are the squares of the triangular numbers; the sums of consecutive *pairs* of triangular numbers are the square numbers, and the sums of consecutive pairs of tetrahedral numbers are the pyramidal numbers.

The *only* number (>1) which is simultaneously square and pyramidal is 4900. This result‡ was conjectured by Lucas in 1875, and proved by Watson in 1918. The proof is by no means elementary.

*See *Sphinx*, 1937, p. 98.

†Cf. H.E. Dudeney, *Amusements in Mathematics*, London, 1917, pp. 26, 167.

‡ E. Lucas, *Nouvelles Annales de Math.* (2), 1875, vol. xiv, p. 336; G.N. Watson, *Messenger of Mathematics* (new series), 1918, vol. xlviii, pp. 1–22.

DIVISIBILITY

If two numbers, x and y, differ by a multiple of p, we say that they are *congruent modulo p*, and write

$$x - y \equiv 0 \text{ (mod } p) \quad \text{or} \quad x \equiv y \text{ (mod } p).$$

Every integer is congruent (mod p) to just one of the p *residues* 0, 1, 2, ..., $p-1$. There is an arithmetic of residues, closely analogous to the arithmetic of ordinary numbers. In it we can add, subtract, or multiply. For instance, working modulo 6 we have

$$3 + 4 \equiv 1, \quad 3 - 4 \equiv 5, \quad 3 \times 4 \equiv 0.$$

The arithmetic becomes more interesting when p is prime (as we shall suppose for the rest of the present chapter). This notion is defined as follows.

A *prime* (or *prime number*) is an integer greater than 1 whose only positive divisors are 1 and itself. (By excluding 1 from the list of primes, we are able to assert that every natural number has a unique decomposition into prime factors; for instance, $504 = 2^3 \, 3^2 \, 7$.) The twenty-five primes less than 100 are 2, 3, 5, 7, 11, 13, 17, 19, 23, 29, 31, 37, 41, 43, 47, 53, 59, 61, 67, 71, 73, 79, 83, 89, 97. Euclid (IX. 20) proved that there are infinitely many primes, arguing somewhat as follows. Consider the product $P = 2.3.5.7...p$ of all the primes up to a particular one. Clearly, $P + 1$ is not divisible by any of these primes. Therefore it has a prime divisor greater than p (including, as a possibility, itself). Thus there is a prime greater than any given prime.

The best attempt that has ever been made towards an explicit formula for primes is Euler's* quadratic form $x^2 + x + 41$, which represents primes for $x = 0, 1, 2, 3, ..., 39$. H.M. Stark† has shown that there is no form $x^2 + x + A$ with $A > 41$ which represents primes for $A - 1$ consecutive values of x.

It is clearly desirable to be able to tell whether a given num-

---

Nouveaux Mémoires de l'Académie royale des Sciences, Berlin, 1772, p. 36.
†*Michigan Mathematical Journal*, 1967, vol. XIV, pp. 1–27.

ber N is prime or composite without having to test every prime less than $\sqrt{N}$ as a possible divisor. I will mention two criteria.

Sir J. Wilson discovered (1770) and Lagrange proved (1773) that N is prime *if, and only if*, $(N-1)!+1$ is divisible by N. For instance, $(7-1)!+1=721$ is divisible by 7, but $(9-1)!+1 = 40321$ is not divisible by 9. Wilson's Theorem is a theoretical rather than practical test for primality, since, if N is large enough for its primality to be in doubt, it is more laborious to find whether $(N-1)!+1$ is or is not divisible by N than to test every prime less than $\sqrt{N}$ as a possible factor of N.

Fermat discovered (1640) and Euler proved (1736) that, if p is prime and a is not divisible by p, then $a^{p-1}-1$ is divisible by p. The case when $a=2$ was known to the Chinese as early as 500 B.C.; they stated also the converse proposition: if N divides $2^{N-1}-1$, then N is a prime. This proposition was rediscovered and "proved" by Leibniz in 1680. However, it is false; it fails for $N=341=11\times31$ and for an infinity of other N's. Modern tests for primality are based on the following converse of Fermat's Theorem, due to Lucas:* if a^x-1 is divisible by N when $x=N-1$, but not when x is a proper divisor of $N-1$, then N is a prime. Modifications of this converse as to both hypothesis and conclusion are necessary before the actual modern tests emerge.†

About 1930, D.H. Lehmer (whose father constructed the first really substantial table of factors and list of primes‡) invented a photo-electric number sieve which factorizes large numbers with astonishing rapidity. During the next forty years, he and his associates improved it until it became an electronic

---

*Théorie des Nombres, Paris, 1891, pp. 423, 441. For an interesting discussion of Fermat's Theorem, see E.T. Bell's Numerology, Baltimore, 1933, pp. 182–185.

†See Bulletin of the American Mathematical Society, 1927, vol. XXXIII, pp. 327–340.

‡D.N. Lehmer, Factor Tables for the First Ten Millions, Washington, 1909; List of Prime Numbers from 1 to 10,006721, Washington, 1914. The latter book has a particularly fascinating Introduction.

sieve* with a speed of a million values per second, which is seven times the speed of the IBM 7090.

It is amusing to ask someone for a quick proof that *every sum of two consecutive odd primes is the product of three integers all greater than* 1; e.g., $7 + 11 = 2 \times 3 \times 3$, $11 + 13 = 2 \times 3 \times 4$. This appears at first to be hard, but is in fact very easy.

THE PRIME NUMBER THEOREM

The fundamental theorem of arithmetic (which says that every natural number can be uniquely factorized) was expressed by Euler (1737) in the elegant form

$$\sum_{n=1}^{\infty} n^{-s} = \prod_{p}(1 + p^{-s} + p^{-2s} + \ldots) = \prod_{p}(1 - p^{-s})^{-1},$$

where the products are over all primes p.

The number of primes not exceeding X is denoted by $\pi(X)$. Thus $\pi(2) = 1$, $\pi(10) = 4$, $\pi(100) = 25$, and so on. After examining a table of the primes less than 400,000, Legendre (1808) asserted that, when X is large, $\pi(X)$ is approximately

$$X/(\log X - B),$$

where B is a constant close to 1. Abel (in a letter of 1823) called this theorem "the most remarkable in the whole of mathematics." Independently, Gauss (1849) noticed that, if X is large while x is comparatively small, the number of primes between X and $X + x$ (or between $X - x$ and X) is approximately $x/(\log X)$, so that $\pi(X)$ is approximately

$$\int_2^X \frac{dt}{\log t}.$$

Both these approximations are consequences of the celebrated *prime number theorem*, which says that *the ratio of $\pi(X)$ to $X/(\log X)$ tends to* 1 *as X tends to infinity.* This was first proved (independently) by Hadamard and de la Vallée

---

*J.D. Brillhart and J.L. Selfridge, *Mathematics of Computation*, 1967, vol. XXI, pp. 87–96. For a readable account of the development of the theory of sieves by Viggo Brun, Atle Selberg, and others, see H. Halberstam and K.F. Roth, *Sequences*, 1966, vol. I, chapter IV.

Poussin* in 1896. Fifty-two years later, Selberg and Erdöst† discovered an entirely elementary (but admittedly long) proof of the prime number theorem. This was surprising for, hitherto, all proofs had depended essentially on techniques from complex variable theory and, in particular, on the idea (going back to B. Riemann‡) of relating the behaviour of the function $\pi(X)$ to the problem of locating the zeros of the Riemann Zeta-function

$$\zeta(s) = \sum_{n=1}^{\infty} n^{-s}$$

in the complex s-plane. In fact, the seemingly intractible Riemann Hypothesis asserts that *all* non-real zeros of $\zeta(s)$ are confined to the line $\mathrm{Re}(s) = \frac{1}{2}$. From this, with $X/(\log X)$ replaced by the more convenient "logarithmic integral"

$$\mathrm{li}\ X = \lim_{\varepsilon \to +0} \left(\int_0^{1-\varepsilon} + \int_{1+\varepsilon}^{X} \right) \frac{dt}{\log t} = \int_2^X \frac{dt}{\log t} + 1 \cdot 04 \ldots,$$

it may be deduced that, for some positive constant c,

$$\left| \mathrm{li}\ X - \pi(X) \right| < cX^{1/2}\log X.$$

This is a far stronger claim than any known established refinement of the prime number theorem.§

As for numerical verification, it has been computed that

$$\mathrm{li}\ 10^9 = 50{,}849235 \quad \text{and} \quad \pi(10^9) = 50{,}847534.$$

(The former is rounded off to the nearest integer; the latter is exact.) Although $\mathrm{li}\ X > \pi(X)$ in every known case, J.E. Littlewood proved (in 1914) that if we go far enough we shall eventually reach a value of X for which the inequality is reversed; moreover, such reversals will occur infinitely often! ||

*Jacques Hadamard, *Bulletin de la Société mathématique de France*, 1896, vol. xxiv, pp. 199–220; C.-J. de la Vallée Poussin, *Annales de la Société scientifique de Bruxelles*, 1896, vol. xx, pp. 183–256.

†*Annals of Mathematics*, 1949, series 2, vol. l, pp. 305–315.

‡*Monatsberichte der Preussischen Akademie der Wissenschaften*, 1859, pp. 671–680.

§A.E. Ingham, *The Distribution of Prime Numbers*, Cambridge, 1932, p. 83.

|| R. Sherman Lehman, *Acta Arithmetica*, 1966, vol. xi, pp. 397–410.

A. de Polignac* has conjectured that every even number is the difference of two consecutive primes in infinitely many ways. Taking the even number to be 2, this means that there are infinitely many pairs of primes that are consecutive odd numbers, such as 5, 7; 11, 13; 17, 19; 29, 31; 41, 43; 59, 61; 71, 73. This conjecture has not been proved or disproved. As evidence in its favour, one finds that there are 36 such pairs in the interval $10^{12} \pm 10^4$. A spectacularly large instance (having 2259 digits) is the prime pair

$$107570463 \times 10^{2250} \pm 1,$$

discovered by Harvey Dubner. †

It can be proved quite easily that the sum of the reciprocals of all the primes less than X increases without limit as X increases. If, however, we consider only the sum of the reciprocals of those primes less than X which differ by 2, then this sum remains bounded as X increases. This fact, due to Brun, shows that there are not "too many" primes which differ by 2.

A somewhat similar conjecture is *Goldbach's Theorem*, that every even number greater than 4 can be expressed as the sum of two odd primes. This has been verified up to 10,000 and for small ranges of very large numbers. Vinogradov‡ proved in 1937 that every sufficiently large odd number is the sum of three primes, and Estermann§ found that *almost all* even numbers are sums of two primes.

MERSENNE NUMBERS

A curious assertion (only partially correct) about the prime or composite character of numbers of the form $2^p - 1$ is to be found in Mersenne's *Cogitata Physico-Mathematica*, published in 1644. In the preface to that work a statement is made about perfect numbers, which implies that the only values of p not

*Nouvelles Annales de Math., 1849, vol. VIII, p. 428.

† *Journal of Recreational Mathematics,* 1986, vol. XVIII, p. 85.

‡I.M. Vinogradov, *The Method of Trigonometrical Sums in the Theory of Numbers,* London, 1954, p. 167.

§*Proceedings of the London Mathematical Society,* ser. 2, 1938, vol. XLIV, pp. 307–314.

greater than 257 which make $2^p - 1$ prime are 2, 3, 5, 7, 13, 17, 19, 31, 67, 127, and 257. This statement is not quite as impressive as it once seemed, for it contains five mistakes. In 1883, I.M. Pervusin discovered that $2^{61} - 1$ is prime.* In 1903, F.N. Cole obtained the factorization

$$2^{67} - 1 = 193707721 \times 761838257287.$$

In 1911 and 1914, R.E. Powers† discovered that $2^{89} - 1$ and $2^{107} - 1$ are prime. In 1922, M. Kraïtchik‡ found $2^{257} - 1$ to be composite.

The modern technique was established by Lucas§ in 1877, and used by him to verify Mersenne's assertion that $2^{127} - 1$ $(= 170{,}141183{,}460469{,}231731{,}687303{,}715884{,}105727)$ is prime. This remained for 75 years (in fact, till 14 July 1951) the largest explicitly known prime. In 1931, D.H. Lehmer ‖ developed Lucas's technique into the following single criterion:

$2^p - 1$ (with $p > 2$) is prime if and only if it divides v_{p-1}, where $v_1 = 4$ and $v_{n+1} = v_n^2 - 2$.

As illustrations, we observe that $2^3 - 1$ divides $v_2 = 14$, $2^4 - 1$ does not divide $v_3 = 194$, and $2^5 - 1$ divides $v_4 = 37{,}634$.

This powerful criterion has been applied by computer to all Mersenne numbers with $p < 132{,}050$. The conclusion is that $2^p - 1$ is prime in just 29 of these cases, namely when

p = 2, 3, 5, 7, 13, 17, 19, 31, 61, 89, 107, 127, 521, 607, 1279, 2203, 2281, 3217, 4253, 4423, 9689, 9941, 11213, 19937, 21701, 23209, 44497, 86243, 132049.

The last seventeen cases were discovered, between 1952 and 1985, by R.M. Robinson, H. Riesel, A. Hurwitz, D.B. Gillies, B. Tuckerman, L. Nickel, C. Noll, H.L. Nelson, and D. Slowinski. For instance, Slowinski (in 1985) used a computer to

*R.C. Archibald, *Scripta Mathematica*, 1935, vol. iii, p. 117.

†*American Mathematical Monthly*, 1911, vol. xviii, pp. 195–197; *Proceedings of the London Mathematical Society*, ser. 2, 1919, vol. xiii, p. 39.

‡See *Sphinx*, 1931, p. 31.

§ *American Journal of Mathematics*, 1878, vol. i, p. 316.

‖ *Sphinx*, 1931, pp. 32, 164.

prove the divisibility of v_{216090} by $2^{216091} - 1$. This much greater Mersenne prime has 65050 digits.

For all the original Mersenne numbers $2^n - 1$ with $n \leq 257$ (n a prime), complete factorization has been achieved (see J.D. Brillhart, D.H. Lehmer, J.L. Selfridge, Bryant Tuckerman, and S.S. Wagstaff, Jr., *Contemporary Mathematics,* vol. XXII, 1983). Here are some remarkable examples:

$$2^{211} - 1 = 15193 \times 60272956433838849161 \times$$
$$3593875704495823757388199894268773153439,$$

$$2^{251} - 1 = 503 \times 54217 \times 178230287214063289511 \times$$
$$61676882198695257501367 \times$$
$$12070396178249893039969681,$$

$$2^{257} - 1 = 535006138814359 \times$$
$$1155685395246619182673033 \times$$
$$374550598501810936581776630096313181393.$$

The first two are due to J. Davis, D. Holdridge, and G. Simmons; the last to M. Penk and R. Baillie.

Harvey Dubner has proved that the numbers $1477! + 1$ and $217833 \times 10^{7150} + 1$ (having 4042 and 7156 digits, respectively) are prime. So too, according to Hugh C. Williams, is $(10^{1031} - 1)/9 = 111 \ldots 1$ (1031 ones).

PERFECT NUMBERS

The theory of *perfect numbers* depends directly on that of Mersenne Numbers. A number is said to be perfect if it is equal to the sum of all its proper divisors. Thus 6 and 28 are perfect numbers, since $6 = 1 + 2 + 3$ and $28 = 1 + 2 + 4 + 7 + 14$. These numbers have had a strong appeal for mystics, since the Creation took 6 days, and there are 28 days in a lunar month.

Euclid proved that $2^{p-1}(2^p - 1)$ is perfect whenever $2^p - 1$ is prime. In fact, the divisors of $2^{p-1}(2^p - 1)$ (including itself) are then 2^n and $2^n(2^p - 1)$, for $n = 0, 1, \ldots, p-1$; and we know that $1 + 2 + 2^2 + \ldots + 2^{p-1} = 2^p - 1$.

Euler showed that this formula includes all *even* perfect numbers. The following simplified proof has been given by

Dickson.* Let $2^n q$ be perfect, where q is odd and $n > 0$. Then $2^{n+1} q = (2^{n+1} - 1)s$, where s is the sum of all the divisors of q. Thus $s = q + d$, where $d = q/(2^{n+1} - 1)$. Hence d is a divisor of q, so that q and d are the only divisors of q. Hence $d = 1$ and $q = 2^{n+1} - 1$ is a prime.

The values $p = 2, 3, 5, 7, 13, 17, 19, 31$ give the Mersenne primes 3, 7, 31, 127, 8191, 131071, 524287, 2147483647, and the perfect numbers 6, 28, 496, 8128, 33550336, 8589869056, 137438691328, 2305843008139952128. It is easy to show that the two final digits of an even perfect number are necessarily either 28 or (apart from 6 itself) 6 preceded by an *odd* digit. Also every even perfect number, except 6, is congruent to 1 modulo 9.

FERMAT NUMBERS

Fermat enriched mathematics with a multitude of new propositions. With one exception all these have been proved or are believed to be true. This exception is his *theorem on binary powers*, in which he asserted that all numbers of the form $2^m + 1$, where $m = 2^n$, are primes,[†] but he added that, though he was convinced of the truth of this proposition, he could not obtain a valid demonstration.

It may be shown that $2^m + 1$ is composite if m is not a power of 2, but of course it does not follow that $2^m + 1$ is a prime if m is a power of 2, say 2^n. As a matter of fact the theorem is not true.

---

*L.E. Dickson, *American Mathematical Monthly*, 1911, vol. XVIII, p. 109. See also his *History of the Theory of Numbers*, vol. I, Washington, 1919. There is reason to believe that every perfect number is even. At any rate, no odd number less than 10^{36} can be perfect; this was proved by Bryant Tuckerman (*Notices of the American Mathematical Society*, 1968, vol. XV, p. 226). It was conjectured by J.J. Sylvester and proved by I.S. Gradshtein (in a Russian paper of 1925) that, if an odd perfect number exists, it must have at least six distinct prime factors. The existence of an odd perfect number would imply the existence of two or more odd numbers whose reciprocals add up to 1. The search for such odd numbers seems more likely to be successful than the search for an odd *perfect* number.

†Letter of Oct. 18, 1640, *Opera*, Toulouse, 1679, p. 162: or Brassinne's *Précis*, p. 143.

In 1732 Euler* showed that if $n = 5$ the formula gives 4294,967297, which is equal to $641 \times 6,700417$. Let us define

$$F_n = 2^{2^n} + 1,$$

so that $F_0 = 3$, $F_1 = 5$, $F_2 = 17$, $F_3 = 257$, and $F_4 = 65,537$. G.T. Bennett† has shown how the divisibility of F_5 by 641 may be verified without actual division: the number

$$641 = 5^4 + 2^4 = 2^7 5 + 1,$$

dividing both $2^{28}(5^4 + 2^4)$ and $(2^7 5)^4 - 1$, divides their difference, which is $2^{32} + 1$.

With the help of Gauss's theory of quadratic residues, it can be proved that any prime factor of F_n (with $n > 1$) must be of the form $2^{n+2}k + 1$, where k is an integer. For instance,

$$F_5 = (2^7 5 + 1)(2^7 52347 + 1)$$

and, as Landry observed in 1880,

$$F_6 = (2^8 1071 + 1)(2^8 262814145745 + 1).$$

In 1909, J.C. Morehead and A.E. Western proved that F_7 and F_8 are composite. This conclusion is most easily checked by means of the following criterion:

F_n (with $n > 0$) is prime if and only if it divides $3^{(F_n - 1)/2} + 1$.

For sixty years after the discovery that F_7 and F_8 are composite, no factors for either of them were found, though it was proved that F_7 is the product of just two primes. At last, in 1970, M.A. Morrison and John Brillhart‡ found the two primes, so now we know that

$$F_7 = (2^9 116503103764643 + 1)(2^9 11141971095088142685 + 1).$$

*Commentarii Academiae Scientiarum Petropolitanae, Leningrad, 1738, vol. VI, p. 104; see also Novi Comm. Acad. Sci. Petrop., Leningrad, 1764, vol. IX, p. 101: or Commentationes Arithmeticae Collectae, Leningrad, 1849, vol. I, pp. 2, 357.

†Compare G.H. Hardy and E.M. Wright, An Introduction to the Theory of Numbers, Oxford, 1954, p. 14.

‡Bulletin of the American Mathematical Society, 1971, vol. LXXVII, p. 264.

Still more remarkably, in 1980, R.P. Brent and J.M. Pollard* found that

$$F_8 = (2^{11}604944512477 + 1)(2^{11}N + 1),$$

where N is an odd number having 59 digits. The smallest Fermat numbers which are unknown to be prime or composite are F_{20}, F_{22}, F_{24}, F_{28}. At least one factor of F_n is known for $n = 9$, 10, $\cdots$, 19 and many larger values, including 23471. In this last case, the factor $2^{23473}5 + 1$ was found by Wilfrid Keller in 1984. F_{23471} is one of the largest numbers that have ever been investigated. To write it down explicitly would be an impossible task, because the number of digits far exceeds the number of particles in the universe (which, according to the late A.S. Eddington, is 51×2^{260}).

All these results support the conjecture that F_n is composite whenever $n > 4$.

FERMAT'S LAST THEOREM

I pass now to another assertion made by Fermat which hitherto has not been proved. This, which is sometimes known as *Fermat's Last Theorem*, is to the effect† that no integral values of x, y, z (with $xyz \neq 0$) can be found to satisfy the equation $x^n + y^n = z^n$ if n is an integer greater than 2. This proposition has acquired extraordinary celebrity from the fact that, while there is no reason to doubt that it is true, no general demonstration of it has ever been given.

Fermat seems to have discovered its truth first‡ for the case $n = 3$, and then for the case $n = 4$. His proof for the former of these cases is lost, but that for the latter is extant,§ and a

---

*See Hans Riesel, *Prime Numbers and Computer Methods for Factorization*, Birkhäuser, 1985.

†Fermat's enunciation will be found in his edition of *Diophantus*, Toulouse, 1670, bk. II, qu. 8, p. 61; or Brassinne's *Précis*, Paris, 1853, p. 53. For bibliographical references, see L.E. Dickson, *History of the Theory of Numbers*, Washington, 1920, vol. II, ch. 26; see also L.J. Mordell, *Fermat's Last Theorem*, Cambridge, 1921, repeated in F. Klein et al., *Famous Problems and Other Monographs*, New York, 1955. Mordell refers to $x^n + y^n = z^n$ as "the most famous of all Diophantine equations."

‡See a letter from Fermat quoted in my *History of Mathematics*, 4th ed., Dover reprint, New York, 1960, chapter xv.

§Fermat's *Diophantus*, note on p. 339; or Brassinne's *Précis*, p. 127.

similar proof for the case $n=3$ was outlined by Euler.* These proofs depend upon showing that, if three integral values of x, y, z can be found which satisfy the equation, then it will be possible to find three other and smaller integers which also satisfy it: in this way finally we show that the equation must be satisfied by three values which obviously do not satisfy it. Thus no integral solution is possible. This method is inapplicable to the general case.

Fermat's discovery of the general theorem was made later. A proof can be given on the assumption that every integer has a unique decomposition into prime factors. While this is true for every *rational* integer, it is not true for every *algebraic* integer, an algebraic integer being defined as a root of a polynomial equation

$$x^n + a_1 x^{n-1} + \ldots + a_n = 0,$$

whose coefficients a_i are rational integers. For instance, in the ring of algebraic integers $a + b\sqrt{10}$, where a and b are rational integers, the number 6 has two distinct decompositions into primes:

$$6 = 2 \times 3 = (4 - \sqrt{10})(4 + \sqrt{10}).$$

Similarly, for some values of n Fermat's equation leads to expressions which can be factorized several ways. It is possible that Fermat's argument rested on the above erroneous supposition, but this is an unsupported conjecture. At any rate he asserted definitely that he had a valid proof – *demonstratio mirabilis sane* – and the fact that no theorem on the subject which he stated he had proved has been subsequently shown to be false must weigh strongly in his favour; the more so because in making the one incorrect statement in his writings (namely, that every F_n is prime) he added that he could not obtain a satisfactory demonstration of it.

*Euler's *Algebra* (English trans. 1797), vol. II, chap. XV, p. 247: one point was overlooked by Euler, but the omission can be supplied. For a full description of the cases $n=4$ and $n=3$, see Hardy and Wright, *An Introduction to the Theory of Numbers*, pp. 191–194.

It must be remembered that Fermat was a mathematician of quite the first rank who had made a special study of the theory of numbers. The subject is in itself one of peculiar interest and elegance, but its conclusions have little practical importance, and for long it was studied by only a few mathematicians. This is the explanation of the fact that it took more than a century before some of the simpler results which Fermat had enunciated were proved, and thus it is not surprising that a proof of the theorem which he succeeded in establishing only towards the close of his life should involve great difficulties.

In 1823 Legendre* obtained a proof for the case $n=5$; in 1832 Lejeune Dirichlet† gave one for $n=14$; and in 1840 Lamé and Lebesgue‡ gave proofs for $n=7$.

To prove the proposition when $n>4$, obviously it is sufficient to confine ourselves to cases where n is a prime. In 1849, Kummer§ proved it for all "regular" primes. (A prime p is said to be regular if it does not divide any of the numerators of the Bernoulli numbers‖ B_1, B_2, ..., $B_{\frac{1}{2}(p-3)}$.) Kummer found that 37, 59, 67, 101, 103, 131, 149, 157 are the only "irregular" primes less than 164. By means of the high-speed calculating machine known as SWAC, J.L. Selfridge and B.W. Pollock¶ have found all the irregular primes less than 25,000, and checked each one, with the conclusion that the theorem is true for $n<25,000$. Other tests have been established; for instance, A. Wieferich** has shown that if the equation is soluble in

*Reprinted in his *Théorie des Nombres*, Paris, 1830, vol. II, pp. 361–368: see also pp. 5, 6.

†*Crelle's Journal*, 1832, vol. IX, pp. 390–393.

‡*Liouville's Journal*, 1841, vol. V, pp. 195–215, 276–279, 348–349.

§References to Kummer's Memoirs are given by H.S. Vandiver, *Transactions of the American Mathematical Society*, 1929, vol. XXXI, pp. 613–642.

‖The Bernoulli numbers appear in the series

$$\frac{x}{2}\cot\frac{x}{2} = 1 - \frac{B_1}{2!}x^2 - \frac{B_2}{4!}x^4 - \frac{B_3}{6!}x^6 - \dots.$$

The values of the first eight are $\frac{1}{6}, \frac{1}{80}, \frac{1}{42}, \frac{1}{30}, \frac{5}{66}, \frac{691}{2730}, \frac{7}{6}, \frac{3617}{510}$.

¶*Proceedings of the National Academy of Sciences* (USA), 1955, vol. XLI, pp. 970–973.

**Crelle's Journal*, 1909, vol. CXXXVI, pp. 293–302.

integers prime to n, where n is an odd prime, then $2^{n-1} - 1$ is divisible by n^2. This restricted problem has been pushed as far as $n = 41{,}000000$ and $250{,}000000$ by J.B. Rosser and D.H. Lehmer,* respectively. A prize † of 100,000 marks was long ago offered for a general proof, to be given before 2007.

Even though Fermat's problem remains unsolved, it has been of great importance for the theory of numbers, because many of the modern methods have been developed in connection with it, and the theories which have arisen in this connection are perhaps more important than a proof of the theorem itself would have been. Naturally there has been much speculation as to how Fermat arrived at the result. Such of his proofs as are extant involve nothing more than elementary geometry and algebra, and indeed some of his arguments do not involve any symbols. This has led some writers to think that Fermat used none but elementary algebraic methods. This may be so, but the following remark, which I believe is not generally known, rather points to the opposite conclusion. He had proposed, as a problem to the English mathematicians, to show that there was only one integral solution of the equation $x^2 + 2 = y^3$: the solution evidently being $x = 5$, $y = 3$. On this he has a note‡ to the effect that there was no difficulty in finding a solution in rational fractions, but that he had discovered an entirely new method – *sane pulcherrima et subtilissima* – which enabled him to solve such questions in integers. It was his intention to write a work § on his researches in the theory of numbers, but it was never completed, and we know but little of his methods of analysis. I venture, however, to add my private suspicion that continued fractions played a not unimportant part in his researches, and as strengthening this conjecture I may note that some of his more recondite results, such as the theorem that a prime of the form $4n + 1$ is expressible as the sum of

*Bulletin of the American Mathematical Society, 1941, vol. XLVII, p. 142.
†L'Intermédiaire des Mathématiciens, vol. XV, pp. 217–218.
‡Fermat's Diophantus, bk. VI, prop. 19, p. 320; or Brassinne's Précis, p. 122.
§Fermat's Diophantus, bk. IV, prop. 31, p. 181; or Brassinne's Précis, p. 82.

two squares,* may be established with comparative ease by properties of such fractions.

GALOIS FIELDS

The familiar properties of addition and multiplication (such as the associative, distributive, and commutative laws, and the possibility of dividing by anything except zero), which hold for the real field, the rational field, and the complex field, also hold for fields in which the number of elements is finite. It can be proved† that this finite number q is always a power of a prime, say $q = p^n$ where p is a prime and n is a positive integer. For each prime power q there is just one field, denoted by GF(q) and called a Galois field, after Evariste Galois (1811–1832) whose brilliant career was cut short by a duel.‡ In particular, GF(p) is the field of residue classes modulo p, and its p elements are denoted by $0, 1, \ldots, p-1$, where (by convention) 0 means the set of all multiples of p, 1 means the set of integers that leave remainder 1 when divided by p, and so on. Of course, $p-1$ can equally well be called -1. For instance, since $641 = 1 + 2^7 5 = 2^4 + 5^4$, the following equations hold in GF(641):

$$2^7 5 = -1, \quad 2^4 = -5^4,$$
$$2^7 = -\tfrac{1}{5}, \quad 2^8 = -\tfrac{2}{5}, \quad 2^{32} = (-\tfrac{2}{5})^4 = 2^4/5^4 = -1.$$

This is how Kraïtchik§ verified that 641 divides $2^{32} + 1$. The statement that $a = b$ in GF(p) has the same meaning as

$$a \equiv b \pmod{p}$$

("a is congruent to b modulo p"), namely that $a - b$ is divisible by p.

The notation $a \equiv b \pmod{m}$ is used also when m is composite, although then the residue classes form only a ring, not a field.

*Fermat's *Diophantus*, bk. III, prop. 22, p. 127; or Brassinne's *Précis*, p. 65.

†Garrett Birkhoff and Saunders MacLane, *A Survey of Modern Algebra* (3rd ed.), New York, 1965, p. 413. See also L.E. Dickson, *History of the Theory of Numbers*, vol. I, Washington, 1919, p. viii.

‡Leopold Infeld, *Whom the Gods Love*, New York, 1948.

§M. Kraïtchik, *Théorie des Nombres*, vol. II, Paris, 1926, p. 221.

For instance, in the ring of residues modulo 4 we have $2 \times 2 \equiv 0$, and 2 has no reciprocal. Thus, although this ring has 4 elements, it is quite different from the field GF(4). To emphasize the distinction, we usually expand the symbol GF(4) to GF(2^2).

Whenever $n > 1$, GF(p^n) can be represented as the field of equivalence classes of polynomials whose coefficients belong to GF(p), two such polynomials being declared equivalent if their difference is divisible by a particular irreducible (or "prime") polynomial of degree n. Thus each of the p^n elements of GF(p^n) can be expressed as a polynomial whose degree is less than n. (Although there are, in general, several irreducible polynomials of degree n, these all yield the same field GF(p^n).)

It is often convenient to use detached coefficients, so that $x^3 + 2x + 1$ appears as 1021 (which has its customary meaning if x stands for ten). For instance, if 1021 is chosen as the modulus for GF(3^3), a typical addition (first in terms of polynomials and then with detached coefficients) is

$$
\begin{array}{rr}
x^2 + 2x + 1 & 121 \\
2x + 2 & 22 \\
\hline
x^2 + x & 110
\end{array}
$$

and a typical multiplication is

$$
\begin{array}{rr}
x^2 + 2x + 1 & 121 \\
x & 10 \\
\hline
x^3 + 2x^2 + x & 1210 \\
x^3 + 2x + 1 & 1021 \\
\hline
2x^2 + 2x + 2 & 222
\end{array}
$$

(The final step consists in subtracting the modulus 1021.) It is known that every element except zero is a power of a suitable "primitive" element. For instance, in the above representation of GF(3^3), every non-zero element is a power of 10 or x:

$$
x^0 = 1, \quad x^1 = 10, \quad x^2 = 100, \quad x^3 = 1000 = 12, \quad \ldots,
$$
$$
x^{13} = 2, \quad \ldots, \quad x^{26} = 1.
$$

The elements 1, x^2, x^4, ... are squares, while (p being odd) x^3, x^5, ... are non-squares. The zero element can be included by making the convention $x^\infty = 0$, which is in reasonable agreement with the rule $x^a x^b = x^{a+b}$.

Over any field and, in particular, a finite field, it is an interesting problem to investigate the solutions of an algebraic equation such as

$$ay^l + bz^m + c = 0.$$

In the cases $(l, m) = (2, 2)$, $(3, 3)$, $(4, 4)$, $(2, 4)$, formulae for the number of solutions were already known to Gauss.* However, the deepest penetration of the general problem appeared much later (with the aid of the theory of algebraic function-fields) through the inspiration of André Weil.† With the function-field defined by such an equation, it is possible to associate a Zeta-function analogous to that of Riemann for the case of the rational field‡ and to formulate a hypothesis concerning the location of its zeros entirely analogous to that of the classical case. It was Weil's signal achievement to provide a proof of this so-called "finite analogue" of the Riemann Hypothesis, which led to far-reaching progress not only in number theory but also in algebraic geometry over finite fields.

*C.F. Gauss, *Werke*, 1900, vol. I, pp. 445–449.
†*Sur les courbes algébriques et les variétés qui s'en déduisent*, Paris, 1948.
‡See M. Eichler, *Algebraic Numbers and Functions*, New York, 1966 (translated from the German edition of 1963), §5.1, for some details of this theory.

CHAPTER III

GEOMETRICAL RECREATIONS

In this chapter and the next one I propose to enumerate certain geometrical questions, puzzles, and games, the discussion of which will not involve necessarily any considerable use of algebra or arithmetic. Most of this chapter is devoted to questions which are of the nature of formal propositions; the next chapter contains a description of various amusements.

In accordance with the rule I laid down for myself in the preface, I exclude the detailed discussion of theorems which involve advanced mathematics. Moreover (with one or two exceptions) I exclude any mention of the numerous geometrical paradoxes which depend merely on the inability of the eye to compare correctly the dimensions of figures when their relative position is changed. This apparent deception does not involve the conscious reasoning powers, but rests on the inaccurate interpretation by the mind of the sensations derived through the eyes, and I do not consider such paradoxes as coming within the domain of mathematics.

GEOMETRICAL FALLACIES

Most educated Englishmen are acquainted with the series of logical propositions in geometry associated with the name of Euclid, but it is not known so generally that these propositions were supplemented originally by certain exercises. Of such exercises Euclid issued three series: two containing easy theorems or problems, and the third consisting of geometrical fallacies, the errors in which the student was required to find.

The collection of fallacies prepared by Euclid is lost, and tradition has not preserved any record as to the nature of the erroneous reasoning or conclusions; but, as an illustration of such questions, I append a few demonstrations, leading to obviously impossible results. Perhaps they may amuse anyone to whom they are new. I leave the discovery of the errors to the ingenuity of my readers.

First fallacy.* *To prove that a right angle is equal to an angle which is greater than a right angle.* Let *ABCD* be a rectangle. From *A* draw a line *AE* outside the rectangle, equal to *AB* or *DC* and making an acute angle with *AB*, as indicated in

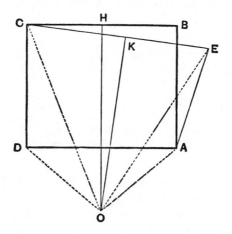

the diagram. Since *CB* and *CE* are not parallel, their perpendicular bisectors *HO* and *KO* will meet, say at *O*. Join *OA*, *OE*, *OC*, and *OD*.

The two triangles *ODC* and *OAE* are clearly congruent. For, since *KO* bisects *CE* and is perpendicular to it, we have *OC*=*OE*. Similarly, since *HO* bisects *CB* and *DA* and is perpendicular to them, we have *OD*=*OA*. Also, by construction, *DC*=*AE*. Therefore the three sides of the triangle *ODC* are equal respectively to the three sides of the triangle *OAE*. Hence, by Euc. I. 8, the triangles are congruent; and therefore the angle *ODC* is equal to the angle *OAE*.

---

*I believe that this and the fourth of these fallacies were first published in this book. They particularly interested Mr. C.L. Dodgson; see the *Lewis Carroll Picture Book*, London, 1899, pp. 264, 266, where they appear in the form in which I originally gave them.

Again, since *HO* bisects *DA* and is perpendicular to it, we have the angle *ODA* equal to the angle *OAD*.

Hence the angle *ADC* (which is the difference of *ODC* and *ODA*) is equal to the angle *DAE* (which is the difference of *OAE* and *OAD*). But *ADC* is a right angle, and *DAE* is necessarily greater than a right angle. Thus the result is impossible.

Second fallacy.* *To prove that a part of a line is equal to the whole line.* Let *ABC* be a triangle; and, to fix our ideas, let us suppose that the triangle is scalene, that the angle *B* is

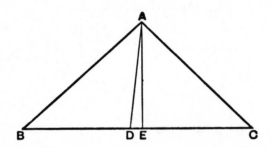

acute, and that the angle *A* is greater than the angle *C*. From *A* draw *AD* making the angle *BAD* equal to the angle *C*, and cutting *BC* in *D*. From *A* draw *AE* perpendicular to *BC*.

The triangles *ABC*, *ABD* are equiangular; hence, by Euc. VI. 19,

$$\triangle ABC : \triangle ABD = AC^2 : AD^2.$$

Also the triangles *ABC*, *ABD* are of equal altitude; hence by Euc. VI. 1,

$$\triangle ABC : \triangle ABD = BC : BD,$$
$$\therefore AC^2 : AD^2 = BC : BD.$$
$$\therefore \frac{AC^2}{BC} = \frac{AD^2}{BD}.$$

Hence, by Euc. II. 13,

$$\frac{AB^2 + BC^2 - 2BC \cdot BE}{BC} = \frac{AB^2 + BD^2 - 2BD \cdot BE}{BD};$$

*See a note by M. Coccoz in *L'Illustration*, Paris, Jan. 12, 1895.

$$\therefore \frac{AB^2}{BC} + BC - 2BE = \frac{AB^2}{BD} + BD - 2BE.$$

$$\therefore \frac{AB^2}{BC} - BD = \frac{AB^2}{BD} - BC.$$

$$\therefore \frac{AB^2 - BC \cdot BD}{BC} = \frac{AB^2 - BC \cdot BD}{BD}.$$

$$\therefore BC = BD,$$

a result which is impossible.

Third fallacy.* *To prove that the sum of the lengths of two sides of any triangle is equal to the length of the third side.* Let

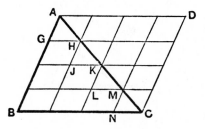

ABC be a triangle. Complete the parallelogram of which AB and BC are sides. Divide AB into $n+1$ equal parts, and through the points so determined draw n lines parallel to BC. Similarly, divide BC into $n+1$ equal parts, and through the points so determined draw n lines parallel to AB. The parallelogram $ABCD$ is thus divided into $(n+1)^2$ equal and similar parallelograms.

I draw the figure for the case in which n is equal to 3; then, taking the parallelograms of which AC is a diagonal, as indicated in the diagram, we have

$$AB + BC = AG + HJ + KL + MN$$
$$+ GH + JK + LM + NC.$$

A similar relation is true however large n may be. Now let n increase indefinitely. Then the lines AG, GH, etc. will get

---

**The Canterbury Puzzles*, by H.E. Dudeney, London, 1919, pp. 51–54.

smaller and smaller. Finally the points $G, J, L, \ldots$ will approach indefinitely near the line AC, and ultimately will lie on it; when this is the case the sum of AG and GH will be equal to AH, and similarly for the other similar pairs of lines. Thus, ultimately,

$$AB + BC = AH + HK + KM + MC$$
$$= AC,$$

a result which is impossible.

Fourth fallacy. *To prove that every triangle is isosceles.* Let ABC be any triangle. Bisect BC in D, and through D draw DO perpendicular to BC. Bisect the angle BAC by AO.

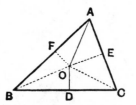

First. If DO and AO do not meet, then they are parallel. Therefore AO is at right angles to BC. Therefore $AB = AC$.

Second. If DO and AO meet, let them meet in O. Draw OE perpendicular to AC. Draw OF perpendicular to AB. Join OB, OC.

Let us begin by taking the case where O is inside the triangle, in which case E falls on AC and F on AB.

The triangles AOF and AOE are congruent, since the side AO is common, angle $OAF =$ angle OAE, and angle $OFA =$ angle OEA. Hence $AF = AE$. Also, the triangles BOF and COE are congruent. For since OD bisects BC at right angles, we have $OB = OC$; also, since the triangles AOF and AOE are congruent, we have $OF = OE$; lastly, the angles at F and E are right angles. Therefore, by Euc. I. 47 and I. 8, the triangles BOF and COE are congruent. Hence $FB = EC$.

Therefore $AF + FB = AE + EC$, that is, $AB = AC$.

The same demonstration will cover the case where DO and AO meet at D, as also the case where they meet outside BC

but so near it that E and F fall on AC and AB and not on AC and AB produced.

Next take the case where DO and AO meet outside the triangle, and E and F fall on AC and AB produced. Draw OE perpendicular to AC produced. Draw OF perpendicular to AB produced. Join OB, OC.

Following the same argument as before, from the congruence of the triangles AOF and AOE, we obtain $AF = AE$; and, from the congruence of the triangles BOF and COE, we obtain $FB = EC$. Therefore $AF - FB = AE - AC$, that is, $AB = AC$.

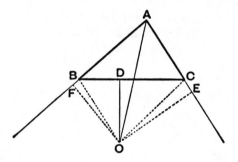

Thus in all cases, whether or not DO and AO meet, and whether they meet inside or outside the triangle, we have $AB = AC$: and therefore every triangle is isosceles, a result which is impossible.

Fifth fallacy.* *To prove that* $\pi/4$ *is equal to* $\pi/3$. On the hypotenuse, BC, of an isosceles right-angled triangle, DBC, describe an equilateral triangle ABC, the vertex A being on the same side of the base as D is. On CA take a point H so that $CH = CD$. Bisect BD in K. Join HK and let it cut CB (produced) in L. Join DL. Bisect DL at M, and through M draw MO perpendicular to DL. Bisect HL at N, and through N draw NO perpendicular to HL. Since DL and HL intersect, therefore MO and NO will also intersect; moreover, since BDC is a right angle, MO and NO both slope away from DC and therefore

---

*This ingenious fallacy is due to Captain Turton: it appeared for the first time in the third edition of this work.

they will meet on the side of *DL* remote from *A*. Join *OC, OD, OH, OL*.

Since the triangles *OMD* and *OML* are congruent, *OD = OL*. Similarly, since the triangles *ONL* and *ONH* are congruent, *OL = OH*. Therefore *OD = OH*. Now in the triangles *OCD* and *OCH*, we have *OD = OH*, *CD = CH* (by construction), and *OC* common, hence (by Euc. I. 8) the angle *OCD* is equal to the angle *OCH*. Hence the angle *BCD* is equal to the angle *BCH*, that is, $\pi/4$ is equal to $\pi/3$, which is absurd.

Sixth fallacy.* *To prove that, if two opposite sides of a quadrilateral are equal, the other two sides must be parallel.* Let *ABCD* be a quadrilateral such that *AB* is equal to *DC*. Bisect *AD* in *M*, and through *M* draw *MO* at right angles to *AD*. Bisect *BC* in *N*, and draw *NO* at right angles to *BC*.

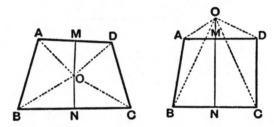

If *MO* and *NO* are parallel, then *AD* and *BC* (which are at right angles to them) are also parallel.

If *MO* and *NO* are not parallel, let them meet in *O*; then *O* must be either inside the quadrilateral as in the left-hand diagram or outside the quadrilateral as in the right-hand diagram. Join *OA, OB, OC, OD*.

Since *OM* bisects *AD* and is perpendicular to it, we have *OA = OD*, and the angle *OAM* equal to the angle *ODM*. Similarly *OB = OC*, and the angle *OBN* is equal to the angle *OCN*. Also by hypothesis *AB = DC*, hence, by Euc. I. 8, the triangles *OAB* and *ODC* are congruent, and therefore the angle *AOB* is equal to the angle *DOC*.

Hence in the left-hand diagram the sum of the angles *AOM*, *AOB* is equal to the sum of the angles *DOM, DOC*; and in the

Mathesis, October 1893, series 2, vol. III, p. 224.

right-hand diagram the difference of the angles *AOM*, *AOB* is equal to the difference of the angles *DOM*, *DOC*; and therefore in both cases the angle *MOB* is equal to the angle *MOC*, i.e. *OM* (or *OM* produced) bisects the angle *BOC*. But the angle *NOB* is equal to the angle *NOC*, i.e. *ON* bisects the angle *BOC*; hence *OM* and *ON* coincide in direction. Therefore *AD* and *BC*, which are perpendicular to this direction, must be parallel. This result is not universally true, and the above demonstration contains a flaw.

Seventh fallacy.* The following argument is taken from a text-book on electricity, published in 1889 by two distinguished mathematicians, in which it was presented as valid. A given vector *OP* of length *l* can be resolved in an infinite number of ways into two vectors *OM*, *MP*, of lengths *l'*, *l''*, and we can make *l'*/*l''* have any value we please from nothing to infinity. Suppose that the system is referred to rectangular axes *Ox*, *Oy*; and that *OP*, *OM*, *MP* make respectively angles θ, θ', θ'' with *Ox*. Hence, by projection on *Oy* and on *Ox*, we have

$$l \sin \theta = l' \sin \theta' + l'' \sin \theta'',$$
$$l \cos \theta = l' \cos \theta' + l'' \cos \theta''.$$
$$\therefore \tan \theta = \frac{n \sin \theta' + \sin \theta''}{n \cos \theta' + \cos \theta''},$$

where $n = l'/l''$. This result is true whatever be the value of n. But n may have any value (e.g., $n = \infty$, or $n = 0$), hence $\tan \theta = \tan \theta' = \tan \theta''$, which obviously is impossible.

Eighth fallacy.* Here is a fallacious investigation of the value of π: it is founded on well-known quadratures. The area of the semi-ellipse bounded by the minor axis is (in the usual notation) equal to $\frac{1}{2}\pi ab$. If the centre is moved off to an indefinitely great distance along the major axis, the ellipse degenerates into a parabola, and therefore in this particular limiting position the area is equal to two-thirds of the circumscribing rectangle. But the first result is true whatever be the dimensions of the curve.

*This was communicated to me by Mr. R. Chartres.

$$\therefore \tfrac{1}{2}\pi ab = \tfrac{2}{3}a \times 2b,$$
$$\therefore \pi = 8/3,$$

a result which obviously is untrue.

Ninth fallacy.   *Every ellipse is a circle.* The focal distance of a point on an ellipse is given (in the usual notation) in terms of the abscissa by the formula $r = a + ex$. Hence $dr/dx = e$. From this it follows that r cannot have a maximum or minimum value. But the only closed curve in which the radius vector has not a maximum or minimum value is a circle. Hence, every ellipse is a circle, a result which obviously is untrue.

GEOMETRICAL PARADOXES

To the above examples I may add the following questions, which, though not exactly fallacious, lead to results which at a hasty glance appear impossible.

First paradox.   The first is a problem sent to me by Mr. W. Renton, to rotate a plane lamina (say, for instance, a sheet of paper) through four right angles so that the effect is equivalent to turning it through only one right angle.

Second paradox.   As in arithmetic, so in geometry, the theory of probability lends itself to numerous paradoxes. Here is a very simple illustration. A stick is broken at random into three pieces. It is possible to put them together into the shape of a triangle provided the length of the longest piece is less than the sum of the other two pieces (cf. Euc. I. 20), that is, provided the length of the longest piece is less than half the length of the stick. But the probability that a fragment of a stick shall be less than half the original length of the stick is 1/2. Hence the probability that a triangle can be constructed out of the three pieces into which the stick is broken would appear to be 1/2. This is not true, for actually the probability is 1/4.

Third paradox.   The following example illustrates how easily the eye may be deceived in demonstrations obtained by actually dissecting the figures and re-arranging the parts. In fact proofs by superposition should be regarded with considerable distrust unless they are supplemented by mathematical

reasoning. The well-known proofs of the propositions Euclid
I. 32 and Euclid I. 47 can be so supplemented and are valid.
On the other hand, as an illustration of how deceptive a non-
mathematical proof may be, I here mention the familiar paradox that a square of paper, subdivided like a chessboard into
64 small squares, can be cut into four pieces which being put
together form a figure containing 65 such small squares.*
This is effected by cutting the original square into four pieces
in the manner indicated by the thick lines in the first figure.
If these four pieces are put together in the shape of a rectangle
in the way shown in the second figure it will appear as if this
rectangle contains 65 of the small squares.

This phenomenon, which in my experience non-mathema-
ticians find perplexing, is due to the fact that the edges of the

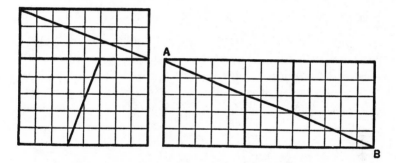

four pieces of paper, which in the second figure lie along the
diagonal AB, do not coincide exactly in direction. In reality
they include a small lozenge or diamond-shaped figure, whose
area is equal to that of one of the 64 small squares in the original
square, but whose length AB is much greater than its breadth.
The diagrams show that the angle between the two sides of
this lozenge which meet at A is $\tan^{-1}\frac{2}{5}-\tan^{-1}\frac{3}{8}$, that is, is
$\tan^{-1}\frac{1}{46}$, which is less than $1\frac{1}{4}°$. To enable the eye to distinguish
so small an angle as this the dividing lines in the first figure

*I do not know who discovered this paradox. It is given in various books,
but I cannot find an earlier reference to it than one in the *Zeitschrift für
Mathematik und Physik*, Leipzig, 1868, vol. XIII, p. 162. Some similar paradoxes
were given by Ozanam, 1803 edition, vol. I, p. 299.

would have to be cut with extreme accuracy and the pieces placed together with great care.

This paradox depends upon the relation $5 \times 13 - 8^2 = 1$. Similar results can be obtained from the formulae

$$13 \times 34 - 21^2 = 1, \quad 34 \times 89 - 55^2 = 1, \; \dots \; ;$$

or from the formulae

$$5^2 - 3 \times 8 = 1, \quad 13^2 - 8 \times 21 = 1, \quad 34^2 - 21 \times 55 = 1, \; \dots .$$

These relations connect sets of three consecutive Fibonacci numbers (see p. 56). The general formula

$$b_n c_{n+1} - c_n b_{n+1} = (-1)^n$$

holds for two adjacent convergents to any continued fraction.

CONTINUED FRACTIONS AND LATTICE POINTS*

Consider a board, divided into a large number of equal squares, with small pegs sticking up at the vertices of all the squares. Let us regard the lines of pegs nearest to two adjacent edges of the board as axes of co-ordinates, so that every peg has co-ordinates which are non-negative integers. If y/x is a fraction

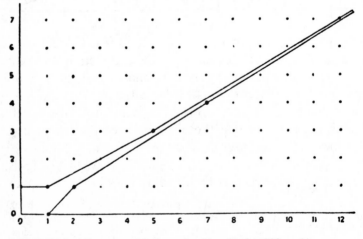

*Cf. F. Klein, *Elementary Mathematics*, New York, 1932, p. 44.

in its lowest terms, a stretched string joining the pegs (0, 0) and (x, y) will not touch any other pegs. This string being fastened at (x, y), let us move it from the other end without allowing it to jump over any pegs. If the string is stretched taut, with its free end at (1, 0), it will, in general, press against certain pegs $(x_1, y_1), (x_2, y_2), \ldots$ between (1, 0) and (x, y). If its free end is held at (0, 1) instead of (1, 0), it will press against other pegs $(x_1', y_1'), (x_2', y_2'), \ldots$. It can be proved that the fractions $y_1/x_1, y_2/x_2, \ldots$ are alternate convergents to y/x, while y_1'/x_1', $y_2'/x_2', \ldots$ are the remaining convergents. (The convergent y_r'/x_r' comes just before or just after y_r/x_r according as y/x is less or greater than 1.)

This construction shows clearly the manner in which the convergents approximate to y/x, alternately by excess and defect. The fraction y/x measures the *gradient* of the string in its original position. There is no difficulty in extending these notions to the case of a string of irrational gradient, firmly attached "at infinity." The case when the gradient is

$$\frac{1}{e-1} = 1/1 + 1/1 + 1/2 + 1/1 + 1/1 + 1/4 + 1/\ldots$$

is shown above. Observe that the string from (1, 1) to (5, 3) touches (without "pressing against") the peg (3, 2). The fraction $\frac{2}{3}$ is one of the *intermediate convergents* which, together with the ordinary or *principal* convergents $(\frac{0}{1}, \frac{1}{1}, \frac{1}{2}, \frac{3}{5}, \frac{4}{7}, \frac{7}{12}, \ldots)$, make up the set of fractions of best approximation.

If $b_{n-1}/c_{n-1}, b_n/c_n, b_{n+1}/c_{n+1}$ are three consecutive (principal) convergents to any continued fraction, we know that

$$b_n/c_n = (b_{n+1} - b_{n-1})/(c_{n+1} - c_{n-1}).$$

Geometrically, this means that the line from (0, 0) to (c_n, b_n) is parallel to the line from (c_{n-1}, b_{n-1}) to (c_{n+1}, b_{n+1}).

GEOMETRICAL DISSECTIONS

Problems requiring the division by straight lines of a given plane rectilinear figure into pieces which can be put together

in some other assigned form are well known. A class of geometrical recreations is concerned with such constructions.

Pythagorean dissection.   Euclid's proof for the theorem of
Pythagoras (Book I, Proposition 47) is clumsy and hard to
memorize. A far more symmetrical proof has been attributed
to the Indian mathematician Arya-Bhata, who was born in
A.D. 466. To prove that $a^2 + b^2 = c^2$ for a right-angled triangle
ABC, he placed four replicas of this triangle inside a square of
side $a + b$, as in figure i. In one arrangement the rest of the area
is made up of squares of sides a and b; in the other (where the
first three triangles have been shifted by translation), it consists
of the single square of side c.

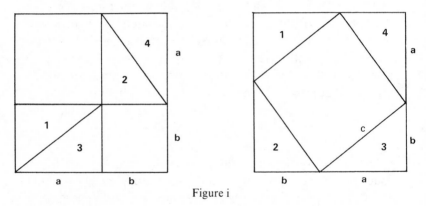

Figure i

In the language of Hilbert's *Foundations of Geometry* (Chicago, 1902, p. 58) the pair of smaller squares and the single
larger square have thus been shown to be "of equal content"
or "equivalent by completion" (*inhaltsgleich*). It would perhaps
be more satisfactory to show that they are "of equal area" or
"equivalent by decomposition" (*zerlegungsgleich*). This refinement was achieved by Perigal,* who placed the squares
AG and DF side by side, as in figure ii (where $AH = HG = a$,
$DE = EF = b$, and $a \geq b$). Cuts BH and BE will divide the combined area into three pieces, two of which are triangles with
sides a, b, c. These triangles ABH and BDE can be translated

*H. Perigal, *Messenger of Mathematics*, 1873, vol. II, N.S., pp. 103–106;
H.E. Dudeney, *Amusements*, London, 1917, p. 32.

to new positions *FEC* and *HGC*, yielding the square *BECH* of side *c*.

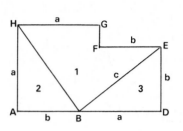
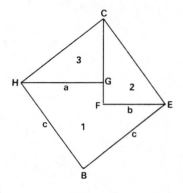

Figure ii

Montucla's dissection.   Demonstrations of a few similar propositions had long been current, but towards the close of the eighteenth century attention was recalled to this kind of solution by Montucla, who proposed and solved the problem of dividing a rectangle so that the parts could be put together in the form of a square; he also solved the converse problem. Later, other solutions of this problem were given by P. Busschop and de Coatpont, who made respectively eight-part and seven-part dissections. The former also had constructions for making a square from a five-part division of a regular hexagon, and from a seven-part division of a regular pentagon.*

Polygonal dissections.   The more general problem of the dissection of a given polygon of any number of sides, and the rearrangement of the parts in the form of another polygon of equal area, had been raised by Bolyai, and a method of solution indicated by Gerwien. The question continued to attract occasional attention. In particular a solution for the case of a polygon and a triangle was given by Euzet in 1854, and the wider problem of two polygons was discussed by E. Guitel in

*Ozanam's *Récréations mathématiques*, 1803. English edition, vol. I, pp. 292–298; 1840 edition, pp. 127–129; Paul Busschop, *Nouvelle Correspondance Mathématique*, Brussels, 1875, vol. II, p. 83; de Coatpont, *ibid.*, 1876, vol. III, p. 116.

1895, by E. Holst in 1896,* and more recently by A. Mineur.†
Consider first the dissection of a given triangle to form a
rectangle of given base. In figure iii, a line *DE*, parallel to the

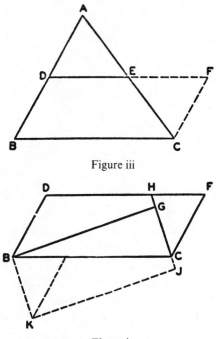

Figure iii

Figure iv

base *BC* of the given triangle *ABC* and bisecting the sides, cuts
the triangle into two parts which may be assembled into a
parallelogram *BCFD*. Now with *B* as centre and with the
given base as radius draw an arc. Make a cut *CH* tangent to
this arc and a cut *BG* perpendicular to *CH*. The point *G* may
fall inside the parallelogram, as in figure iv, or outside, as in
figure v. In the latter case, locate *J* (on *CH*) so that *CJ* = *HG*,
and make the cut *JL* parallel to *BG*. Then the pieces may be

*P. Gerwien, *Crelle's Journal*, 1833, p. 228; M. Euzet, *Nouvelles Annales de
Mathématiques*, 1854, vol. XIII, pp. 114–115; E. Guitel, *Association Française
pour l'Avancement des Sciences*, 1895, pp. 264–267; E. Holst, *L'Intermédiaire
des Mathématiciens*, 1896, vol. III, pp. 91–92.

†See *Mathesis*, 1931, pp. 150–152. The following treatment is due to Michael
Goldberg.

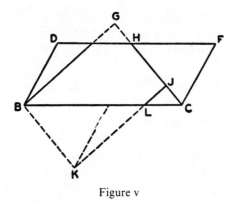

Figure v

assembled into the desired rectangle *BGJK*. If the base *BG* is too long or too short to follow this procedure directly, it will be necessary first to change the proportions of the parallelogram *BCFD* by cutting it into three pieces and rearranging these as in figure vi (where $E'F = BC' \geqq \frac{1}{2}BC$). Of course, if the base of

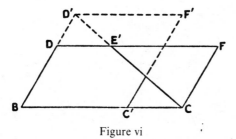

Figure vi

the rectangle had been chosen to be the side of the square equivalent to the triangle, the given triangle would have been converted into a square.

Every polygon can be cut into triangles by drawing a sufficient number of diagonals. Therefore every polygon can be transformed into a square or rectangle by transforming the component triangles into rectangles all having the same base and then stacking the rectangles into a column.

We can always use a rectangle as an intermediate stage in the transformation of one polygon into another. Given the dissections which transform the given polygon and the desired polygon each into the same rectangle, we go from the given polygon to the rectangle by using the first transformation and then to the desired polygon by reversing the second transformation.

Minimum dissections.   The above-mentioned writers aimed only at finding a solution, and did not in general trouble themselves with considering the smallest number of pieces required. In 1905, the special cases of a four-part dissection of a pair of triangles, a four-part dissection of a triangle and parallelogram, and dissections of a pair of parallelograms were given by H.M. Taylor,* and he recognized the desirability of finding the smallest necessary number of cuts.

Puzzle dissections.   More   recently   H.E. Dudeney   has propounded various ingenious puzzles of this kind, in all cases the number of parts being specified.† His reputation served to attract attention to this class of problems. As illustrating his results and as geometrical recreations of this type, I pick out

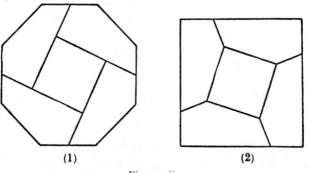

(1)   (2)

Figure vii

his problems, (a) to divide, by two straight lines, a Greek cross, that is, one made up of five equal squares, into four pieces all of the same shape and size which can be put together to make a square; (b) to divide an isosceles right-angled triangle into four pieces which can be put together to make a Greek cross;

*Messenger of Mathematics, vol. XXXV, pp. 81–101.
†For instance, see his *Amusements in Mathematics*, London, 1917, p. 27 et seq.

(c) to divide a regular pentagon into six pieces which can be put together to make a square; (d) to divide an equilateral triangle into four pieces which can be put together to make a square. The reader interested in the subject will like to compare Dudeney's solution of the last question with that given by Taylor already mentioned, and that by Macaulay referred to below.

A. H. Wheeler and M. Goldberg* have divided a regular pentagon into six pieces which can be put together to make an equilateral triangle; and J. Travers has divided a regular octagon into five pieces which can be put together to make a square. (See figure vii.)

Macaulay's four-part dissections.   The theory of four-part dissections of pairs of rectilinear figures of equal area has been discussed by W.H. Macaulay.† He has treated four-part dissections of pairs of triangles, of a triangle and a parallelogram, of pairs of quadrilaterals, of pairs of pentagons each with two sides equal and parallel, and of pairs of certain related hexagons. His results are projective and all of them are deducible from his hexagon dissections. This is an interesting generalization.

Volume dissections.   It is natural to ask whether every polyhedral solid can be divided, by a finite number of plane cuts, into pieces which will fit together to form any other polyhedral solid of the same volume. The answer is No; it was proved by M. Dehn‡ that not every tetrahedron can be transformed by dissection into a prism. This annihilates any hopes for a general method of volume dissections analogous to polygonal dissections, although special volume dissections are quite possible.

The duplication of the cube.   Here is a variant of the "Delian Problem" (which we shall consider in chapter XII): given a line of length $2^{1/3}$ (or $\sqrt[3]{2}$), divide two cubes of unit edge, by plane

*American Mathematical Monthly, 1952, vol. LIX, pp. 106–107.

†Mathematical Gazette, 1914, vol. VII, p. 381; vol. VIII, 1915, pp. 72, 109; Messenger of Mathematics, vol. XLVIII, 1919, p. 159; vol. XLIX, 1919, p. 111.

‡See H.C. Lenhard, Elemente der Mathematik, 1962, vol. XVII, pp. 108–109.

cuts, into pieces which can be assembled into a single cube. This problem can be solved by two applications of the transformation of a given rectangle into a rectangle of given base. The "minimum" dissection is obtained by using the method of figure vi, rather than that of figure iv.

First place the two equal cubes together to form a square prism $2 \times 1 \times 1$. Then mark on one of the rectangular faces (2×1) the cuts necessary to transform it into a rectangle of base $2^{1/3}$, which is equal to the edge of the duplicated cube. Through these lines pass planes perpendicular to this face, cutting the prism into three parts.

Assemble these parts to form the new prism $2^{1/3} \times 2^{2/3} \times 1$. Then, on either of the faces $2^{2/3} \times 1$, mark the cuts necessary to transform this rectangle into a square of side $2^{1/3}$. Through these lines pass planes perpendicular to this face, and assemble the three parts into the desired cube. The original prism $2 \times 1 \times 1$ has been divided into seven irregular solids by these cuts. If the pieces are disarranged, the task of assembling them to form the cube or the prism is not an easy puzzle.

This problem was proposed by W.F. Cheney, Jr., and solved thus by A.H. Wheeler.*

CYCLOTOMY

At the age of nineteen, Gauss† proved that the solution of the *cyclotomic equation* $x^p = 1$, where p is a prime, can be reduced to the solution of a succession of quadratic equations whenever the prime p is one of the numbers F_m (in the notation of page 68). Later,‡ P.L. Wantzel strengthened this "whenever" to "if and only if." Since the roots of the cyclotomic equation are

$$\cos(2r\pi/p) + i \sin(2r\pi/p) \quad (r = 0, 1, \ldots, p-1),$$

it follows that a regular n-gon, where n is odd, can be drawn by a Euclidean construction (i.e. with ruler and compasses) if, and only if, n is a *Fermat prime* F_m, or the product of several dif-

*American Mathematical Monthly, 1935, vol. XLII, p. 509.
†Disquisitiones Arithmeticae, 1801.
‡Liouville's Journal de Mathématiques, 1837, vol. II, pp. 366–372.

ferent Fermat primes. It suffices to consider odd values of n, since the construction of a $(2^k n)$-gon involves that of an n-gon followed by k angle-bisections.

If we assume that all Fermat Numbers after F_4 are composite, it follows* that n must be a divisor of $3 . 5 . 17 . 257 . 65537$ $= 2^{32} - 1 = 4294{,}967295$. If a polygon with a larger odd number of sides can be drawn with ruler and compasses, this larger number must have at least 39457 digits, since the first Fermat Number whose composite character is in doubt is F_{17} $(= 2^{131072} + 1)$.

Constructions for the triangle and pentagon are familiar. A construction for the 15-gon follows immediately, since $\frac{4}{5}\pi - \frac{2}{3}\pi = \frac{2}{15}\pi$; you merely have to inscribe in one circle both a triangle and a pentagon. The following construction for the 17-gon is due to H.W. Richmond, who gave also an analogous construction for the pentagon to compare with it.

The problem is to inscribe, in a given circle, a regular 17-gon with one vertex at P_0 (see figure viii). Let OB be a radius, per-

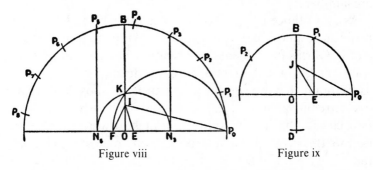

Figure viii   Figure ix

pendicular to the diameter through P_0. Find a point I on OB so that $OI = \frac{1}{4}OB$. Join IP_0, and find points E, F on OP_0 so that angle $OIE = \frac{1}{4}OIP_0$ and angle $FIE = \frac{1}{4}\pi$. Let the circle on FP_0 as diameter cut OB in K, and let the circle with centre E and radius EK cut OP_0 in N_3 (between O and P_0) and N_5. Let lines N_3P_3, N_5P_5, parallel to OB, cut the original circle in P_3, P_5. Then the arcs P_0P_3, P_0P_5 will be $\frac{3}{17}$ and $\frac{5}{17}$ of the circumference.

*Kraïtchik, *La Mathématique des Jeux*, Brussels, 1930, p. 99.

The proof* involves repeated application of the principle that the roots of the equation $x^2 + 2x \cot 2C = 1$ are $\tan C$ and $-\cot C$.

For the pentagon† (figure ix), define B as before, bisect OB at J, and find E on OP_0 so that JE bisects the angle OJP_0. If the line through E parallel to OB cuts the circle in P_1 the arc P_0P_1 will be $\frac{1}{5}$ of the circumference.

Constructions for polygons of 51, 85, and 255 sides follow immediately; so do constructions for polygons whose numbers of sides are these numbers multiplied by any power of 2.

COMPASS PROBLEMS

It is well known that Euclid in his *Elements* confined his constructions to those which could be made with ungraduated rulers and compasses. The use of a ruler is, however, unnecessary.‡ Mascheroni§ established a connected series of propositions by constructions made with compasses alone. Of course, the logical sequence is very different from that with which we are familiar.

As an instance which will illustrate the subject, I select the problem to find a point midway between two given points A and B. Of this fundamental proposition Mascheroni gave five solutions (prop. 66). Here are two of them: they rest on the assumption that we can draw a semicircle whose centre and one extremity are given, a result he had previously established. In each case the demonstration is straightforward. Other solutions of this proposition have been given, some of which evade the use of a semicircle.

One of his constructions is as follows. With B as centre,

Quarterly Journal of Mathematics, 1893, vol. XXVI, p. 206.

†A slightly simpler construction for the pentagon has been suggested by H.E. Dudeney, *Amusements in Mathematics*, p. 38. With centre J and radius JP_0, draw an arc to cut BO (produced) in D. Then an arc with centre P_0 and radius P_0D will cut the original circle in P_1.

‡Michael Goldberg, *School Science and Mathematics*, 1925, vol. XXV, pp. 961–965.

§His work was published at Pavia in 1797. However, it has recently become known that he was largely anticipated by G. Mohr. whose *Euclides Danicus* was published at Amsterdam in 1672 (and at Copenhagen in 1928).

and a radius *BA*, describe a semicircle of which *A* and *C* are the extremities. With centres *A* and *C*, and radii *AB* and *CA*, describe circles which cut in *P* and *Q*. With centres *P* and *Q*, and a radius equal to *AB*, describe circles. These will cut in a point midway between *A* and *B*.

Here is another of his solutions, which for some purposes he preferred. With *B* as centre, and radius *BA*, describe a semicircle of which *A* and *C* are the extremities. With *A* and *C* as centres, and a radius equal to *AB*, describe circles which cut the semicircle in *H* and *K* respectively. With *A* and *C* as centres, and a radius equal to *AC*, describe circles which cut the last-mentioned circles (above *AC*) in *Q* and *P* respectively. With centres *P* and *C*, and radii *PA* and *PQ*, describe circles. These will cut in a point midway between *A* and *B*.

Numerous geometrical recreations of this kind can be made by any one, for all that is necessary is to select at random one of Euclid's propositions, and see how it can be established by using only circles. As instances, I select the construction on a given line of a triangle similar to a given triangle (prop. 125); and the construction of a regular pentagon of given dimensions (prop. 137). Whatever be the solution obtained, it is always interesting to turn to Mascheroni's book, and see how the question is tackled there.

THE FIVE-DISC PROBLEM

The problem of completely covering a fixed red circular space by placing over it, one at a time, five smaller equal circular tin discs is familiar to frequenters of English fairs. Its effectiveness depends on making the tin discs as small as possible, and therefore leads to the interesting geometrical question of finding the size of the smallest tin discs which can be used for the purpose.

The problem is soluble if the radius of each tin disc is just greater than three-fifths of the radius of the red circle. Of course, in a show the visitor is not allowed to move the discs when once he has put them down, and it is only rarely, very rarely, that he succeeds empirically in placing them correctly.

The rule works out thus. If O is the centre of the red circle, a its radius, and AOB a diameter of it, we take on OA a point P such that approximately $OP = a/35$. We then place the first tin disc with its centre on OB, and so that P is on its edge: suppose that its edge meets the edge of the red circle in C and C', points on opposite sides of AB. Place the next two discs so that in each case AP is a chord of the disc. Suppose that these discs meet the edge of the red circle in D and D' respectively, C and D being on the same side of AB. Then if we place the two remaining discs so that CD and $C'D'$ are chords of them, the problem will solved. The minimum discs are not obtained, as one might at first guess, by the overlapping of five discs placed at the vertices of an inscribed pentagon.

In practice and for simplicity it is desirable to make the tin discs a trifle larger than the theory requires, and to treat P as coincident with O.

The mathematical discussion on this problem is too technical and long to insert here. Probably a bare statement of the results such as is given above is all that most readers will want. Should closer approximations be desired, here they are.* If the radius of the red circle be taken as one foot, the critical radius for the discs, below which the problem is impossible, is ·609383 foot. Also $OP = $·028547 foot; hence O lies very near but not quite on the circumference of the first disc which is put down. If three of the discs are placed so that their edges pass through O, the radius of each of them must exceed ·6099579 foot, a length practically indistinguishable from that of the radius of the minimum disc. If the discs are put with their centres at the vertices of an inscribed pentagon and their edges passing through O, the radius of each of them must exceed ·6180340 foot. It follows that unless the discs are cut with extreme accuracy the problem may be solved by making the circumference of each disc pass through O; the possibility of using this inaccurate rule is a serious defect in the problem when used as the foundation of a puzzle.

*See E.H. Neville, *Proceedings of the London Mathematical Society*, 1915, second series, vol. XIV, pp. 308–326.

I believe that the discs used in fairs are generally large enough to allow the employment of the inaccurate rule, though even then it is safer to use the correct method. In an example made for myself I put a minute faint mark near the centre of the red circle but just far enough away to ensure failure for those who make that point lie on the edge of each disc. Notwithstanding their neglect or ignorance of the correct rule, showmen seem to find that the game is profitable, and obviously this is an excellent test of its merits from their point of view.

LEBESGUE'S MINIMAL PROBLEM

Here is an unsolved problem about covering an area. The *diameter* of any figure may be defined as the longest straight line that can be drawn so as to join two points of the figure. Lebesgue's problem* is the determination of *the plane lamina of least area which, when suitably oriented, will cover any given plane figure of unit diameter.*

A circle of unit diameter is too small, since, although it will cover a square of unit diagonal, it will not cover an equilateral triangle of unit side. On the other hand, a regular hexagon circumscribed about the circle is unnecessarily large. Thus the required area lies between $\frac{1}{4}\pi$ and $\frac{1}{2}\sqrt{3}$. Its precise value, and the shape, have never been determined.

KAKEYA'S MINIMAL PROBLEM

A somewhat similar problem is the determination of the least area swept over by a straight line of unit length which is reversed in direction by a continuous motion during which it takes every possible orientation in the plane. Kakeya's problem had assumed the proportions of a famous unsolved problem although its complete solution was published only ten years after it was proposed. Osgood and Kubota suggested the deltoid (or three-cusped hypocycloid) as a possible solu-

---

*Julius Pál, K. Danske videnskabernes selskab, Copenhagen. Mathematisk-fysiske meddelelser, 1920, vol. III, no. 2, pp. 1–35; S. Kakeya, Tôhoku Science Reports, 1917, vol. VI, pp. 71–78; Coxeter, Eureka, 1958, vol. XXI, p. 13.

tion, the area in this case being just half that of the circle of unit diameter. (See figure x.)

Besicovitch showed* that there is no smallest area, that the area can be made as small as we please! This astounding fact is difficult to believe, so I proceed to give a very brief outline of his elegant demonstration.

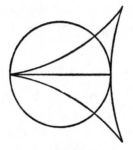

Figure x

It is enough to describe a figure in which a unit line segment can be turned through a right angle, because then by putting two such figures together we can make one with Kakeya's property. Begin with a triangle ABC in which $AC = BC$, C is a right angle, and the altitude from C to AB equals 1. A unit segment initially on AC can obviously be turned in the triangle so as to end up on BC. The idea of Besicovitch was to do this turning bit by bit. First cut the triangle into a large number of thin triangles by lines joining C to points evenly spaced on AB. Then slide each of the pieces a different distance along AB, not changing its size, shape, or orientation, so as to make them all overlap as much as possible. Besicovitch proved the surprising fact that by taking the number of pieces large enough and sliding each one the right amount you can get in the end a figure having area as small as you please. Unfortunately, while a unit segment can make lots of small changes of direction in this figure, it cannot turn continuously through a whole right angle for lack of a way to get from each small triangle to the next. This defect is corrected by enlarging the

*A.S. Besicovitch, *Mathematische Zeitschrift*, 1928, vol. XXVII, p. 312.

figure with so-called "joins," namely paths connecting successive triangles. To keep down the area contributed by these joins, they are made rather circuitous; each one consists of two long, almost parallel, line segments, one starting from each of the two triangles to be joined, and where they meet an additional thin triangle of height 1. The area of the long region enclosed between the two lines of a join is not counted in the area of the figure. This means that the finished figure is a rather lacy, spidery thing, with many narrow loops radiating out long distances from it.

Mathematicians continued to wonder for some years whether the perforated character of Besicovitch's figure, and its tendency to reach out long distances, can be avoided. Technically the remaining problem was: how small an area is possible for a *simply connected* (and/or bounded) set with Kakeya's property? (A set is called simply connected if it has no holes.) Incidentally, the corresponding problem for *convex* sets had already been solved by J. Pál* before Besicovitch's solution of the general problem; the answer is the equilateral triangle of area $1/\sqrt{3}$. Not until 1965 did Bloom and Schoenberg† independently discover simply connected sets with the Kakeya property smaller than Kakeya's own deltoid; the areas of their examples approach $(5 - 2\sqrt{2})\pi/24$, which is approximately $\pi/11$. Then Cunningham found that actually restricting the figure to be simply connected and contained in a circle of radius 1 does not change the answer as compared with Besicovitch's: you can still make the area as small as you please.

A problem nevertheless remains, which is still unsolved. A set is called *star-shaped* if there is a point in it which can be joined to every other point of the set by a line segment contained in the set. How small can the area of a star-shaped figure with Kakeya's property be? The examples of Bloom and Schoenberg are star-shaped, showing that such sets can

---

*Julius Pál, *Mathematische Annalen*, 1921, vol. LXXXIII, pp. 311–319.

†F. Cunningham, Jr. and I.J. Schoenberg, *Canadian Journal of Mathematics*, 1965, vol. XVII, pp. 946–956.

have areas as near as you please to the number (nearly $\pi/11$) mentioned above. (Schoenberg picturesquely describes a typical one of his figures as resembling the locus of the end of a Foucault pendulum swinging ten thousand times.) On the other hand, Cunningham's examples are not star-shaped, and he has proved that the area of a star-shaped set with the Kakeya property cannot be less than $\pi/108$. Thus, the lower bound for areas of star-shaped Kakeya sets has some value (between $\pi/108$ and $\pi/11$) which remains to be found.

ADDENDUM

Note. Page 84. The required rotation of the lamina can be effected thus. Suppose that the result is to be equivalent to turning it through a right angle about a point O. Describe on the lamina a square $OABC$. Rotate the lamina successively through two right angles about the diagonal OB as axis and through two right angles about the side OA as axis, and the required result will be attained.

CHAPTER IV

GEOMETRICAL RECREATIONS (*continued*)

Leaving now the question of formal geometrical propositions, I proceed to enumerate a few games or puzzles which depend mainly on the relative position of things, but I postpone to chapter x the discussion of such amusements of this kind as necessitate any considerable use of arithmetic or algebra. Some writers regard draughts, solitaire, chess, and such-like games as subjects for geometrical treatment in the same way as they treat dominoes, backgammon, and games with dice in connection with arithmetic: but these discussions require too many artificial assumptions to correspond with the games as actually played or to be interesting.

The amusements to which I refer are of a more trivial description, and it is possible that a mathematician may like to omit this chapter. In some cases it is difficult to say whether they should be classified as mainly arithmetical or geometrical, but the point is of no importance.

STATICAL GAMES OF POSITION

Of the innumerable statical games involving geometry of position I shall mention only three or four.

Three-in-a-row. First, I may mention the games of three-in-a-row, of which noughts and crosses, one form of merrillees, and go-bang are well-known examples. These games are played on a board – generally in the form of a square containing n^2 small squares or cells. The common practice is for one player to place a white counter or piece or to make a cross on each small square or cell which he occupies: his opponent similarly uses black counters or pieces or makes a nought on each cell which he occupies. Whoever first gets three (or any other assigned number) of his pieces in three adjacent cells and in a straight line wins. There is no difficulty in giving the complete analysis for boards of 9 cells and of 16 cells: but it is lengthy, and not particularly interesting. Most of these games were known to the ancients,* and it is for that reason I mention them here.

Three-in-a-row. Extension. I may, however, add an elegant but difficult extension which has not previously found its way, so far as I am aware, into any book of mathematical recreations. The problem is to place n counters on a plane so as to form as many rows as possible, each of which shall contain three, and only three, counters.†

It is easy to arrange the counters in a number of rows equal to the integral part of $(n-1)^2/8$. This can be effected by the following construction. Let P be any point on a cubic. Let the tangent at P cut the curve again in Q. Let the tangent at Q cut the curve in A. Let PA cut the curve in B, QB cut it in C, PC cut it in D, QD cut it in E, and so on. Then the counters must be placed at the points P, Q, A, B, ... Thus 9 counters can be placed in 8 such rows; 10 counters in 10 rows; 15 counters in 24 rows; 81 counters in 800 rows; and so on.

Sylvester found that, by a suitable choice of the initial point P, the number of rows can be increased to the integral part of $(n-1)(n-2)/6$. Thus 9 counters can be arranged in 9 rows; 10 counters in 12 rows; 15 counters in 30 rows; 81 counters in 1053 rows; and so on.

These, however, are inferior limits, and may be exceeded; for instance, Sylvester stated that 9 counters can be placed in 10 rows, each containing three counters; I do not know how he placed them, but one way of so arranging them is to put marks 1 and 8 at the centres of two rectangles (or parallelograms) 2365 and 4367 (having a common side 36). Then the lines 18, 27, 36, 45 all pass through one point, which

*Becq de Fouquières, *Les Jeux des Anciens*, second edition, Paris, 1873, chap. XVIII.

†*Educational Times Reprints*, 1868, vol. VIII, p. 106; *ibid.*, 1886, vol. XLV, pp. 127–128.

may be marked 9. The ten rows of three are all the sets of three different digits a, b, c that satisfy $a + b + c \equiv 0 \pmod 9$.

Sylvester raised the question whether it is possible to arrange n counters (not all in one straight row) so that each pair of them is in line with at least one other. He did not live to see a proof that the answer is No.*

Extension to p-in-a-row. The problem mentioned above at once suggests the extension of placing n counters so as to form as many rows as possible, each of which shall contain p and only p counters. Such problems can often be solved immediately by placing at infinity the points of intersection of some of the lines, and (if it is so desired) subsequently projecting the diagram thus formed so as to bring these points to a finite distance. One instance of such a solution is given above.

As examples I may give the arrangement of 10 counters in 5 rows, each containing 4 counters; the arrangement of 16 counters in 15 rows, each containing 4 counters; the arrangement of 18 counters in 9 rows, each containing 5 counters; and the arrangement of 19 counters in 10 rows, each containing 5 counters. These problems I leave to the ingenuity of my readers (see p. 129).

Tessellation. Another of these statical recreations is known as tessellation, and consists in the formation of geometrical designs or mosaics covering a plane area by the use of tiles of given geometrical forms.

If the tiles are regular polygons and two adjacent tiles have in common a whole side (or just a vertex), the resulting forms are easily enumerated. For instance, if we confine ourselves to the use of like tiles, each being a regular p-gon, we are restricted to the use of equilateral triangles, squares, or hexagons. For suppose that to fill the space round a vertex we require q polygons. Each internal angle of the p-gon is equal to $(p-2)\pi/p$. Hence $q(p-2)\pi/p = 2\pi$, that is,

$$(p-2)(q-2) = 4.$$

Since p and q are both greater than 2, we merely have to con-

*See, for instance, Coxeter, *Introduction to Geometry*, p. 65.

sider the possible ways of expressing 4 as the product of two positive integers. Letting p^q denote such a regular tessellation (of p-gons, q at each vertex), we deduce that the only possibilities* are 6^3, 4^4, 3^6.

If, however, we allow the use of unlike regular tiles (triangles, squares, etc.), we can construct numerous geometrical designs covering a plane area. If at each point the same number and kind of polygons are used, in the same (or the opposite) cyclic

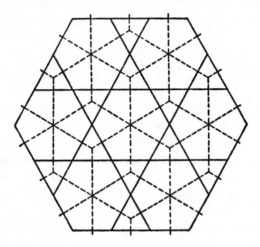

order, analysis similar to the above† shows that we can get eight possible arrangements – namely,

$$3.12^2, \ 4.6.12, \ 4.8^2, \ (3.6)^2, \ 3.4.6.4, \ 3^3.4^2, \ 3^2.4.3.4, \ 3^4.6.$$

The tessellation $3^4.6$ (with four triangles and a hexagon at each vertex) has the peculiarity of existing in two enantiomorphous forms; i.e., it is not superposable with its image in a mirror (unless we allow the whole plane to be turned over).

When every edge of such a tessellation is replaced by a perpendicular line, joining the centres of two adjacent tiles,

*The analogous problem for polygons covering the surface of a sphere is equivalent to the construction of regular polyhedra (see page 131). For an application of 6^3 to the voting system known as Proportional Representation, see G. Pólya, *L'Enseignement mathématique*, 1918, vol. xx, p. 367.

†See Kraïtchik, pp. 272–282, figs. 421–423, 425, 426, 432, 433, 440.

we obtain a *reciprocal* tessellation, whose tiles are all alike (though not necessarily regular). In this sense, 6^3 is reciprocal to 3^6 (and vice versa), while 4^4 is self-reciprocal (or rather, reciprocal to a congruent 4^4). The tessellation $(3.6)^2$ and its reciprocal are drawn on page 106.

Anallagmatic pavements. The use of colours introduces new considerations. One formation of a pavement by the employment of square tiles of two colours is illustrated by the common chess-board; in this the cells are coloured alternately white and black. Another variety of a pavement made with square tiles of two colours was invented by Sylvester,* who termed it anallagmatic. In the ordinary chess-board, if any two rows or any two columns are placed in juxtaposition, cell to cell, the cells which are side by side are either all of the same colour or all of different colours. In an anallagmatic arrangement the cells are so coloured (with two colours) that when any two columns or any two rows are placed together side by side, half the cells next to one another are of the same colour and half are of different colours.

Anallagmatic pavements composed of m^2 cells or square tiles cannot be constructed if m is odd or oddly-even. It has been conjectured that they exist whenever m is a multiple of 4. The first doubtful case is when $m = 188$.

If solutions when $m = a$ and when $m = b$ are known, a solution when $m = ab$ can be deduced at once; we merely have to replace every black cell of the a-pavement by the whole b-pavement, and every white cell of the a-pavement by the b-pavement with its colours reversed. Repeated application of this principle gives a solution whenever m is a power of 2. The case when $m = 8$ is shown in the first of the following drawings.

When p is a prime of the form $4k - 1$ and n is odd, an anallagmatic pavement with $m = p^n + 1$ can be derived from the addition table for $GF(p^n)$ (see p. 73) as follows. In the bottom row

---

*See *Mathematical Questions from the Educational Times*, London, vol. x, 1868, pp. 74–76; vol. LVI, 1892, pp. 97–99. The results are closely connected with theorems in the theory of equations.

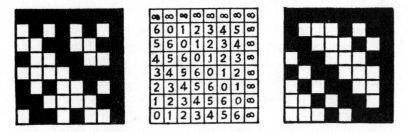

and left column of the table we write the $m-1$ elements of the finite field and an extra mth "element" ∞ which is defined so as to be unchanged when any element (including itself) is added to it. The rest of the table is completed by adding the pairs of elements that thus determine the columns and rows. (The case when $m=8$, so that $p=7$ and $n=1$, is illustrated above.) Each cell is coloured white or black according as its entry is a square or a non-square, with the (somewhat artificial*) convention that ∞ is a non-square. (Thus, when $m=8$, the elements 0, 1, 2, 4 are white, while 3, 5, 6, ∞ are black.)

The case when $m=28$ (so that $p=n=3$) is illustrated on page 109, the elements appearing in their "natural" order 0, 1, 2, 10, 11, 12, 20, ..., 221, 222, ∞. The squares are 0, 1, 20, 21, 22, 100, 102, 110, 111, 120, 121, 202, 211, 221. These (apart from 0) are most easily calculated as being alternate terms in the sequence of powers of 10 (namely 1, 10, 100, 12, 120, ..., 201).

The underlying theory is due to H. Davenport and R.E.A.C. Paley.† The latter gave also a more complicated rule to cover the case when $m=2(p^n+1)$ where p^n is of the form $4k+1$. By combining these methods, he showed how an anallagmatic tessellation of m^2 cells can be constructed whenever m is

---

*One would naturally expect ∞ to be its own square, like 0. I avoid using the words "quadratic residues," because 0 is undeniably a square, although it is not included among the quadratic residues.

†*Journal of Mathematics and Physics* (Cambridge, Mass.), 1933, vol. XII, pp. 311–320. Actually, Paley used subtraction instead of addition, but the consequent changes are quite trivial. An interesting application of this theory (to m-dimensional geometry) has been made by J.A. Barrau, *Nieuw Archief voor Wiskunde*, 1906, series 2, vol. VII.

divisible by 4 and of the form $2^k(p^n + 1)$, where p is an odd prime. (The case $m = 92$ was settled another way in 1961, with the aid of an electronic computer.*)

An Anallagmatic Pavement, $m = 28$

In all these cases the pavement has one completely black row and column; consequently every other row or column is half black and half white. When m is a power of 4, it is possible to construct an anallagmatic pavement which is "isochromatic" in the sense that half the rows (and columns) have $\sqrt{m}$ more black than white tiles, while the rest have $\sqrt{m}$ more white than black. The tenth edition of this book shows such a pavement with $m = 16$.

Polyominoes. A novel form of tessellation problem was suggested by Golomb† in 1953, but first became known to the

*S.W. Golomb and L.D. Baumert, *American Mathematical Monthly*, 1963, vol. LXX, pp. 12–17; Marshall Hall Jr., *Combinatorial Theory*, Waltham, Mass., 1967, p. 207.

†S.W. Golomb, *Polyominoes*, New York, 1965.

general public in the May 1957 issue of the *Scientific American*.*
A *polyomino* is a "super-domino" consisting of a complex of
edge-connected unit squares. Polyominoes are usually classified by the number of their constituent squares. Thus, while
a *monomino* consists merely of one unit square, a *domino* is the
rectangle formed by two unit squares, and a *tromino* is a shape
formed by three successively adjacent squares. This can be
extended to *tetrominoes, pentominoes, hexominoes, hepto-
minoes, octominoes, enneominoes, decominoes,* etc., each
consisting of four, five, six, seven, eight, nine, and ten unit
squares, respectively. Two *n-ominoes* are considered to be
distinct if there does not exist a rotation or reflection which
superimposes them.

Pentominoes† are of special interest. There are exactly
twelve such figures, each of which resembles (more or less) a
letter of the alphabet:

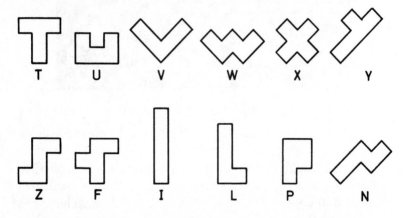

Combinatorial problems include forming for each pentomino
a scale model three times its size out of nine other pieces,
arranging replicas of any piece into a rectangle (this is only

*M. Gardner, *Scientific American*, 1957, vol. CXCVI, no. 5, pp. 154–156; 1957,
vol. CXCVII, no. 6, pp. 126–129; 1960, vol. CCIII, no. 5, pp. 186–194; 1962, vol.
CCVII, no. 5, pp. 151–159; 1969, vol. CCXXI, no. 6, pp. 122–127.

†A puzzle on pentominoes appeared in Dudeney's book, *The Canterbury
Puzzles*, London, 1919, as problem no. 74, "The Broken Chessboard," pp. 119–
120.

possible for the *L*, *I*, *P*, *Y* pentominoes), and placing the twelve pieces on a chess-board so that any given 2 by 2 square is left vacant. It is also possible to fit the twelve pentominoes into a 3 by 20, 4 by 15, 5 by 12, or 6 by 10 rectangle. There are two solutions for the 3 by 20 rectangle, sufficiently clearly indicated by *V Z Y W T F N L I P X U* and *V L N F T W Y Z I P X U*. On the other hand, 368 basically different solutions exist for the 4 by 15 rectangle, 1010 for the 5 by 12, and 2339 for the 6 by 10. The accompanying figure shows the only way to assemble the 6 by 10 by combining two 5 by 6 rectangles.

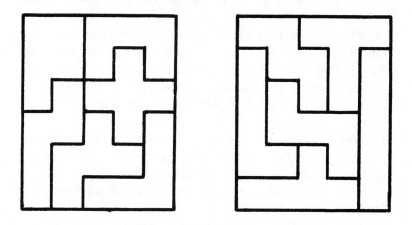

Pentominoes may also be used in constructing several interesting board games; of these the following has been described by Golomb.* Two players alternately place one pentomino at a time on an eight by eight board; play continues until one player (the loser) is unable to move. The game is intriguing in that great skill and insight are required for good play. A variation of the game is to distribute the pentominoes between the players before the game begins; this changes the strategy somewhat.

An unsolved problem is to find a pleasing formula for $p(n)$,

*In the book cited above. See also M. Gardner, *Scientific American*, 1965, vol. CCXIII, no. 4, pp. 96–104. A set of pieces for such a game, called *Pan-Kai*, has been manufactured by Phillips Publishers (1961).

the number of n-ominoes. Computers have been used to determine $P(n)$ for small values of n; for example, the following table has been compiled independently by several people:

n	1	2	3	4	5	6	7	8	9	10	11	12	13	14	15	16
$P(n)$	1	1	2	5	12	35	108	369	1285	4655	17073	63600	238591	901971	3426576	13079255

M. Eden,[*] the first to consider the n-omino enumeration problem, studied the rate of growth of the function $P(n)$. He showed that $(3 \cdot 14)^n < P(n) < (6 \cdot 75)^n$ for all sufficiently large n. D.A. Klarner[†] showed that $(P(n))^{1/n}$ tends to a limit θ, and $3 \cdot 72 < \theta$. Later, Klarner and Rivest[‡] showed that $\theta < 4 \cdot 65$. Thus, to date (1972), the best bounds on $P(n)$ are $(3 \cdot 72)^n < P(n) < (4 \cdot 65)^n$ for all sufficiently large n. Evidence computed from the known values of $P(n)$ supports the conjecture that the ratio $P(n+1)/P(n)$ increases with n. If this is true, then $P(n+1)/P(n)$ is a lower bound for θ for all n. In particular, $n = 15$ would give $3 \cdot 817 < \theta$, a substantial improvement over Klarner's lower bound.

Higher-dimensional n-ominoes have been considered also. For example, $P_3(n)$, the number of 3-dimensional n-ominoes, has been calculated for $n \leq 7$ and these values are given in the following table. (Here two congruent solid n-ominoes are considered different if a reflection is required to transform one into the other. This is the convention with gloves and shoes: we distinguish between right and left copies of the "same" thing.)

n	1	2	3	4	5	6	7	...
$P_3(n)$	1	1	2	8	29	166	1023	...

Various 3-dimensional puzzles have been proposed having such solid n-ominoes as the pieces. *Soma*, the most popular of these puzzles (thanks to Martin Gardner's article on the

[*]*Proceedings of the Fourth Berkeley Symposium on Mathematical Statistics and Probability*, 1961, vol. IV, pp. 223–239.

[†]*Canadian Journal of Mathematics*, 1967, vol. XIX, pp. 851–863.

[‡]"A procedure for improving the upper bound for the number of n-ominoes," *Canadian Journal of Mathematics*, 1973, vol. XXV, pp. 585–602.

subject*), was invented by Piet Hein. Soma consists of all solid n-ominoes with $n \leq 4$ which are not bricks. The volume of these pieces is 27, and one of the problems is to pack them into a cube of edge 3.

Many of the readers of Gardner's Soma cube article were inspired to make sets of 3-dimensional n-ominoes. Dubbing these figures n-cubes, Klarner made sets of tetracubes, pentacubes, and hexacubes, and invented many packing problems involving these pieces. The set of eight tetracubes may be used to pack each figure formed by magnifying one of the tetracubes by a factor of 2. For example, this includes boxes with dimensions $2 \times 2 \times 8$, and $2 \times 4 \times 4$. By leaving out any one of the twenty-nine pentacubes, the remaining set of twenty-eight may be used to pack boxes with dimensions $4 \times 5 \times 7$, $2 \times 5 \times 14$, and $2 \times 7 \times 10$. C.J. Bouwkamp showed that the $4 \times 5 \times 7$ box can be cut into sub-boxes and packed with twenty-eight of the pentacubes in over 84000,000000 ways. One of the sub-boxes with dimensions $3 \times 4 \times 5$ is packed with the set of 12 "planar" pentacubes, that is, the 12 solid pentominoes. Bouwkamp† has also prepared a catalogue of all solutions of the $3 \times 4 \times 5$ planar pentacube packing problem; the total number of essentially different solutions is 3940. Klarner has packed the set of 166 hexacubes into a $2 \times 6 \times 83$ box, and he packed the set of hexacubes with the $1 \times 1 \times 6$ piece removed into five boxes with dimensions $2 \times 9 \times 11$. These sub-boxes may be assembled in various ways to form boxes with dimensions $9 \times 10 \times 11$, $2 \times 9 \times 55$, and $2 \times 11 \times 45$. It is also possible to pack four cubes with side 6 using 144 of the hexacubes.

Colour-cube problem. As an example of a recreation analogous to tessellation I will mention the colour-cube problem.‡ Stripped of mathematical technicalities the problem may be enunciated as follows. A cube has six faces, and if six colours

*Scientific American, 1958, vol. CXCIX, no. 3. pp. 182–188.

†Catalogue of Solutions of the Rectangular $3 \times 4 \times 5$ Solid Pentomino Problem, Technische Hogeschool, Eindhoven, The Netherlands, 1967.

‡P.A. MacMahon, London Mathematical Society Proceedings, vol. XXIV, 1893, pp. 145–155; and New Mathematical Pastimes, Cambridge, 1921, pp. 42–46. See also F. Winter, Die Spiele der 30 bunten Würfel, Leipzig, 1934.

are chosen we can paint each face with a different colour. By permuting the order of the colours we can obtain thirty such cubes, no two of which are coloured alike. Take any one of these cubes, K; then it is desired to select eight out of the remaining twenty-nine cubes, such that they can be arranged in the form of a cube (whose linear dimensions are double those of any of the separate cubes) coloured like the cube K, and placed so that where any two cubes touch each other the faces in contact are coloured alike.

Only one collection of eight cubes can be found to satisfy these conditions. These eight cubes can be determined by the following rule. Take any face of the cube K: it has four angles, and at each angle three colours meet. By permuting the colours cyclically we can obtain from each angle two other cubes, and the eight cubes so obtained are those required. A little consideration will show that these are the required cubes, and that the solution is unique.

For instance, suppose that the six colours are indicated by the letters a, b, c, d, e, f. Let the cube K be put on a table, and to fix our ideas suppose that the face coloured f is at the bottom, the face coloured a is at the top, and the faces coloured b, c, d, and e front respectively the east, north, west, and south points of the compass. I may denote such an arrangement by $(f; a; b, c, d, e)$. One cyclical permutation of the colours which meet at the northeast corner of the top face gives the cube $(f; c; a, b, d, e)$, and a second cyclical permutation gives the cube $(f; b; c, a, d, e)$. Similarly cyclical permutations of the colours which meet at the northwest corner of the top face of K give the cubes $(f; d; b, a, c, e)$ and $(f; c; b, d, a, e)$. Similarly from the top southwest corner of K we get the cubes $(f; e; b, c, a, d)$ and $(f; d; b, c, e, a)$; and from the top southeast corner we get the cubes $(f; e; a, c, d, b)$ and $(f; b; e, c, d, a)$.

The eight cubes being thus determined, it is not difficult to arrange them in the form of a cube coloured similarly to K, and subject to the condition that faces in contact are coloured alike; in fact they can be arranged in two ways to satisfy these conditions. One such way, taking the cubes in the numerical order given above, is to put the cubes 3, 6, 8, and 2 at the SE, NE, NW, and SW corners of the bottom face; of course, each placed with the colour f at the bottom, while 3 and 6 have the colour b to the east, and 2 and 8 have the colour d to the west. The cubes 7, 1, 4, and 5 will then form the SE, NE, NW and SW corners of the top face; of course, each placed with the colour a at the top, while 7 and 1 have the colour b to the east, and 5 and 4 have the colour d to the west. If K is not given, the difficulty of the problem is increased. Similar puzzles in two dimensions can be made.

Dissecting a square into different squares. Z. Moroń* observed that a certain set of nine unequal squares can be fitted together to make a rectangle 32×33. This was the beginning of a fascinating investigation that is sometimes referred to as "squaring the square." Let a square or rectangle be called *perfect* when it is dissected into unequal squares. The constituent squares are its *elements* and their number is its *order*. It is *simple* if it contains no smaller perfect rectangle, and *compound* otherwise. Many perfect squares have been obtained by the "empirical method" of making a catalogue of perfect rectangles and trying to fit together some of its members, with possible deletion of corner squares. In this way, R. Sprague† obtained a perfect square of the 55th order. R.L. Brooks, C.A.B. Smith, A.H. Stone, and W.T. Tutte‡ similarly obtained one of the 26th order. These authors also described a "theoretical method" whereby perfect squares of orders 39 and higher could be constructed. A.J.W. Duijvestijn§ has described the unique simple perfect square of minimal order, namely 21, with elements 2, 4, 6, 7, 8, 9, 11, 15, 16, 17, 18, 19, 24, 25, 27, 29, 33, 35, 37, 42, 50. It now appears on the cover of each issue of the *Journal of Combinatorial Theory*.

---

*Przeglad matematyczno-fizyczny, Warsaw, 1925, vol. III, pp. 152–153.
†Mathematische Zeitschrift, 1939, vol. XLV, pp. 607–608.
‡Duke Mathematical Journal, 1940, vol. VII, pp. 312–340.
§ Journal of Combinatorial Theory, 1978, vol. B XXV, pp. 240–243.

In 1968, R.L. Brooks found the first simple perfect rectangle with sides in the ratio 2:1. Its order is 1323. Since then, Federico has developed an "empirical method" which gives simple perfect 2 by 1 rectangles of much smaller order, in some cases as low as 23.

C.J. Bouwkamp, A.J.W. Duijvestijn, and P. Medema have published *Tables relating to simple perfect rectangles of orders nine through fifteen*.* Duijvestijn † has found by a computer search that the smallest order for a simple perfect 2 by 1 rectangle is 22, and that one such rectangle has elements 2, 5, 6, 7, 11, 13, 17, 20, 22, 23, 24, 27, 29, 30, 35, 49, 53, 65, 67, 69, 71, 83.

DYNAMICAL GAMES OF POSITION

Games which are played by moving pieces on boards of various shapes – such as merrilees, fox and geese, solitaire, backgammon, draughts, and chess – present more interest. In general, possible movements of the pieces are so numerous that mathematical analysis is not practicable, but in a few games the possible movements are sufficiently limited as to permit of mathematical treatment; one or two of these are given later: here I shall confine myself mainly to puzzles and simple amusements.

Shunting problems. The first I will mention is a little puzzle which I bought some years ago, and which was described as the "Great Northern Puzzle." It is typical of a good many problems connected with the shunting of trains, and though it rests on a most improbable hypothesis, I give it as a specimen of its kind.

The puzzle shows a railway, *DEF*, with two sidings, *DBA* and *FCA*, connected at *A*. The portion of the rails at *A* which is common to the two sidings is long enough to permit of a single wagon, like *P* or *Q*, running in or out of it; but is too short to contain the whole of an engine, like *R*. Hence, if an

*Technische Hogeschool, Eindhoven, 1960.
† *Journal of Combinatorial Theory,* 1979, vol. xxvi, pp. 372–373.

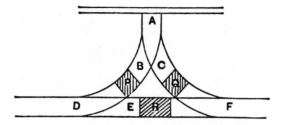

engine runs up one siding, such as *DBA*, it must come back the same way.

Initially a small block of wood, *P*, coloured to represent a wagon, is placed at *B*; a similar block, *Q*, is placed at *C*; and a longer block of wood, *R*, representing an engine, is placed at *E*. The problem is to use the engine *R* to interchange the wagons *P* and *Q*, without allowing any flying shunts.

Another shunting puzzle, on sale in the streets in 1905, under the name of the "Chifu-Chemulpo Puzzle," is made as follows. A loop-line *BGE* connects two points *B* and *E* on a railway track *AF*, which is supposed blocked at both ends, as shown in the diagram. In the model, the track *AF* is 9 inches long, $AB = EF = 1\frac{5}{6}$ inches, and $AH = FK = BC = DE = \frac{1}{4}$ inch. On the

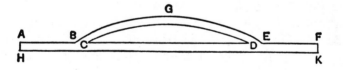

track and loop are eight wagons, numbered successively 1 to 8, each 1 inch long and one-quarter of an inch broad, and an engine, *e*, of the same dimensions. Originally the wagons are on the track from *A* to *F* and in the order 1, 2, 3, 4, 5, 6, 7, 8, and the engine is on the loop. The construction and the initial arrangement ensure that at any one time there cannot be more than eight vehicles on the track. Also if eight vehicles are on it, only the penultimate vehicle at either end can be moved on to the loop, but if less than eight are on the track, then the last two vehicles at either end can be moved on to the loop. If the points at each end of the loop-line are clear, the main track will hold four, but not more than four, vehicles; but the loop-line will hold five cars, or four and the engine. The object is to reverse the order of the wagons on the track, so that from A to F they will be numbered successively 8 to 1; and to do this by means which will involve as few transferences of the engine or a wagon to or from the loop as is possible. Twenty-six moves are required, and there is more than one solution in 26 moves (see p. 129).

Other shunting problems are not uncommon, but these two examples will suffice.

Ferry-boat problems. Everybody is familiar with the story of the showman who was travelling with a wolf, a goat, and a basket of cabbages; and for obvious reasons was unable to leave the wolf alone with the goat, or the goat alone with the cabbages. The only means of transporting them across a river was a boat so small that he could take in it only one of them at a time. The problem is to show how the passage could be effected.*

A somewhat similar problem is to arrange for the passage of a river by three men and three boys who have the use of a boat which will not carry at one time more than one man or two boys. Fifteen passages are required.†

Problems like these were proposed by Alcuin, Tartaglia, and other medieval writers. The following is a common type of such questions. Three‡ beautiful ladies have for husbands three men, who are young, gallant, and jealous. The party is travelling, and finds on the bank of a river over which it has to pass, a small boat which can hold no more than two persons. How can they cross the river, it being agreed that, in order to avoid scandal, no woman shall be left in the society of a man unless her husband is present? Eleven passages are required. With two married couples five passages are required. The similar problem with four married couples is insoluble.

Another similar problem is the case of n married couples

---

*Ozanam, 1803 edition, vol. I, p. 171; 1840 edition, p. 77.
†H.E. Dudeney, *The Tribune*, October 4, 1906.
‡Bachet, Appendix, problem IV, p. 212.

who have 'to cross a river by means of a boat which can be rowed by one person and will carry $n-1$ people, but not more, with the condition that no woman is to be in the society of a man unless her husband is present. Alcuin's problem given above is the case of $n=3$. Let y denote the number of passages from one bank to the other which will be necessary. Then it has been shown that if $n=3$, $y=11$; if $n=4$, $y=9$; and if $n>4$, $y=7$.

The following analogous problem is due to E. Lucas.* To find the smallest number x of persons that a boat must be able to carry in order that n married couples may by its aid cross a river in such a manner that no woman shall remain in the company of any man unless her husband is present; it being assumed that the boat can be rowed by one person only. Also to find the least number of passages, say y, from one bank to the other which will be required. M. Delannoy has shown that if $n=2$, then $x=2$ and $y=5$; if $n=3$, then $x=2$ and $y=11$; if $n=4$, then $x=3$ and $y=9$; if $n=5$, then $x=3$ and $y=11$; and finally if $n>5$, then $x=4$ and $y=2n-3$.

M. De Fonteney has remarked that, if there was an island in the middle of the river, the passage might be always effected by the aid of a boat which could carry only two persons. If there are only two or only three couples, the island is unnecessary, and the case is covered by the preceding method. His solution, involving $8n-6$ passages, is as follows. The first nine passages will be the same, no matter how many couples there may be: the result is to transfer one couple to the island and one couple to the second bank. The result of the next eight passages is to transfer one couple from the first bank to the second bank; this series of eight operations must be repeated as often as necessary until there is left only one couple on the first bank, only one couple on the island, and all the rest on the second bank. The result of the last seven passages is to transfer all the couples to the second bank. It would, however,

---

*Récréations mathématiques, Paris, 1883, vol. i, pp. 15–18, 237–238; hereafter I shall refer to this work by the name of the author.

seem that if *n* is greater than 3, we need not require more than $6n - 7$ passages from land to land.*

M.G. Tarry has suggested an extension of the problem, which still further complicates its solution. He supposes that each husband travels with a harem of *m* wives or concubines; moreover, as Moslem women are brought up in seclusion, it is reasonable to suppose that they would be unable to row a boat by themselves without the aid of a man. But perhaps the difficulties attendant on the travels of one wife may be deemed sufficient for Christians, and I content myself with merely mentioning the increased anxieties experienced by Moslems in similar circumstances.

Geodesics.   Geometrical problems connected with finding the shortest routes from one point to another on a curved surface are often difficult, but geodesics on a flat surface or flat surfaces are in general readily determinable.

I append one instance,† though I should have hesitated to do so had not experience shown that some readers do not readily see the solution. It is as follows. A room is 30 feet long, 12 feet wide, and 12 feet high. On the middle line of one of the smaller side-walls and 1 foot from the ceiling is a wasp. On the middle line of the opposite wall and 11 feet from the ceiling is a fly. The wasp catches the fly by crawling all the way to it: the fly, paralysed by fear, remaining still. The problem is to find the shortest route that the wasp can follow.

To obtain a solution we observe that we can cut a sheet of paper so that, when folded properly, it will make a model to scale of the room. This can be done in several ways. If, when the paper is again spread out flat, we can join the points representing the wasp and the fly by a straight line lying wholly on the paper, we shall obtain a geodesic route between them. Thus the problem is reduced to finding the way of cutting out the paper which gives the shortest route of the kind.

*See H.E. Dudeney, *Amusements in Mathematics*, London, 1917, p. 237.

†This is due to Mr. H.E. Dudeney. I heard a similar question propounded at Cambridge in 1903, but I first saw it in print in the *Daily Mail*, London, February 1, 1905.

Here is the diagram corresponding to a solution of the above question, where *A* represents the floor, *B* and *D* the longer side-walls, *C* the ceiling, and *W* and *F* the positions on the two smaller side-walls occupied initially by the wasp and fly. In the diagram the square of the distance between *W* and *F* is $(32)^2 + (24)^2$; hence the distance is 40 feet.

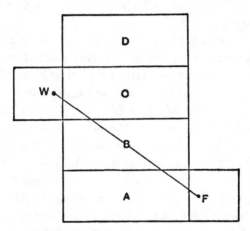

Problems with counters placed in a row. Numerous dynamical problems and puzzles may be illustrated with a box of counters, especially if there are counters of two colours. Of course, coins or pawns or cards will serve equally well. I proceed to enumerate a few of these played with counters placed in a row.

First problem with counters. The following problem must be familiar to many of my readers. Ten counters (or coins) are placed in a row. Any counter may be moved over two of those adjacent to it on the counter next beyond them. It is required to move the counters according to the above rule so that they shall be arranged in five equidistant couples.

If we denote the counters in their initial positions by the numbers 1, 2, 3, 4, 5, 6, 7, 8, 9, 10, we proceed as follows. Put 7 on 10, then 5 on 2, then 3 on 8, then 1 on 4, and lastly 9 on 6. Thus they are arranged in pairs on the places originally occupied by the counters 2, 4, 6, 8, 10.

Similarly by putting 4 on 1, then 6 on 9, then 8 on 3, then 10 on 7, and lastly 2 on 5, they are arranged in pairs on the places originally occupied by the counters 1, 3, 5, 7, 9.

If two superposed counters are reckoned as only one, solutions analogous to those given above will be obtained by putting 7 on 10, then 5 on 2, then 3 on 8, then 1 on 6, and lastly 9 on 4; or by putting 4 on 1, then 6 on 9, then 8 on 3, then 10 on 5, and lastly 2 on 7.*

There is a somewhat similar game played with eight counters, but in this case the four couples finally formed are not equidistant. Here the transformation will be effected if we move 5 on 2, then 3 on 7, then 4 on 1, and lastly 6 on 8. This form of the game is applicable equally to $8 + 2n$ counters, for if we move 4 on 1, we have left on one side of this couple a row of $8 + 2n - 2$ counters. This again can be reduced to one of $8 + 2n - 4$ counters, and in this way finally we have left eight counters which can be moved in the way explained above.

A more complete generalization would be the case of n counters, where each counter might be moved over the m counters adjacent to it on to the one beyond them. For instance, we may place twelve counters in a row and allow the moving of a counter over three adjacent counters. By such movements we can obtain four piles, each pile containing three counters. Thus, if the counters be numbered consecutively, one solution can be obtained by moving 7 on 3, then 5 on 10, then 9 on 7, then 12 on 8, then 4 on 5, then 11 on 12, then 2 on 6, and then 1 on 2. Or again we may place sixteen counters in a row and allow the moving of a counter over four adjacent counters on to the next counter available. By such movements we can get four piles, each pile containing four counters. Thus, if the counters be numbered consecutively, one solution can be obtained by moving 8 on 3, then 9 on 14, then 1 on 5, then 16 on 12, then 7 on 8, then 10 on 7, then 6 on 9, then 15 on 16, then 13 on 1, then 4 on 15, then 2 on 13, and then 11 on 6.

---

*Note by J. Fitzpatrick to a French translation of the third edition of this work, Paris, 1898.

Second problem with counters. Another problem,† of a some-what similar kind, is of Japanese origin. Place four florins (or white counters) and four halfpence (or black counters) alternately in a line in contact with one another. It is required in four moves, each of a pair of two contiguous pieces, without altering the relative position of the pair, to form a continuous line of four halfpence followed by four florins.

This can be solved as follows. Let a florin be denoted by *a* and a halfpenny by *b*, and let xx denote two contiguous blank spaces. Then the successive positions of the pieces may be represented thus:

Initially	xx*abababab*
After the first move	*baababaxxb*
After the second move	*baabxxaabb*
After the third move	*bxxbaaaabb*
After the fourth move	*bbbbaaaaxx*

The operation is conducted according to the following rule. Suppose the pieces to be arranged originally in circular order, with two contiguous blank spaces, then we always move to the blank space for the time being that pair of coins which occupies the places next but one and next but two to the blank space on one assigned side of it.

A similar problem with $2n$ counters, n of them being white and n black, will at once suggest itself, and, if n is greater than 4, it can be solved in n moves. I have, however, failed to find a simple rule which covers all cases alike, but solutions, due to M. Delannoy, have been given* for the four cases where n is of the form $4m$, $4m+2$, $4m+1$, or $4m+3$; in the first two cases the first $\frac{1}{2}n$ moves are of pairs of dissimilar counters and the last $\frac{1}{2}n$ moves are of pairs of similar counters; in the last two cases the first move is similar to that given above (namely, of the penultimate and antepenultimate counters to the be-

*Bibliotheca Mathematica, 1896, series 3, vol. VI, p. 323; P.G. Tait, *Philosophical Magazine*, London, January, 1884, series 5, vol. XVII, p. 39; or *Collected Scientific Papers*, Cambridge, vol. II, 1890, p. 93.
†*La Nature*, June 1887, p. 10.

ginning of the row), the next $\frac{1}{2}(n-1)$ moves are of pairs of dissimilar counters, and the final $\frac{1}{2}(n-1)$ moves are of similar counters.

The problem is also capable of solution if we substitute the restriction that at each move the pair of counters taken up must be moved to one of the two ends of the row instead of the condition that the final arrangement is to be continuous.

Tait suggested a variation of the problem by making it a condition that the two coins to be moved shall also be made to interchange places; in this form it would seem that five moves are required; or, in the general case, $n+1$ moves are required.

Problems on a chess-board with counters or pawns. The following three problems require the use of a chess-board as well as of counters or pieces of two colours. It is more convenient to move a pawn than a counter, and if therefore I describe them as played with pawns, it is only as a matter of convenience, and not that they have any connection with chess. The first is characterized by the fact that in every position not more than two moves are permitted; in the second and third problems not more than four moves are permitted in any position. With these limitations, analysis is possible. I shall not discuss the similar problems in which more moves are permitted.

*First problem with pawns.** On a row of seven squares on a chess-board 3 white pawns (or counters), denoted in the diagram by a's, are placed on the 3 squares at one end, and 3 black pawns (or counters), denoted by b's, are placed on the 3 squares at the other end – the middle square being left vacant.

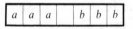

Each piece can move only in one direction; the a pieces can move from left to right, and the b pieces from right to left. If the square next to a piece is unoccupied, it can move on to that; or if the square next to it is occupied by a piece of the opposite

*Lucas, vol. ii, part 5, pp. 141–143.

colour and the square beyond that is unoccupied, then it can leap over that piece on to the unoccupied square beyond it. The object is to get all the white pawns in the places occupied initially by the black pawns and vice versa.

The solution requires 15 moves. It may be effected by moving first a white pawn, then successively two black pawns, then three white pawns, then three black pawns, then three white pawns, then two black pawns, and then one white pawn. We can express this solution by saying that if we number the cells (a term used to describe each of the small squares on a chess-board) consecutively, then initially the vacant space occupies the cell 4, and in the successive moves it will occupy the cells 3, 5, 6, 4, 2, 1, 3, 5, 7, 6, 4, 2, 3, 5, 4. Of these moves, six are simple and nine are leaps.

More generally, if we have m white pawns at one end of a row of $m+n+1$ cells, and n black pawns at the other end, the arrangement can be reversed in $mn+m+n$ moves, of which $m+n$ are simple and mn are leaps.

*Second problem with pawns.** A similar game may be played on a rectangular or square board. The case of a square board containing 49 cells, or small squares, will illustrate this sufficiently: in this case the initial position is shown in the annexed diagram where the a's denote the pawns or pieces of one colour, and the b's those of the other colour. The a pieces can move horizontally from left to right or vertically down, and the b pieces can move horizontally from right to left or vertically up, according to the same rules as before.

a	a	a	a	b	b	b
a	a	a	a	b	b	b
a	a	a	a	b	b	b
a	a	a		b	b	b
a	a	a	b	b	b	b
a	a	a	b	b	b	b
a	a	a	b	b	b	b

*Lucas, vol. II, part 5, p. 144.

The solution reduces to the preceding case. The pieces in the middle column can be interchanged in 15 moves. In the course of these moves every one of the seven cells in that column is at some time or other vacant, and whenever that is the case, the pieces in the row containing the vacant cell can be interchanged. To interchange the pieces in each of the seven rows will require 15 moves. Hence to interchange all the pieces will require $15 + (7 \times 15)$ moves – that is, 120 moves.

If we place $2n(n+1)$ white pawns and $2n(n+1)$ black pawns in a similar way on a square board of $(2n+1)^2$ cells, we can transpose them in $2n(n+1)(n+2)$ moves: of these $4n(n+1)$ are simple and $2n^2(n+1)$ are leaps.

Third problem with pawns. The following analogous problem is somewhat more complicated. On a square board of 25 cells, place eight white pawns or counters on the cells de-

a	b	c		
d	e	f		
g	h	*	H	G
		F	E	D
		C	B	A

noted by small letters in the annexed diagram, and eight black pawns or counters on the cells denoted by capital letters, the cell marked with an asterisk (*) being left blank. Each pawn can move according to the laws already explained, the white pawns being able to move only horizontally from left to right or vertically downwards, and the black pawns being able to move only horizontally from right to left or vertically upwards. The object is to get all the white pawns in the places initially occupied by the black pawns and vice versa. No moves outside the dark line are permitted.

Since there is only one cell on the board which is unoccupied, and since no diagonal moves and no backward moves are permitted, it follows that at each move not more than two pieces of either colour are capable of moving. There are, however, a very large number of empirical solutions. The following, due to Mr. H.E. Dudeney, is effected in 46 moves:

$$H\,h\,g^*\,F\,f\,c^*\,C\,B\,H\,h^*\,G\,D\,F\,f\,e\,h\,b\,a\,g^*\,G\,A\,B\,H\,E\,F\,f\,d\,g^*\,H\,h\,b\,c^*\,C\,F\,f^*\,G\,H\,h^*$$

the letters indicating the cells from which the pieces are successively moved. It will be noticed that the first twenty-three moves lead to a symmetrical position, and that the next twenty-two moves can be at once obtained by writing the first twenty-two moves in reverse order and interchanging small and capital letters. Similar problems with boards of various shapes can be easily constructed.

Probably, were it worth the trouble, the mathematical theory of games such as that just described might be worked out by the use of Vandermonde's notation, described later in chapter vi, or by the analogous method employed in the theory of the game of solitaire.*

Problems on a chess-board with chess-pieces. There are several mathematical recreations with chess-pieces, other than pawns. Some of these are given later in chapter vi.

Paradromic rings. - The difficulty of mentally realizing the effect of geometrical alterations in certain simple figures is illustrated by the familiar experiment of making *paradromic rings* by cutting a paper ring prepared in the following manner.

Take a strip of paper or piece of tape (say, for convenience, an inch or two wide, and at least 9 or 10 inches long), rule a line in the middle down the length AB of the strip, gum one end A over the other end B, and we get a ring like a section of a cylinder. If this ring is cut by a pair of scissors along the ruled line, we obtain two rings exactly like the first, except that they are only half the width. Next suppose that the end A is twisted through two right angles before it is gummed to B (the result of which is that the back of the strip at A is gummed over the

---

*On the theory of solitaire, see Reiss, "Beiträge zur Theorie des Solitär-Spiels," *Crelle's Journal*, Berlin, 1858, vol. liv, pp. 344–379; and Lucas, vol. i, part v, pp. 89–141.

front of the strip at *B*), then a cut along the line will produce only one ring. Next suppose that the end *A* is twisted once completely round (i.e. through four right angles) before it is gummed to *B*, then a similar cut produces two interlaced rings. If any of my readers think that these results could be predicted off-hand, it may be interesting to them to see if they can predict correctly the effect of again cutting the rings formed in the second and third experiments down their middle lines in a manner similar to that above described.

The theory is due to Listing and Tait,* who discussed the case when the end *A* receives *m* half-twists (that is, is twisted through *m*π) before it is gummed to *B*.

If *m* is even, we obtain a surface which has two sides and two edges. If the ring is cut along a line midway between the edges, we obtain two rings, each of which has *m* half-twists, and which are linked together $\frac{1}{2}m$ times.

If *m* is odd, we obtain a surface having only one side and one edge. If this ring is cut along its mid-line, we obtain only one ring, but it has $2m+2$ half-twists, and if *m* is greater than unity it is knotted. If the ring, istead of being bisected, is trisected,† we obtain two interlocked rings: one like the original ring (coming from the middle third), and one like the bisected ring. The manner in which these two rings are interlocked is illustrated by the following diagrams (of the cases $m=3$ and $m=5$). The ring with $m=1$ is called a *Möbius strip*.

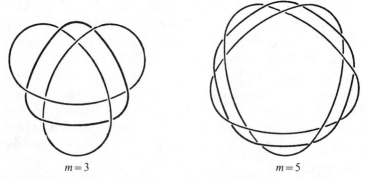

$m=3$ $m=5$

*Vorstudien zur Topologie, Göttinger Studien, 1847, part x.
†This remark is due to J.M. Andreas.

ADDENDUM

Note. Page 105. One method of arranging 16 counters in 15 lines, as stated in the text, is as follows. The five diagonals of a regular pentagon cross one another to form the vertices of a second pentagon. Its diagonals cross one another to form the vertices of a third pentagon. The fifteen vertices of these three pentagons, along with their common centre, form the desired configuration. (Dudeney's *Canterbury Puzzles*, 1919, p. 175)

An arrangement of 18 counters in 9 rows, each containing 5 counters, can be obtained thus. From one angle, A, of an equilateral triangle $AA'A''$, draw lines AD, AE inside the triangle making any angles with AA'. Draw from A' and A'' lines similarly placed in regard to $A'A''$ and $A''A$. Let $A'D'$ cut $A''E''$ in F, and $A'E'$ cut $A''D''$ in G. Then AFG is a straight line. The 3 vertices of the triangle and the 15 points of intersection of AD, AE, AF, with the similar pencils of lines drawn from A', A'', will give an arrangement as required.

An arrangement of 19 counters in 10 rows, each containing 5 counters, can be obtained by placing counters at the 19 points of intersection of the 10 lines $x = \pm a, x = \pm b, y = \pm a, y = \pm b, y = \pm x$: of these points two are at infinity.

As a further example, consider the problem of arranging 28 counters in 36 rows, each containing 4 counters. Such an arrangement can be obtained by joining certain vertices of a regular enneagon $A_1A_2 \ldots A_9$. We determine two other enneagons, $B_1B_2 \ldots B_9$, $C_1C_2 \ldots C_9$, concentric with the first. B_1 is the point of intersection of A_3A_6 with A_5A_8, while C_1 is the point of intersection of A_2A_6 with A_5A_9. The line B_4B_7 contains C_5 and C_6. The twenty-eighth point is the centre, which lies on nine lines such as $A_1B_1C_1$.

Note. Page 116. The Great Northern Shunting Problem is effected thus. (i) R pushes P into A. (ii) R returns, pushes Q up to P in A, couples Q to P, draws them both out to F, and then pushes them to E. (iii) P is now uncoupled, R takes Q back to A, and leaves it there. (iv) R returns to P, takes P back to C, and leaves it there. (v) R running successively through F, D, B comes to A, draws Q out, and leaves it at B.

Note. Page 117. One solution of the Chifu-Chemulpo Puzzles is as follows. Move successively wagons 2, 3, 4 up, i.e., on to the loop line. [Then push 1 along the straight track close to 5; this is not a "move."] Next, move 4 down, i.e. on to the straight track and push it along to 1. Next, move 8 up, 3 down to the end of the track and keep it there temporarily, 6 up, 2 down, e down, 3 up, 7 up. [Then push 5 to the end of the track and keep it there temporarily.] Next, move 7 down, 6 down, 2 up, 4 up. [Then push e along to 1.] Next, move 4 down to the end of the track and keep it there temporarily, 2 down, 5 up, 3 down, 6 up, 7 up, 8 down to the end of the track, e up, 5 down, 6 down, 7 down. In this solution we moved e down to the track at one end, then shifted it along the track, and finally moved it up to the loop from the other end of the track. We might equally well move e down to the track at one end, and finally move it back to the loop from the same end. In this solution the pieces successively moved are 2, 3, 4, 4, e, 8, 7, 3, 2, 6, 5, 5, 6, 3, 2, 7, 2, 5, 6, 3, 7, e, 8, 5, 6, 7.

CHAPTER V

POLYHEDRA

"Although a Discourse of Solid Bodies be an uncommon and neglected Part of Geometry, yet that it is no inconsiderable or unprofitable Improvement of the Science will (no doubt) be readily granted by such, whose Genius tends as well to the Practical as Speculative Parts of it, for whom this is chiefly intended."*

A polyhedron is a solid figure† with plane faces and straight edges, so arranged that every edge is both the join of two vertices and a common side of two faces. Familiar instances are the pyramids and prisms. (A pentagonal pyramid has six vertices, ten edges, and six faces; a pentagonal prism has ten, fifteen, and seven. See figure 7 on plate ɪ.) I would mention also the *antiprism*,‡ whose two bases, though parallel, are not similarly situated, but each vertex of either corresponds to a side of the other, so that the lateral edges form a zigzag. (Thus a pentagonal antiprism has ten vertices, twenty edges, and twelve faces. See figure 9 on plate ɪ.)

The tessellations described on page 106 may be regarded as infinite polyhedra.

SYMMETRY AND SYMMETRIES

It is convenient to say that a figure is *reflexible* § if it is superposable with its image in a plane mirror (i.e. if it is, in the most elementary sense, "symmetrical"). A figure which is not reflexible forms, with its mirror-image, an *enantiomorphous* pair.

---

*Abraham Sharp, *Geometry Improv'd*, London, 1717, p. 65.
†More precisely, it is the *surface* of such a solid figure.
‡Or "prismoid." See the *Encyclopaedia Britannica* (xɪvth edition), article "Solids."
§Or "self-reflexible."

Plate I

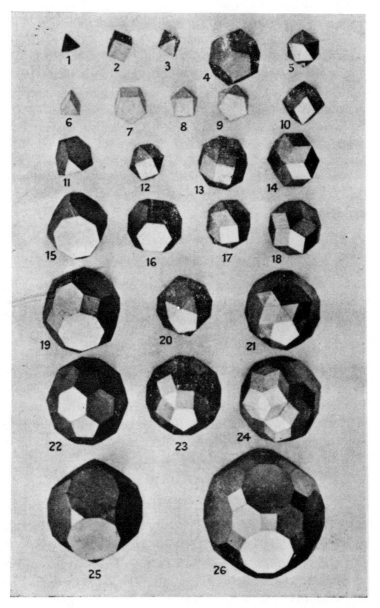

(The obvious example is a pair of shoes.) A reflexible figure has at least one *plane of symmetry*; the operation of reflecting in such a plane leaves the figure unchanged as a whole. A figure may also be symmetrical by rotation about an *axis of symmetry*. The vague statement that a figure has a certain amount of "symmetry" can be made precise by saying that the figure has a certain number of *symmetries*, a symmetry* being defined as any combination of motions and reflections which leaves the figure unchanged as a whole.

For a regular polygon $ABC...X$, there is a symmetry (in fact, a rotation) which cyclically permutes the vertices, changing A into B, B into $C, ...$, and X into A.

THE FIVE PLATONIC SOLIDS

Let A be a vertex belonging to a face α of a polyhedron. The polyhedron is said to be *regular* if it admits two particular symmetries: one which cyclically permutes the vertices of α, and one which cyclically permutes the faces surrounding A. It follows that all the faces are regular and equal, all the edges are equal, and all the vertices are surrounded alike. If each vertex is surrounded by q p-gons, we may denote the polyhedron by the symbol p^q (as on page 106) or $\{p, q\}$.

If the polyhedron is finite, the faces at one vertex form a solid angle. The internal angle of each face being $(p-2)\pi/p$, we now have $q(p-2)\pi/p < 2\pi$, that is,

$$(p-2)(q-2) < 4.$$

Since p and q are both greater than 2, we merely have to consider the possible ways of expressing 1 or 2 or 3 as the product of two positive integers, and then in each case the polyhedron can be built up, face by face. Letting V, E, F denote the number of vertices, edges, faces, the results are as in the following table. (See plate I, figure 1, 2, 3, 4, 5.)

*Or "symmetry operation." (Any rotation or translation may be regarded as a combination of two reflections.)

$\{p, q\}$	V	E	F	Name
$\{3, 3\}$	4	6	4	Regular tetrahedron
$\{4, 3\}$	8	12	6	Cube
$\{3, 4\}$	6	12	8	Octahedron
$\{5, 3\}$	20	30	12	Dodecahedron
$\{3, 5\}$	12	30	20	Icosahedron

Clearly, $qV = 2E = pF$. A less obvious relation is

$$E^{-1} = p^{-1} + q^{-1} - \tfrac{1}{2}.$$

This follows easily from Euler's Formula $F + V - E = 2$, which will be proved in chapter VIII, on pages 232–233.

In four ways the tetrahedron can be regarded as a triangular pyramid, and the octahedron as a triangular antiprism. In three ways the octahedron can be regarded as a square double-pyramid, and the cube as a square prism. In six ways the icosahedron can be regarded as a pentagonal antiprism with two pentagonal pyramids stuck on to its bases. The faces of the dodecahedron consist of two opposite pentagons (in parallel planes), each surrounded by five other pentagons.

An icosahedron can be inscribed in an octahedron, so that each vertex of the icosahedron divides an edge of the octahedron according to the "golden section."* A cube can be inscribed in a dodecahedron so that each edge of the cube lies in a face of the dodecahedron (and joins two alternate vertices of that face).

These five figures have been known since ancient times. The earliest thorough investigation of them is probably that of Theaetetus.† It has been suggested that Euclid's *Elements* was originally written, not as a general treatise on geometry, but in order to supply the necessary steps for a full appreciation of the five regular solids. At any rate, Euclid begins by constructing an equilateral triangle, and ends by constructing a dodecahedron.

*See page 39.
†T. Heath, *A History of Greek Mathematics*, Oxford, 1921, vol. I, p. 162.

Mystically minded Greeks associated the regular polyhedra with the four Elements and the Universe. Kepler* justified the correspondence as follows. Of the five solids, the tetrahedron has the smallest volume for its surface, the icosahedron the largest; these therefore exhibit the qualities of dryness and wetness, respectively, and correspond to Fire and Water. The cube, standing firmly on its base, corresponds to the stable Earth; but the octahedron, which rotates freely when held by two opposite corners, corresponds to the mobile Air. Finally, the dodecahedron corresponds to the Universe, because the zodiac has twelve signs. He illustrated this correspondence by drawing a bonfire on his tetrahedron; a lobster and fishes on his icosahedron; a tree, a carrot, and gardening tools on his cube; birds and clouds on his octahedron; and the sun, moon, and stars on his dodecahedron.

With each of these polyhedra we may associate three concentric spheres: one (the "circum-sphere") through all the vertices, one touching all the edges, and one (the "insphere") touching all the faces. Consider the second of these spheres. If we replace each edge by a perpendicular line touching this sphere at the same point, the edges at a vertex lead to the sides

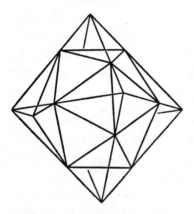

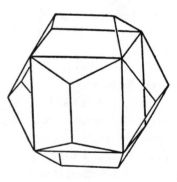

Icosahedron and Octahedron   Cube and Dodecahedron

*Opera Omnia, Frankfort, 1864, vol. v, p. 121.

of a polygon. Such polygons are the V faces of another "reciprocal" polyhedron, which has F vertices. The reciprocal of $\{p, q\}$ is $\{q, p\}$. Thus the cube and the octahedron are reciprocal, likewise the dodecahedron and the icosahedron. The tetrahedron is self-reciprocal or, rather, reciprocal to another tetrahedron. The diagonals of the faces of a cube are the edges of two reciprocal tetrahedra. (See figure 27 on plate II, facing this page.) The term *reciprocal* arises from the existence of a reciprocating sphere, with respect to which the vertices of $\{q, p\}$ are the poles of the face-planes of $\{p, q\}$, and vice versa. The ratio of circum-radius R to in-radius r is exactly the same for the cube as for the octahedron, and for the dodecahedron as for the icosahedron. In fact, if the reciprocating sphere has radius ρ, the reciprocal of a given polyhedron has circum-radius ρ^2/r and in-radius ρ^2/R. Thus the relative size of two reciprocal polyhedra may be adjusted so as to make them have the same circum-sphere and the same in-sphere. (In general, their corresponding edges will no longer intersect.)

If two reciprocal regular solids of the same in-radius (and therefore the same circum-radius) stand side by side on a horizontal plane (such as a table top), the distribution of vertices in horizontal planes is the same for both – i.e. the planes are the same, and the numbers of vertices in each plane are proportional. This fact was noticed by Pappus,* but has only recently been adequately explained, although its various extensions indicated that it was no mere accident. One of these extensions is to the Kepler-Poinsot polyhedra, which will be described later. Another is to tessellations of a plane. Consider the tessellation $\{6, 3\}$ (i.e. hexagons, three round each vertex). By picking out alternate vertices of each hexagon in a consistent manner, we derive the triangular tessellation $\{3, 6\}$ (which, in a different position, is the reciprocal tessellation). We then find that every circle concentric with a face of the $\{6, 3\}$ contains twice as many vertices of the $\{6, 3\}$ as of the $\{3, 6\}$. (This, however, is obvious, since the omitted vertices of the $\{6, 3\}$ belong to another $\{3, 6\}$, congruent to the first.)

*T. Heath, *A History of Greek Mathematics*, vol. II, pp. 368–369.

Plate II

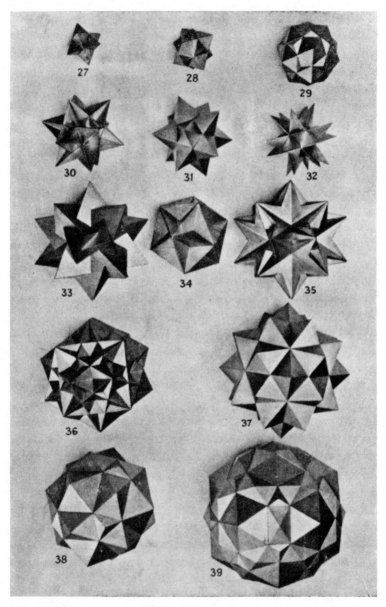

The fact that the vertices of a hexagonal tessellation belong also to two triangular tessellations is analogous to our observation (p. 134) that the vertices of a cube belong also to two regular tetrahedra. These two tetrahedra may be said to form a *compound* – Kepler's *stella octangula*; their eight faces lie in the facial planes of an octahedron. There is also a compound of *five tetrahedra* having the vertices of a dodecahedron and the facial planes of an icosahedron; this occurs in two enantiomorphous varieties. By putting the two varieties together, so as to have the same twenty vertices, we obtain a compound of *ten tetrahedra*, oppositely situated pairs of which can be replaced by *five cubes* (having the twenty vertices of the dodecahedron, each taken twice). It is quite easy to visualize one such cube in a given dodecahedron (as in the second drawing on page 133); the whole set of five makes a very pretty model. Finally, by reciprocating the five cubes we obtain a compound of *five octahedra* having the facial planes of an icosahedron, each taken twice. This icosahedron is inscribed in each one of the octahedra as in the first drawing on page 133. (See also plate II, figures 27, 33, 35, 36, 37.)

Among the edges of a regular polyhedron, we easily pick out a skew polygon or zigzag, in which the first and second edges are sides of one face, the second and third are sides of another face, and so on. This zigzag is known as a *Petrie polygon*, and has many applications. Each finite polyhedron can be orthogonally projected on to a plane in such a way that one Petrie polygon becomes a regular polygon with the rest of the projection inside it. It can be shown in various simple ways that the Petrie polygon of $\{p, q\}$ has h sides, where

$$\cos^2(\pi/h) = \cos^2(\pi/p) + \cos^2(\pi/q).$$

The h sides of the Petrie polygon of $\{p, q\}$ are crossed by h edges of the reciprocal polyhedron $\{q, p\}$; these form a Petrie polygon for $\{q, p\}$.

The regular polyhedra are symmetrical in many different ways. There is an *axis* of symmetry through the centre of every face, through the mid-point of every edge, and through

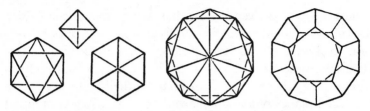

The Platonic Solids and their Petrie Polygons

every vertex: $E+1$ axes altogether. There are also $3h/2$ *planes* of symmetry.

THE ARCHIMEDEAN SOLIDS

A polyhedron is said to be *uniform* if it has regular faces and admits symmetries which will transform a given vertex into every other vertex in turn. The Platonic polyhedra are uniform; so are the right regular prisms and antiprisms, of suitable height – namely, when their lateral faces are squares and equilateral triangles, respectively. Such a polyhedron may be denoted by a symbol giving the numbers of sides of the faces around one vertex (in their proper cyclic order); thus the n-gonal prism and antiprism are $4^2 . n$ and $3^3 . n$. It is quite easy to prove* that, apart from these, there are just thirteen (finite, convex) uniform polyhedra:

$$3.6^2, \quad 4.6^2, \quad 3.8^2, \quad 5.6^2, \quad 3.10^2, \quad 4.6.8, \quad 4.6.10,$$
$$(3.4)^2, \quad (3.5)^2, \quad 3.4^3, \quad 3.4.5.4, \quad 3^4.4, \quad 3^4.5.$$

These are the *Archimedean solids*.

Let σ denote the sum of the face-angles at a vertex. (This must be less than 2π in order to make a solid angle.) Then the number of vertices is given by the formula† $(2\pi - \sigma)V = 4\pi$. For instance, $3^4 . 5$ has 60 vertices, since $\sigma = (\frac{4}{3} + \frac{3}{5})\pi$.

If we regard the *stella octangula* as consisting of two interpenetrating solid tetrahedra, we may say that their common part is an octahedron. Also, as we have already observed,

---

*See, for instance, T.R.S. Walsh, *Geometriae Dedicata*, 1972, vol. I, pp. 117–123.

†E. Steinitz and H. Rademacher, *Vorlesungen über die Theorie der Polyeder*, Berlin, 1934, p. 11.

their edges are diagonals of the faces of a cube. Analogously, the common part of a cube and an octahedron, in the properly reciprocal position (with corresponding edges perpendicularly bisecting each other), is the *cuboctahedron* $(3.4)^2$. Each pair of corresponding edges (of the cube and octahedron) are the diagonals of a rhomb, and the twelve such rhombs are the faces of a "semi-regular" polyhedron known as the *rhombic dodecahedron*. (The latter is not uniform, but "isohedral." See the first drawing on page 151.) After suitable magnification, the edges of the cuboctahedron intersect those of the rhombic dodecahedron (at right angles); in fact, these two polyhedra are reciprocal, just as the octahedron and cube are reciprocal. The icosahedron and dodecahedron lead similarly to the *icosidodecahedron* $(3.5)^2$, and to its reciprocal, the *triaconta-hedron*. (See plate II, figures 28, 29, and plate I, figures 12, 10, 20, 18. Compare the tessellations drawn on page 106.) The compound of five cubes has the 30 facial planes of a triacontahed-ron. Reciprocally, the compound of five octahedra has the 30 vertices of an icosidodecahedron.

The faces of the icosidodecahedron consist of 20 triangles and 12 pentagons (corresponding to the faces of the two parent regulars). Its 60 edges are perpendicularly bisected by those of the reciprocal triacontahedron (although the latter edges are not bisected by the former: see plate II, figure 39). The 60 points where these pairs of edges cross one another are the vertices of a polyhedron whose faces consist of 20 triangles, 12 pentagons, and 30 rectangles. By slightly displacing these points (towards the mid-points of the edges of the triaconta-hedron), the rectangles can be distorted into squares, and we have another Archimedean solid, the *rhombicosidodecahedron*, 3.4.5.4. (Plate I, figure 23; compare the tessellation 3.4.6.4.) An analogous construction leads to the *rhombicuboctahedron** 3.4^3, whose faces consist of 8 triangles and $6+12$ squares. (See plate II, figure 38, and plate I, figure 13.) In attempting

---

*I.e., "rhombi-cub-octahedron"; but the other is "rhomb-icosi-dodeca-hedron."

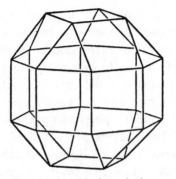

Pseudo-rhombicuboctahedron

to make a model of this polyhedron, J.C.P. Miller* accidentally discovered a "pseudo-rhombicuboctahedron," bounded likewise by 8 triangles and 18 squares, and isogonal in the loose or "local" sense (each vertex being surrounded by one triangle and three squares), but not in the strict sense (which implies that the appearance of the solid as a whole must remain the same when viewed from the direction of each vertex in turn).

On cutting off the corners of a cube, by planes parallel to the faces of the reciprocal octahedron, we leave small triangles, and reduce the square faces to octagons. For suitable positions of the cutting planes these octagons will be regular, and we have another Archimedean solid, the *truncated cube*, 3.8^2. (Cf. the tessellations 4.8^2 and 3.12^2.) Each of the five Platonic solids has its truncated variety;† so have the cuboctahedron and the icosidodecahedron, but in these last cases ($4.6.8$ and $4.6.10$) a distortion is again required, to convert rectangles into squares.‡ (Cf. the tessellation $4.6.12$.)

All the Archimedean solids so far discussed are reflexible (by reflection in the plane that perpendicularly bisects any edge). The remaining two, however, are not reflexible: the

Philosophical Transactions of the Royal Society, 1930, series A, vol. CCXXIX, p. 336.

†The "truncated $\{p, q\}$" is $q.(2p)^2$. See plate I, figures 11, 15, 16, 25, 22.

‡On account of this distortion, the truncated cuboctahedron ($4.6.8$) is sometimes called the "great rhombicuboctahedron," and then 3.4^3 is called the *small* rhombicuboctahedron; similarly for the truncated icosidodecahedron and rhombicosidodecahedron.

snub cube $3^4.4$, and the *snub dodecahedron* $3^4.5$ (plate I, figures 17 and 21). Let us draw one diagonal in each of the 30 squares of the rhombicosidodecahedron, choosing between the two possible diagonals in such a way that just one of these new lines shall pass through each of the 60 vertices. (The choice in the first squares determines that in all the rest.) Each square has now been divided into two right-angled isosceles triangles; by distorting these into equilateral triangles we obtain the snub dodecahedron.* The snub cube is similarly derivable from the rhombicuboctahedron, provided we remember to operate only on the 12 squares that correspond to the edges of the cube (and not on the 6 squares that correspond to its faces). The tessellation $3^4.6$ may be regarded as a "snub $\{6, 3\}$," and $3^2.4.3.4$ as a "snub $\{4, 4\}$." Moreover, the "snub tetrahedron" is the icosahedron $\{3, 5\}$, derived as above from the cuboctahedron (or "rhombi-tetra-tetrahedron").

The snub cube and the snub dodecahedron both occur in two enantiomorphous varieties. Their metrical properties involve the solution of cubic equations, whereas those of the reflexible Archimedeans (and of the regulars) involve nothing worse than square roots; in other words, the reflexibles are capable of Euclidean construction, but the two proper snubs are not.

MRS. STOTT'S CONSTRUCTION

The above description of the Archimedean solids is essentially Kepler's. A far more elegant construction for the reflexible figures has been devised by Alicia Boole Stott.† Her method is free from any employment of distortion, and the final edge-length is the same as that of the regular solid from which we start. In the process called *expansion*, certain sets of elements (viz., edges or faces) are moved directly away from the centre, retaining their size and orientation, until the consequent in-

*This name is unfortunate, since the figure is related to the icosahedron just as closely as to the dodecahedron. "Snub icosidodecahedron" would be far better.

†*Verhandelingen der Koninklijke Akademie van Wetenschappen*, Amsterdam, 1910, vol. XI, no. 1.

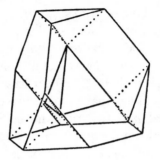

(1) Tetrahedron and Truncated Tetrahedron

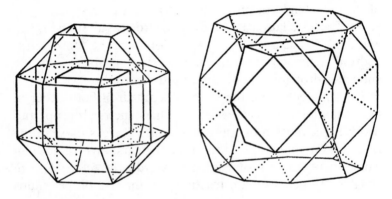

(2) Cube and Rhombicuboctahedron   (3) Truncated Cube and Cuboctahedron

terstices can be filled with new regular faces. The reverse process is called *contraction*. By expanding any regular solid according to its edges, we derive the "truncated" variety. By expanding the cube (or the octahedron) according to its faces, we derive the rhombicuboctahedron, 3.4^3. By expanding this according to its 12 squares which correspond to the edges of the cube, or by expanding the truncated cube according to its octagons, we derive the truncated cuboctahedron, $4.6.8$. By contracting the truncated cube according to its triangles, we derive the cuboctahedron. And so on. Mrs. Stott has represented these processes by a compact symbolism, and extended them to spaces of more than three dimensions, where they are extraordinarily fruitful.

EQUILATERAL ZONOHEDRA

The solids that I am about to describe were first investigated by E.S. Fedorov.* Their interest has been enhanced by P.S. Donchian's observation that they may be regarded as three-dimensional projections of n-dimensional *hyper-cubes* (or *measure-polytopes*, or *regular orthotopes*†). Their edges are all equal, and their faces are generally rhombs, but sometimes higher "parallel-sided $2m$-gons," i.e., equilateral $2m$-gons which are symmetrical by a half-turn. The subject begins with the following theorem on polygonal dissection.

Every parallel-sided $2m$-gon (and, in particular, every regular $2m$-gon) can be dissected‡ into $\frac{1}{2}m(m-1)$ rhombs of the same length of side. This is easily proved by induction, since every parallel-sided $2(m+1)$-gon can be derived from a parallel-sided $2m$-gon by adding a "ribbon" of m rhombs. In fact, the pairs of parallel sides of such a $2m$-gon can take any m different directions, and there is a component rhomb for every pair of these directions; hence the number $\frac{1}{2}m(m-1)$. For two perpendicular directions, the rhomb is a square.

Consider now any sheaf of n lines through one point of space,§ and suppose first that no three of the lines are coplanar. Then there is a polyhedron whose faces consist of $n(n-1)$ rhombs, and whose edges, in sets of $2(n-1)$, are parallel to the n given lines. In fact, for every pair of the n lines, there is a pair of opposite faces whose sides lie in those directions. To construct this *equilateral zonohedron*, imagine a plane through any one of the n lines, gradually rotating through a complete turn. Each time that this plane passes through one

---

Zeitschrift für Krystallographie und Mineralogie, 1893, vol. XXI, p. 689; *Nachala Ucheniya o Figurakh*, Leningrad, 1953. See also Coxeter, *Twelve Geometric Essays*, Carbondale, Illinois, 1968, chap. 4.

†L. Schläfli, *Quarterly Journal of Mathematics*, 1860, vol. III, p. 66: "(4, 3, 3, ..., 3)"; C.H. Hinton, *The Fourth Dimension*, London, 1906; P.H. Schoute, *Mehrdimensionale Geometrie*, Leipzig, 1905, vol. II, pp. 243–246; D.M.Y. Sommerville, *An Introduction to the Geometry of* n *Dimensions*, London, 1929, pp. 49, 171, 182, 190.

‡In how many ways?

§This construction is due to P.S. Donchian.

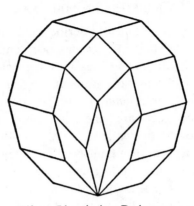

Fifteen Rhombs in a Dodecagon

of the other $n-1$ lines, take a rhomb whose edges are parallel to the two lines, and juxtapose it to the rhomb previously found (without changing its orientation). This process leads eventually to a closed ribbon of $2(n-1)$ rhombs. By fixing our attention on another of the n lines, we obtain another such ribbon, having two parallel faces in common with the first. When a sufficient number of these ribbons (or *zones*) have been added, the polyhedron is complete.

If m of the n lines are coplanar, we have a pair of opposite parallel-sided $2m$-gons, to replace $\frac{1}{2}m(m-1)$ pairs of opposite rhombs. If these m lines are symmetrically disposed, the $2m$-gons will be regular.

In this manner, three perpendicular lines lead to a cube, and three lines of general direction to a parallelepiped with rhombic faces. (This is called a *rhombohedron* only if the six faces are congruent.) More generally, m coplanar lines and one other line lead to a parallel-sided $2m$-gonal prism (a *right* prism if the "other" line is perpendicular to the plane of the first m).

The four "diameters" of the cube (joining pairs of opposite vertices) lead to the rhombic dodecahedron, the six diameters of the icosahedron lead to the triacontahedron, and the ten diameters of the (pentagonal) dodecahedron lead to an *ennea-*

*contahedron** whose faces are 30 rhombs of one kind and 60 of another. The six diameters of the cuboctahedron lead to the truncated octahedron, whose faces are 6 squares and 8 hexagons (the equivalent of 8×3 rhombs), and the fifteen diameters of the icosidodecahedron lead to the truncated icosidodecahedron, whose faces are 30 squares, 20 hexagons ($= 20 \times 3$ rhombs), and 12 decagons ($= 12 \times 10$ rhombs). As a final example, the nine diameters of the octahedron and cuboctahedron (taken together in corresponding positions)[†] lead to the truncated cuboctahedron, whose faces are 12 squares, 8 hexagons ($= 24$ rhombs), and 6 octagons ($= 36$ rhombs). (See plate I, figures 10, 18, 24, 16, 26, 19.)

As in these examples, so in general, the solid has the same type of symmetry as the given sheaf of lines. A rhombic $n(n-1)$-hedron having a centre of symmetry and an n-gonal axis occurs for every value of n, being given by a sheaf of n lines symmetrically disposed around a cone.[‡] The faces are all alike when n is 3; they can be all alike when n is 4 or 5, if the lines are suitably chosen, viz., if the angle between alternate lines is supplementary to the angle between consecutive lines. Then $n = 4$ gives the rhombic dodecahedron; $n = 5$ gives a *rhombic icosahedron*[§] (plate I, figure 14) which can be derived from the triacontahedron by removing any one of the zones and bringing together the two pieces into which the remainder of the surface is thereby divided. By removing a suitable zone from the rhombic icosahedron we obtain Bilinski's new rhombic dodecahedron with faces all alike but different from those of the classical rhombic dodecahedron.

Fedorov's general zonohedron can be derived from the equilateral zonohedron by lengthening or shortening all the edges that lie in each particular direction. Thus rhombic

---

*This somewhat resembles a figure described by A. Sharp, *Geometry Improv'd*, p. 87.

[†]I.e., perpendiculars to the nine planes of symmetry of the cube (or of the octahedron).

[‡]B.L. Chilton and H.S.M. Coxeter, *American Mathematical Monthly*, 1963, vol. LXX, pp. 946–951.

[§]Stanko Bilinski, *Glasnik*, 1960, vol. XV, pp. 252–262.

faces become parallelograms, and "parallel-sided $2m$-gons" cease to be equilateral. When each higher face is replaced by its proper number of parallelograms, we have $F = n(n-1)$, $E = 2F$, and $V = F + 2$. In fact, *every convex polyhedron bounded solely by parallelograms is a zonohedron.**

One final remark on this subject: there is a three-dimensional analogue for the theorem that a parallel-sided $2m$-gon can be dissected into $\frac{1}{2}m(m-1)$ parallelograms. The zonohedron can be dissected into $\frac{1}{6}n(n-1)(n-2)$ parallelepipeds (viz., one for every three of the n directions).

THE KEPLER-POINSOT POLYHEDRA

By extending the sides of a regular pentagon till they meet again, we derive the star-pentagon or pentacle or *pentagram*, which has long been used as a mystic symbol. We may regard the pentagram $\{\frac{5}{2}\}$ as a generalized polygon, having five sides which enclose the centre twice. Each side subtends an angle $\frac{4}{5}\pi$ at the centre, whereas each side of an ordinary n-gon subtends an angle $2\pi/n$. Thus the pentagram behaves as if it were an n-gon with $n = \frac{5}{2}$. Analogously, any rational number n (> 2) leads to a polygon $\{n\}$, the numerator giving the number of sides, and the denominator the "density" (or "species").

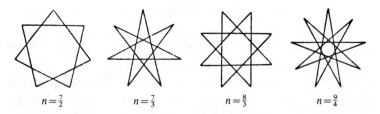

$$n = \tfrac{7}{2} \qquad n = \tfrac{7}{3} \qquad n = \tfrac{8}{3} \qquad n = \tfrac{9}{4}$$

This process of "stellation" may also be applied in space. The stellated faces of the regular dodecahedron meet by fives at twelve new vertices, forming the *small stellated dodecahedron* $\{\frac{5}{2}, 5\}$. These new vertices belong also to an icosahedron. By

---

*Coxeter, *Regular Polytopes*, New York, 1973, p. 27.

inserting the edges of this icosahedron, but keeping the original twelve facial planes, we obtain a polyhedron whose faces are twelve ordinary pentagons, while the section near a vertex is a pentagram; this is the *great dodecahedron,** $\{5, \frac{5}{2}\}$. It is reciprocal to $\{\frac{5}{2}, 5\}$, as its symbol implies. By stellating the faces of $\{5, \frac{5}{2}\}$, we derive the *great stellated dodecahedron,* $\{\frac{5}{2}, 3\}$, which has the twenty vertices of an ordinary dodecahedron. Its reciprocal, the *great icosahedron*† $\{3, \frac{5}{2}\}$, has twenty triangular faces, and its vertices are those of an ordinary icosahedron. (See plate II, figures 31, 34, 32, 30.)

Thus we increase the number of finite regular polyhedra from five to nine. One way to see that these exhaust all the possibilities‡ is by observing that the "Petrie polygon" of $\{p, q\}$ is still characterized by the number h, where

$$\cos^2(\pi/h) = \cos^2(\pi/p) + \cos^2(\pi/q),$$

even when p and q are not integers. Writing this equation in the symmetrical form

$$\cos^2(\pi/p) + \cos^2(\pi/q) + \cos^2(\pi/k) = 1$$

(where $1/k = \frac{1}{2} - 1/h$), we find its rational solutions to be the three permutations of 3, 3, 4, and the six permutations of 3, 5, $\frac{5}{2}$, making nine in all, as required.

$\{p, q\}$	V	E	F	D	Name	Discoverer
$\{\frac{5}{2}, 5\}$	12	30	12	3	Small stellated dodecahedron	Kepler (1619)
$\{\frac{5}{2}, 3\}$	20	30	12	7	Great stellated dodecahedron	,,
$\{5, \frac{5}{2}\}$	12	30	12	3	Great dodecahedron	Poinsot (1809)
$\{3, \frac{5}{2}\}$	12	30	20	7	Great icosahedron	,,

*The *Encyclopaedia Britannica* (XIVth edition, article "Solids") unhappily calls this the "small stellated dodecahedron," and vice versa. (Cf. the XIth edition, article "Polyhedron.")

†Good drawings of these figures are given by Lucas (in his *Récréations mathématiques*), vol. II, pp. 206–208, 224.

‡This was first proved (another way) by Cauchy, *Journal de l'École Polytechnique*, 1813, vol. IX, pp. 68–86.

The polyhedra $\{\frac{5}{2}, 5\}$ and $\{5, \frac{5}{2}\}$ fail to satisfy Euler's Formula $V - E + F = 2$, which holds for all ordinary polyhedra. The reason for this failure (which apparently induced Schläfli* to deny the existence of these two figures) will appear in chapter VIII. However, all the nine finite regular polyhedra satisfy the following extended theorem, due to Cayley:

$$d_V V - E + d_F F = 2D,$$

where d_F is the "density" of a face (viz., 1 for an ordinary polygon, 2 for a pentagram), d_V is the density of a vertex (or rather, of the section near a vertex), and D is the density of the whole polyhedron (i.e. the number of times the faces enclose the centre).

"Archimedean" star polyhedra have been investigated,† but are beyond the scope of this book.

THE 59 ICOSAHEDRA

Imagine a large block of wood with a small tetrahedron or cube (somehow) drawn in the middle. If we make saw-cuts along all the facial planes of the small solid, and throw away all the pieces that extend to the surface of the block, nothing remains but the small solid itself. But if, instead of a tetrahedron or cube, we start with an octahedron, we shall be left with nine pieces: the octahedron itself, and a tetrahedron on each face, converting it into a *stella octangula* which has the appearance of two interpenetrating tetrahedra (the regular compound mentioned above). Similarly, a dodecahedron leads to $1 + 12 + 30 + 20$ pieces: the dodecahedron itself, twelve pentagonal pyramids which convert this into the small stellated dodecahedron, thirty wedge-shaped tetrahedra which convert the latter into the great dodecahedron, and twenty triangular double-pyramids which convert this last into the great stellated dodecahedron.

*Quarterly Journal of Mathematics, 1860, vol. III, pp. 66, 67. He defined "$(\frac{5}{2}, 3), (3, \frac{5}{2})$,".but not "$(\frac{5}{2}, 5), (5, \frac{5}{2})$."

†Coxeter, Longuet-Higgins, and Miller, Philosophical Transactions of the Royal Society, 1954, series A, vol. CCXLVI, pp. 401–450.

Finally, the icosahedron* leads to $1 + 20 + 30 + 60 + 20 + 60 + 120 + 12 + 30 + 60 + 60$ pieces, which can be put together to form 32 different reflexible solids, all having the full icosahedral symmetry, and 27 pairs of enantiomorphous solids, having only the symmetry of rotation. The former set of solids includes the original icosahedron, the compound of five octahedra (made of the first $1 + 20 + 30$ pieces), the compound of ten tetrahedra (made of the first $1 + 20 + 30 + 60 + 20 + 60 + 120$ pieces), and the great icosahedron (made of all save the last 60 pieces). The latter set includes the compound of five tetrahedra, and a number of more complicated figures having the same attractively "twisted" appearance.†

SOLID TESSELLATIONS

Just as there are many symmetrical ways of filling a plane with regular polygons, so there are many symmetrical ways of filling space with regular and Archimedean solids. For the sake of brevity, let us limit our discussion to those ways in which all the edges (as well as all the vertices) are surrounded alike. Of such "solid tessellations" there are just five,‡ an edge being surrounded by (i) four cubes, or (ii) two tetrahedra and two octahedra, arranged alternately, or (iii) a tetrahedron and three truncated tetrahedra, or (iv) three truncated octahedra, or (v) an octahedron and two cuboctahedra. Let us denote these by the symbols $[4^4]$, $[3^4]$, $[3^2 . 6^2]$, $[4 . 6^2]$, $[3^2 . 4]$, which indicate the polygons (interfaces) that meet at an edge.

The "regular" space-filling $[4^4]$ is familiar. It is "self-reciprocal" in the sense that the centres of all the cubes are the vertices of an identical space-filling. Its alternate vertices

*A.H. Wheeler, *Proceedings of the International Mathematical Congress*, Toronto, 1924, vol. I, pp. 701–708; M. Brückner, *Vielecke und Vielflache*, Leipzig, 1900 (plate VIII, nos. 2, 26; plate IX, nos. 3, 6, 11, 17, 20; plate X, no. 3; plate XI, nos. 14, 24).

†For J.F. Petrie's exquisite drawings of all these figures, see "The 59 Icosahedra," *University of Toronto Studies* (Mathematical Series), no. 6, 1938.

‡A. Andreini, *Memorie della Società italiana delle Scienze*, 1905, series 2, vol. XIV, pp. 75–129, figs. 12, 15, 14, 18, 33.

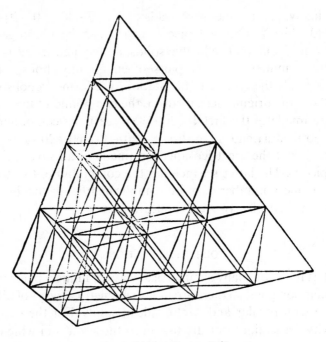

The Solid Tessellation [3⁴]

give the space-filling $[3^4]$, one tetrahedron being inscribed in each cube, and one octahedron surrounding each omitted vertex. This has a particularly high degree of regularity (although its solids are of two kinds, unlike those of $[4^4]$); for, not merely the vertices and edges, but also the triangular interfaces, are all surrounded alike; in fact, each triangle belongs to one solid of either kind. If we join the centres of adjacent solids, by lines perpendicular to the interfaces, and by planes perpendicular to the edges, we obtain the "reciprocal" space-filling, say $[3^4]'$; this consists of rhombic dodecahedra, of which four surround some vertices (originally centres of tetrahedra), while six surround others (originally centres of octahedra).

The space-filling $[3^2.6^2]$ can be derived from $[3^4]$ by making each of a certain set of tetrahedra of the latter adhere to its four adjacent octahedra and to six other tetrahedra which

connect these in pairs, so as to form a truncated tetrahedron.*
Thus $[3^2 . 6^2]$ has half the vertices of $[3^4]$, which in turn has
half the vertices of $[4^4]$.

The space-filling of truncated octahedra, $[4 . 6^2]$, is reciprocal to a space-filling of "isosceles" tetrahedra (or tetragonal
bisphenoids) whose vertices belong to two reciprocal $[4^4]$'s
(the "body-centred cubic lattice" of crystallography). The
vertices of $[3^2 . 4]$ are the mid-points of the edges (*or* the centres
of the squares) of $[4^4]$.

BALL-PILING OR CLOSE-PACKING

A large box can be filled with a number of small equal spheres
arranged in horizontal layers, one on top of another, in various
ways, of which I will describe three. It might be filled so that
each sphere rests on the top of the sphere immediately below it
in the next layer, touches each of four adjacent spheres in the
same layer, and touches one sphere in the layer above it; thus
each sphere is in contact with six others. Or we might slightly
spread out the spheres of each layer, so as to be not quite in
contact, and let each sphere rest on four in the layer below and
help to support four in the layer above, the "spreading out"
being adjusted so that the points of contact are at the vertices
of a cube. We might also fill the box with spheres arranged so
that each of them is in contact with four spheres in the next
lower layer, with four in the same layer, and with four in the
next higher layer. This last arrangement is known as *normal
piling* or *spherical close-packing*; it gives the greatest number
of spheres with which the box can be filled. (Although it is
impossible for one sphere to touch more than twelve others of
the same size, we shall see later that there are many different
ways of packing equal spheres so that each touches exactly
twelve others.)

These three arrangements may be described as follows. In
the first, the centres of the spheres are at the vertices of the
space-filling $[4^4]$, and the spheres themselves are inscribed in

*Analogously, any of the plane tessellations 6^3, $(3 . 6)^2$, $3^4 . 6$ may be derived
from 3^6 by making certain sets of six triangles coalesce to form hexagons.

the cubes of the reciprocal $[4^4]$. In the second, the spheres are inscribed in the truncated octahedra of $[4.6^2]$ (touching the hexagons, but just missing the squares). In the third, the spheres are inscribed in the rhombic dodecahedra of $[3^4]'$, and their centres are at the vertices of $[3^4]$.

Now, the vertices of $[3^4]$ form triangular tessellations 3^6 in a series of parallel planes.

A		A		A		A	
	B		B		B		
C		C		C		C	
	A		A		A		
B		B		B		B	
	C		C		C		
A		A		A		A	
	B		B		B		
C		C		C		C	

Our figure shows a "plan" of this arrangement of points, projected orthogonally on one of the planes, which we take to be horizontal. The points A are projected from one plane, the points B from the next, C from the next, A again from the next, and so on, in cyclic order. Now imagine solid spheres centred at all these points. The points A give a layer of close-packed spheres, each touching six others. The points B represent another such layer, resting on the first; each sphere of either touches three of the other. The points C represent a third layer, resting on the second; but an equally "economical" piling of spheres is obtained if the centres of the third layer lie above the points A again, instead of lying above the points C. And so, at every stage, the new layer may or may not lie vertically above the last but one.

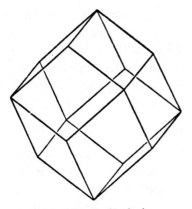

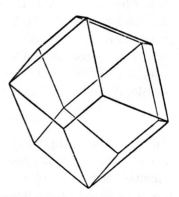

(1) Rhombic Dodecahedron   (2) Trapezo-rhombic Dodecahedron

The arrangement *A B C A B C*... represents spherical close-packing; on the other hand, the arrangement *A B A B A B*... is known as *hexagonal* close-packing. In both cases space is filled to the extent of 74 per cent. If a large number of equal balls of "plasticene" or modelling clay are rolled in chalk, packed in either fashion, and squeezed into a solid lump, those near the middle tend to form rhombic dodecahedra or trapezorhombic dodecahedra* respectively. If, instead of being carefully stacked, the balls are shaken into a random arrangement as dense as possible, and are then squeezed as before, the resulting shapes are irregular polyhedra of various kinds. The average number of faces† is not 12 but about 13·3. Equal spheres arranged in such a *random* piling have not been proved to occupy less space than the same spheres in normal piling; but it is clear that any small displacement will increase the total volume by enlarging the interstices.

If you stand on wet sand, near the sea-shore, it is very noticeable that the sand gets comparatively dry around your feet, whereas the footprints that you leave contain free water. The following explanation is due, I believe, to Osborne Reynolds. The grains of sand, rolled into approximately spherical shape

*Cf. Steinhaus, *Mathematical Snapshots*, New York, 1938, p. 88.
†J.D. Bernal, *Nature*, 1959, vol. CLXXXIII, pp. 141–147; Coxeter, *Introduction to Geometry*, pp. 410–412.

by the motion of the sea, have been deposited in something like random piling. The pressure of your feet disturbs this piling, increasing the interstices between the grains. Water is sucked in from around about, to fill up these enlarged interstices. When you remove your feet, the random piling is partially restored, and the water is left above.

REGULAR SPONGES

The definition of regularity on page 131 depends on two symmetries, which, in every case so far discussed, are *rotations*. By allowing the number of vertices, edges, and faces to be infinite, this definition includes the plane tessellations $\{3, 6\}$, $\{6, 3\}$, $\{4, 4\}$. It would be absurd to allow each face to have infinitely many sides, or to allow infinitely many faces to surround one vertex; therefore the special symmetries must be periodic. However, they need not be rotations; they may be rotatory-reflections. (A rotatory-reflection is the combination of a rotation and a reflection, which may always be chosen so that the axis of the rotation is perpendicular to the reflecting plane.) Such an operation interchanges the "inside" and "outside" of the polyhedron; consequently the inside and outside are identical, and the polyhedron (dividing space into two equal parts) must be infinite. The dihedral angles at the edges of a given face are alternately positive and negative, and the edges at a vertex lie alternately on the two sides of a certain plane. This allows the sum of the face-angles at a vertex to exceed 2π.

It can be proved that the polyhedra $\{p, q\}$ of this type are given by the integral solutions of the equation

$$2 \sin (\pi/p) \sin (\pi/q) = \cos (\pi/k),$$

namely $\{6, 6\}$ $(k=3)$, $\{6, 4\}$ and $\{4, 6\}$ $(k=4)$, $\{3, 6\}$ $(k=6)$, and $\{4, 4\}$ $(k=\infty)$. The three plane tessellations occur because a plane rotation may be regarded indifferently as a rotation in space or as a rotatory-reflection. The three new ffgures are "sponges" with k-gonal holes.*

*For photographs of models, see Coxeter, *Twelve Geometric Essays*, p. 77, where the three sponges are denoted by $\{6,6|3\}$, $\{6,4|4\}$, $\{4,6|4\}$.

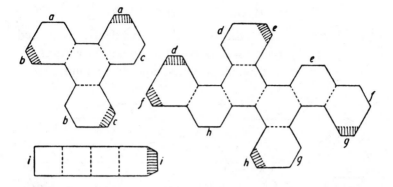

The faces of $\{6, 6\}$ are the hexagons of the solid tessellation $[3^2 . 6^2]$; those of $\{6, 4\}$ are the hexagons of $[4 . 6^2]$; and those of $\{4, 6\}$ are half the squares of $[4^4]$. The remaining interfaces of the solid tessellations appear as holes. The last two sponges (discovered by J.F. Petrie in 1926) are *reciprocal*, in the sense that the vertices of each are the face centres of the other;* $\{6, 6\}$ is self-reciprocal or, rather, reciprocal to another congruent $\{6, 6\}$.

To make a model of $\{6, 6\}$, cut out sets of four hexagons (of thin cardboard), stick each set together in the form of the hexagonal faces of a truncated tetrahedron ($3 . 6^2$), and then stick the sets together, hexagon on hexagon, taking care that no edge shall belong to more than two faces. (In the finished model, the faces are double, which makes for greater strength besides facilitating the construction.) Similarly, to make $\{6, 4\}$, use sets of eight hexagons, forming the hexagonal faces of truncated octahedra ($4 . 6^2$). Finally, to make $\{4, 6\}$, use rings of four squares. This last model, however, is not rigid; it can gradually collapse, the square holes becoming rhombic. (In fact, J.C.P. Miller once made an extensive model and mailed it, flat in an envelope).

---

*Plane tessellations can be reciprocal in this sense, but finite polyhedra cannot. The centres of the faces of an octahedron are the vertices of a cube, while the vertices of the octahedron are the centres of the faces of another (larger) cube.

ROTATING RINGS OF TETRAHEDRA

J.M. Andreas and R.M. Stalker have independently discovered a family of non-rigid finite polyhedra having $2n$ vertices, $6n$ edges (of which $2n$ coincide in pairs), and $4n$ triangular faces,

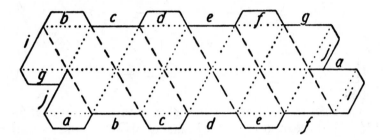

for $n=6$ or 8 or any greater integer. The faces are those of n tetrahedra, joined together in cyclic order at a certain pair of opposite edges of each, so as to form a kind of ring. When $n=6$, the range of mobility is quite small, but when $n=8$, the ring can turn round indefinitely, like a smoke-ring. When n is even, the figure tends to take up a symmetrical position; it is particularly pretty when $n=10$.* When n is odd, the entire lack of symmetry seems to make the motion still more fascinating. When $n \geq 22$, the ring can occur in a knotted form.

A model of any such ring may be made from a single sheet of paper. For the case when $n=6$, copy the above diagram, cut it out, bend the paper along the inner lines, upwards or downwards according as these lines are broken or dotted, and stick the flaps in the manner indicated by the lettering. The ends have to be joined somewhat differently when n is a multiple of 4 (see figure XXXIV, page 216). When n is odd, either method of joining can be used at will.

Since there are two types of edge, such a polyhedron is not regular, and no symmetry is lost by making the triangles isosceles instead of equilateral. If the doubled edges are sufficiently

*One of the "Stephanoids" described by M. Brückner in his *Vielecke und Vielflache*, p. 216 (and plate VIII, no. 4) consists of a ring of ten irregular tetrahedra.

short compared to the others, the ring with $n = 6^*$ can be made to turn completely, like the rings with $n \geq 8$.

THE KALEIDOSCOPE [†]

The ordinary kaleidoscope consists essentially of two plane mirrors, inclined at $\pi/3$ or $\pi/4$, and an object (or set of objects) placed in the angle between them so as to be reflected in both. The result is that the object is seen 6 or 8 times (according to the angle), in an attractively symmetrical arrangement. By making a hinge to connect two (unframed) mirrors, the angle between them can be varied at will, and it is clear that an angle π/n gives $2n$ images (including the object itself). As a limiting case, we have two parallel mirrors and a theoretically infinite number of images (restricted in practice only by the brightness of the illumination and the quality of the mirrors). If the object is a point on the bisector of the angle between the mirrors, the images are the vertices of a regular $2n$-gon. If the object is a point on one of the mirrors, the images coincide in pairs at the vertices of a regular n-gon. The point may be represented in practice by a candle, or by a little ball of plastic clay or putty.

Regarding the two mirrors as being vertical, let us introduce a third vertical mirror in such a way that each pair of the three mirrors makes an angle of the form π/n. In other words, any horizontal section is to be a triangle of angles π/l, π/m, π/n, where l, m, n are integers. The solutions of the consequent equation

$$\frac{1}{l} + \frac{1}{m} + \frac{1}{n} = 1$$

are 3, 3, 3; 2, 3, 6; 2, 4, 4. In each case the number of images is infinite. By varying the position of a point-object in the triangle, we obtain the vertices of certain isogonal tessella-

---

*Such a ring (of six tetragonal bisphenoids) has been on sale in the United States as a child's toy, with letters of the alphabet on its 24 faces. (Patent No. 1,997,022, issued in 1935.) See also M. Goldberg, *Journal of Mathematics and Physics*, 1947, vol. xxvi, pp. 10–21.

[†]E. Hess, *Neues Jahrbuch für Mineralogie, Geologie und Palaeontologie*, 1889, vol. i, pp. 54–65.

tions.* In particular, if the point is taken at a vertex of the triangle, or where an angle-bisector meets the opposite side, or at the in-centre (where all three angle-bisectors concur), then the tiles of the tessellation are regular polygons. In the notation of page 106, the results of putting the point in these various positions are as indicated in the following diagrams.

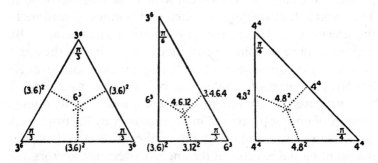

The network of triangles, which the mirrors appear to create, may be coloured alternately white and black. By taking a suitable point within every triangle of one colour (but ignoring the corresponding point within every triangle of the other colour), we obtain the vertices of 3^6 (again), $3^4.6$, and $3^2.4.3.4$, respectively. (The remaining uniform tessellation, $3^3.4^2$, is not derivable by any such method.) The above diagrams reveal many relationships between the various tessellations: that the vertices of 3^6 occur among the vertices of 6^3, that the vertices of 6^3 trisect the edges of (another) 3^6, that the vertices of one 4^4 bisect the edges of another, and so on.

If the third mirror is placed horizontally instead of vertically – i.e. if the two hinged mirrors stand upon it – the number of images is no longer infinite; in fact, it is $4n$, where π/n is the angle between the two vertical mirrors. For a point on one of the vertical mirrors, the images coincide in pairs at the vertices of an n-gonal prism. Two of the three dihedral angles

*Placing a lighted candle between three (unframed) mirrors, the reader will see an extraordinarily pretty effect. The University of Minnesota has made use of this idea in two short films: *Dihedral Kaleidoscopes* and *The Symmetries of the Cube.*

between pairs of the three mirrors are now right angles. A natural generalization is the case where these three angles are π/l, π/m, π/n.

Since, for any reflection in a plane mirror, object and image are equidistant from the plane, we easily see that all the images of a point in this generalized kaleidoscope lie on a sphere, whose centre is the point of intersection of the planes of the three mirrors. On the sphere, these planes cut out a spherical triangle, of angles π/l, π/m, π/n. The resulting image-planes divide the whole sphere into a network (or "map") of such triangles, each containing one image of any object placed within the first triangle. The number of images is therefore equal to the number of such triangles that will suffice to fill the whole spherical surface. Taking the radius as unity, the area of the whole sphere is 4π, while that of each triangle is $(\pi/l) + (\pi/m) + (\pi/n) - \pi$. Hence the required number is

$$4\bigg/\left(\frac{1}{l}+\frac{1}{m}+\frac{1}{n}-1\right).$$

Since this must be positive, the numbers l, m, n have to be chosen so as to satisfy

$$\frac{1}{l}+\frac{1}{m}+\frac{1}{n}>1.$$

This inequality has the solutions $2, 2, n$; $2, 3, 3$; $2, 3, 4$; $2, 3, 5$. The first case has already been mentioned; the rest are depicted on page 158 (by J.F. Petrie).

For a practical demonstration, the mirrors should in each case be cut as circular sectors (of the same fairly large radius), whose angles* are equal to the sides of a spherical triangle of angles π/l, π/m, π/n.

By varying the position of a point-object in the spherical triangle (or in the solid angle between the three mirrors) we

---

*In the three cases, these angles are respectively: 54°44′, 54°44′, 70°32′; 35°16′, 45°, 54°44′; 20°54′, 31°43′, 37°23′.

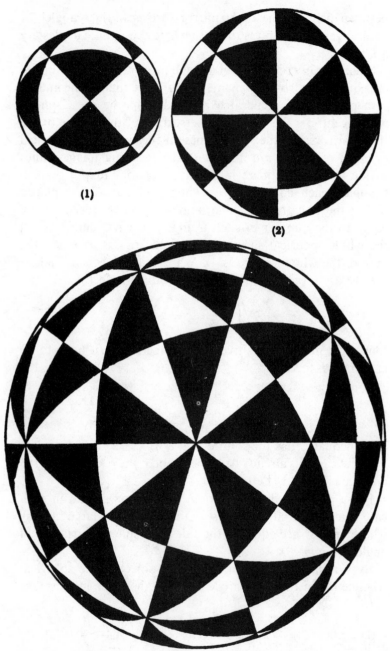

(1)

(2)

(3)

obtain the vertices of certain isogonal polyhedra. In particular, if the point is on one of the edges where two mirrors meet, or on one of the mirrors and equidistant from the other two, or at the centre of a sphere which touches all three, then the faces of the polyhedra are regular polygons. The manner in which the various uniform polyhedra arise* is indicated in the following diagrams, analogous to those given for tessellations on page 156.

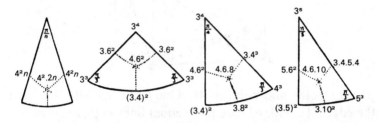

By taking a suitable point within each white (or black) triangle,† we obtain the vertices of $3^3 . n$, 3^5, $3^4 . 4$, or $3^4 . 5$, respectively. It has already been remarked that the snub cube $3^4 . 4$ exists in two enantiomorphous forms; the vertices of one form lie in the white triangles, those of the other in the black. The same thing happens in the case of the snub dodecahedron, $3^4 . 5$.

By introducing a fourth mirror, we obtain solid tessellations. Tetrahedra of three different shapes can be formed by four planes inclined at angles that are submultiples of π. These three shapes can conveniently be cut out from a rectangular block of dimensions $1 \times \sqrt{2} \times \sqrt{2}$. Suppose $ABCD$ to be a horizontal square of side $\sqrt{2}$, at height 1 above an equal square $A'B'C'D'$. After cutting off the alternate corners A', B, C', D, by planes through sets of three other vertices, we are left with the tetragonal bisphenoid $AB'CD'$, which is one of the required shapes.

*See Möbius, *Gesammelte Werke*, 1861, vol. ii, p. 656, figs. 47, 51, 54; W.A. Wythoff, *Proceedings of the Royal Academy of Sciences, Amsterdam*, 1918, vol. xx, pp. 966–970; G. de B. Robinson, *Journal of the London Mathematical Society*, 1931, vol. vi, pp. 70–75; H.S.M. Coxeter, *Proceedings of the London Mathematical Society*, 1935, series 2, vol. xxxviii, pp. 327–339.

†Möbius, *loc. cit.*, figs. 46, 49, 53.

Another is any one of the corner pieces that were cut off, such as *ABCB'*. (Two such pieces can be fitted together to make a shape just like *AB'CD'*.) The third is obtained by cutting *ABCB'* in half along its plane of symmetry, which is the plane *BB'E*,

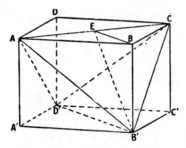

where *E* is the mid-point of *AC*. One half is *AEBB'*. Note that the edges *AE*, *EB*, *BB'* are three equal lines in three perpendicular directions.

A point-object in such a tetrahedron will give rise to the vertices of a solid tessellation in various ways,* some of which are indicated in the following diagrams (which show *AB'CD'*, *ABCB'*, *AEBB'*, in the same orientation as before).

Five mirrors can be arranged in the form of certain triangular prisms; these lead to solid tessellations of prisms. Six can be arranged rectangularly in three pairs of parallels, as when we

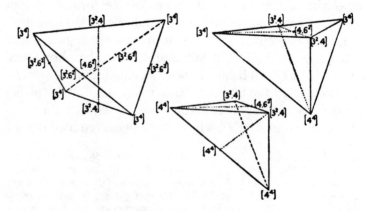

*Andreini, *loc. cit.* (p. 147 above), figs. 17–24 bis.

have a mirror in the ceiling and floor as well as all four walls of an ordinary room; these give a solid tessellation of rectangular blocks. G. Pólya has proved* that any kaleidoscope is effectively equivalent to one having at most six mirrors.

*Annals of Mathematics, 1934, vol. xxxv, p. 594.

ADDENDUM

Note. Page 143. A *rhombohedron* is a parallelepiped bounded by six congruent rhombs. It has two opposite vertices at which the three face-angles are equal; it is said to be *acute* or *obtuse* according to the nature of these angles. A *golden* rhombohedron has faces whose diagonals are in the "golden" ratio τ : 1 (see page 56). The Japanese architect Koji Miyazaki observed in 1977 that ten golden rhombohedra, five acute (A_6) and five obtuse (O_6), can be fitted together to form a rhombic icosahedron (F_{20}), and that two rhombic icosahedra, symmetrically placed with a common "obtuse" vertex, can be surrounded by further rhombohedra (thirty acute and thirty obtuse) to form a large rhombic icosahedron whose edges are twice as long; symbolically

$$30 \, A_6 + 30 \, O_6 + 2 \, F_{20} = F_{20} \, 2.$$

Consequently the whole space can be filled with golden rhombohedra so as to form a honeycomb having a pentagonal axis of symmetry.

A different space-filling of A_6's and O_6's was discovered independently, at about the same time, by Robert Ammann in Massachusetts. He was seeking a 3-dimensional analogue for Roger Penrose's non-periodic tilings of the plane (see the *Mathematical Intelligencer,* 1979, vol. II, p. 36). Ammann's procedure was extended by the Japanese physicist Tohru Ogawa, whose idea is to iterate the construction

$$55 \, A_6 + 34 \, O_6 = A_6 \, \tau^3, \qquad 34 \, A_6 + 21 \, O_6 = O_6 \, \tau^3.$$

Since the volumes of A_6 and O_6 are in the ratio τ to 1, Ogawa's hierarchic rule illustrates the identities

$$55 \, \tau + 34 = \tau^{10}, \qquad 34 \, \tau + 21 = \tau^9$$

(which involve the same consecutive Fibonacci numbers 21, 34, 55 as the paradox on page 86, although now everything is exact). Since all the edges of this *quasilattice* are parallel to the six diameters joining pairs of opposite vertices of the regular icosahedron, icosahedral symmetry occurs in a statistical sense, agreeing with the apparent icosahedral symmetry of certain alloys of aluminium and manganese (see A.L. Mackay, *Nature,* 1986, vol. CCCXIX, pp. 102–104).

CHAPTER VI

CHESS-BOARD RECREATIONS

A chess-board and chess-men lend themselves to recreations, many of which are geometrical. The problems are, however, of a distinct type, and sufficiently numerous to deserve a chapter to themselves. A few problems which might be included in this chapter have been already considered in chapter IV.

The ordinary chess-board consists of 64 small squares, known as cells, arranged as shown below in 8 rows and 8 columns. Usually the cells are coloured alternately white and black, or white and red. The cells may be defined by the numbers 11, 12, etc., where the first digit denotes the number of

18	28	38	48	58	68	78	88
17	27	37	47	57	67	77	87
16	26	36	46	56	66	76	86
15	25	35	45	55	65	75	85
14	24	34	44	54	64	74	84
13	23	33	43	53	63	73	83
12	22	32	42	52	62	72	82
11	21	31	41	51	61	71	81

the column and the second digit the number of the row, the two digits representing respectively the abscissa and ordinate of the mid-points of the cells. I use this notation in the following pages. A generalized board consists of n^2 cells arranged in n rows and n columns. Most of the problems which I shall describe can be extended to meet the case of a board of n^2 cells.

The usual chess-pieces are Kings, Queens, Bishops, Knights, and Rooks or Castles; there are also Pawns. I assume that the moves of these pieces are known to the reader.

With the game itself and with chess problems of the usual type I do not concern myself. Particular positions of the pieces may be subject to mathematical analysis, but in general the moves open to a player are so numerous as to make it impossible to see far ahead. Probably this is obvious, but it may emphasize how impossible it is to discuss the theory of the game effectively if I add that it has been shown that there may be as many as 197299 ways of playing the first four moves, and 71782 different positions at the end of the first four moves (two on each side), of which 16556 arise when the players move pawns only.*

RELATIVE VALUE OF PIECES

The first question to which I will address myself is the determination of the relative values of the different chess-pieces.†

If a piece is placed on a cell, the number of cells it commands depends in general on its position. We may estimate the value of the piece by the average number of cells which it commands when placed in succession on every cell of the board. This is equivalent to saying that the value of a piece may be estimated by the chance that if it and a king are put at random on the board, the king will be in check: if no other restriction is imposed, this is called a simple check. On whatever cell the piece is originally placed there will remain 63 other cells on which the king may be placed. It is equally probable that it may be put on any one of them. Hence the chance that it will be in check is 1/63 of the average number of cells commanded by the piece.

A rook put on any cell commands 14 other cells. Wherever the rook is placed there will remain 63 cells on which the king

---

*L'Intermédiaire des Mathématiciens, Paris, December 1903, vol. x, pp. 305–308; also Royal Engineers Journal, London, August–November 1889; or British Association Transactions, 1890, p. 745.

†H.M. Taylor, Philosophical Magazine, March 1876, series 5, vol. I, pp. 221–229.

may be placed, and on which it is equally likely that it will be placed. Hence the chance of a simple check is 14/63, that is, 2/9. Similarly, on a board of n^2 cells the chance is $2(n-1)/(n^2-1)$, that is, $2/(n+1)$.

A knight, when placed on any of the 4 corner cells like 11, commands 2 cells. When placed on any of the 8 cells like 12 and 21, it commands 3 cells. When placed on any of the 4 cells like 22 or any of the 16 boundary cells like 13, 14, 15, 16, it commands 4 cells. When placed on any of the 16 cells like 23, 24, 25, 26, it commands 6 cells. And when placed on any of the remaining 16 middle cells, it commands 8 cells. Hence the average number of cells commanded by a knight put on a chessboard is $(4 \times 2 + 8 \times 3 + 20 \times 4 + 16 \times 6 + 16 \times 8)/64$, that is, 336/64. Accordingly, if a king and a knight are put on the board the chance that the king will be in simple check is $336/(64 \times 63)$, that is, 1/12. Similarly, on a board of n^2 cells the chance is $8(n-2)/n^2(n+1)$.

A bishop when placed on any of the ring of 28 boundary cells commands 7 cells. When placed on any ring of the 20 cells next to the boundary cells, it commands 9 cells. When placed on any of the 12 cells forming the next ring, it commands 11 cells. When placed on the 4 middle cells, it commands 13 cells. Hence, if a king and a bishop are put on the board, the chance that the king will be in simple check is $(28 \times 7 + 20 \times 9 + 12 \times 11 + 4 \times 13)/(64 \times 63)$, that is, 5/36. Similarly, on a board of n^2 cells, when n is even, the chance is $2(2n-1)/3n(n+1)$. When n is odd, the analysis is longer, owing to the fact that in this case the number of white cells on the board differs from the number of black cells. I do not give the work, which presents no special difficulty.

A queen when placed on any cell of a board commands all the cells which a bishop and a rook when placed on that cell would do. Hence, if a king and a queen are put on the board, the chance that the king will be in simple check is $(2/9) + 5/36$, that is, 13/36. Similarly, on a board of n^2 cells, when n is even, the chance is $2(5n-1)/3n(n+1)$.

On the above assumptions the relative values of the rook,

knight, bishop, and queen are 8, 3, 5, 13. According to Staunton's *Chess-Player's Handbook*, the actual values, estimated empirically, are in the ratio of 548, 305, 350, 994; according to Von Bilguer, the ratios are 540, 350, 360, 1000, the value of a pawn being taken as 100.

There is considerable discrepancy between the above results as given by theory and practice. It has been, however, suggested that a better test of the value of a piece would be the chance that when it and a king were put at random on the board, it would check the king without giving the king the opportunity of taking it. This is called a safe check, as distinguished from a simple check.

Applying the same method as above, the chances of a safe check work out as follows. For a rook the chance of a safe check is $(4 \times 12 + 24 \times 11 + 36 \times 10)/(64 \times 63)$, that is, $1/6$, or on a board of n^2 cells is $2(n-2)/n(n+1)$. For a knight all checks are safe, and therefore the chance of a safe check is $1/12$; or on a board of n^2 cells is $8(n-2)/n^2(n+1)$. For a bishop the chance of a safe check is $364/(64 \times 63)$, that is, $13/144$, or on a board of n^2 cells, when n is even, is $2(n-2)(2n-3)/3n^2(n+1)$. For a queen the chance of a safe check is $1036/(64 \times 63)$, that is, $37/144$, or on a board of n^2 cells $2(n-2)(5n-3)/3n^2(n+1)$, when n is even.

On this view the relative values of the rook, knight, bishop, and queen are 24, 12, 13, 37; while, according to Staunton, experience shows that they are approximately 22, 12, 14, 40, and according to Von Bilguer, 18, 12, 12, 33.

The same method can be applied to compare the values of combinations of pieces. For instance, the value of two bishops (one restricted to white cells and the other to black cells) and two rooks, estimated by the chance of a simple check, are respectively $35/124$ and $37/93$. Hence on this view a queen in general should be more valuable than two bishops, but less valuable than two rooks. This agrees with experience.

An analogous problem consists in finding the chance that two kings, put at random on the board, will not occupy adjoining cells – that is, that neither would (were such a move possible) check the other. The chance is 43/48, and therefore the chance that they will occupy adjoining cells is 5/48. If three kings are put on the board, the chance that no two of them occupy adjoining cells is 1061/1488. The corresponding chances* for a board of n^2 cells are

$$(n-1)(n-2)(n^2+3n-2)/n^2(n^2-1)$$

and $(n-1)(n-2)(n^4+3n^3-20n^2-30n+132)/n^2(n^2-1)(n^2-2)$.

THE EIGHT QUEENS PROBLEM[†]

One of the classical problems connected with a chess-board is the determination of the number of ways in which eight queens can be placed on a chess-board (or, more generally, in which n queens can be placed on a board of n^2 cells) so that no queen can take any other. This was proposed originally by Franz Nauck in 1850.

In 1874 Dr. S. Günther[‡] suggested a method of solution by means of determinants. For, if each symbol represents the corresponding cell of the board, the possible solutions for a board of n^2 cells are given by those terms, if any, of the determinant

$$\begin{vmatrix} a_1 & b_2 & c_3 & d_4 & \cdots \\ \beta_2 & a_3 & b_4 & c_5 & \cdots \\ \gamma_3 & \beta_4 & a_5 & b_6 & \cdots \\ \delta_4 & \gamma_5 & \beta_6 & a_7 & \cdots \\ \cdots & & & & \\ \cdots & & a_{2n-3} & b_{2n-2} \\ \cdots & & \beta_{2n-2} & a_{2n-1} \end{vmatrix}$$

in which no letter and no suffix appears more than once.

The reason is obvious. Every term in a determinant con-

---

*L'Intermédiaire des Mathématiciens, Paris, 1897, vol. IV, p. 6, and 1901, vol. VIII, p. 140.

†On the history of this problem see W. Ahrens, Mathematische Unterhaltungen und Spiele, Leipzig, 1901, chap. IX. For later developments see Kraïtchik, La Mathématique des Jeux, Brussels, 1930, pp. 300–356.

‡Grunert's Archiv der Mathematik und Physik, 1874, vol. LVI, pp. 281–292.

tains one, and only one, element out of every row and out of every column: hence any term will indicate a position on the board in which the queens cannot take one another by moves rook-wise. Again, in the above determinant the letters and suffixes are so arranged that all the same letters and all the same suffixes lie along bishop's paths: hence, if we retain only those terms in each of which all the letters and all the suffixes are different, they will denote positions in which the queens cannot take one another by moves bishop-wise. It is clear that the signs of the terms are immaterial.

In the case of an ordinary chess-board the determinant is of the 8th order, and therefore contains 8!, that is, 40320 terms, so that it would be out of the question to use this method for the usual chess-board of 64 cells or for a board of larger size unless some way of picking out the required terms could be discovered.

A way of effecting this was suggested by Dr. J.W.L. Glaisher* in 1874, and so far as I am aware the theory remains as he left it. He showed that if all the solutions of n queens on a board of n^2 cells were known, then all the solutions of a certain type for $n+1$ queens on a board of $(n+1)^2$ cells could be deduced, and that all the other solutions of $n+1$ queens on a board of $(n+1)^2$ cells could be obtained without difficulty. The method will be sufficiently illustrated by one instance of its application.

It is easily seen that there are no solutions when $n=2$ or $n=3$. If $n=4$ there are two terms in the determinant which give solutions, namely, $b_2 c_5 \gamma_3 \beta_6$ and $c_3 \beta_2 b_6 \gamma_5$. To find the solutions when $n=5$, Glaisher proceeded thus. In this case, Günther's determinant is

$$\begin{vmatrix} a_1 & b_2 & c_3 & d_4 & e_5 \\ \beta_2 & a_3 & b_4 & c_5 & d_6 \\ \gamma_3 & \beta_4 & a_5 & b_6 & c_7 \\ \delta_4 & \gamma_5 & \beta_6 & a_7 & b_8 \\ \varepsilon_5 & \delta_6 & \gamma_7 & \beta_8 & a_9 \end{vmatrix}$$

*Philosophical Magazine, London, December 1874, series 4, vol. XLVIII, pp. 457–467.

To obtain those solutions (if any) which involve a_9, it is sufficient to append a_9 to such of the solutions for a board of 16 cells as do not involve a. As neither of those given above involves an a, we thus get two solutions, namely, $b_2c_5\gamma_3\beta_6a_9$ and $c_3\beta_2b_6\gamma_5a_9$. The solutions which involve a_1, e_5, and ε_5 can be written down by symmetry. The eight solutions thus obtained are all distinct; we may call them of the first type.

The above are the only solutions which can involve elements in the corner squares of the determinant. Hence the remaining solutions are obtainable from the determinant

$$\begin{vmatrix} 0 & b_2 & c_3 & d_4 & 0 \\ \beta_2 & a_3 & b_4 & c_5 & d_6 \\ \gamma_3 & \beta_4 & a_5 & b_6 & c_7 \\ \delta_4 & \gamma_5 & \beta_6 & a_7 & b_8 \\ 0 & \delta_6 & \gamma_7 & \beta_8 & 0 \end{vmatrix}$$

If, in this, we take the minor of b_2 and in it replace by zero every term involving the letter b or the suffix 2, we shall get all solutions involving b_2. But in this case the minor at once reduces to $d_6a_5\delta_4\beta_8$. We thus get one solution, namely, $b_2d_6a_5\delta_4\beta_8$. The solutions which involve β_2, δ_4, δ_6, β_8, b_8, d_6, and d_4 can be obtained by symmetry. Of these eight solutions it is easily seen that only two are distinct: these may be called solutions of the second type.

Similarly, the remaining solutions must be obtained from the determinant

$$\begin{vmatrix} 0 & 0 & c_3 & 0 & 0 \\ 0 & a_3 & b_4 & c_5 & 0 \\ \gamma_3 & \beta_4 & a_5 & b_6 & c_7 \\ 0 & \gamma_5 & \beta_6 & a_7 & 0 \\ 0 & 0 & \gamma_7 & 0 & 0 \end{vmatrix}$$

If, in this, we take the minor of c_3, and in it replace by zero every term involving the letter c or the suffix 3, we shall get all the solutions which involve c_3. But in this case the minor vanishes. Hence there is no solution involving c_3, and therefore by symmetry no solutions which involve γ_3, γ_7, or c_7. Had there been any solutions involving the third element in the first or last row or column of the determinant, we should have described them as of the third type.

Thus in all there are ten, and only ten, solutions: namely, eight of the first type, two of the second type, and none of the third type.

Similarly, if $n = 6$, we obtain no solutions of the first type, four solutions of the second type, and no solutions of the third type, that is, four solutions in all. If $n = 7$, we obtain sixteen solutions of the first type, twenty-four solutions of the second type, no solutions of the third type, and no solutions of the fourth type, that is, forty solutions in all. If $n = 8$, we obtain sixteen solutions of the first type, fifty-six solutions of the second type, and twenty solutions of the third type, that is, ninety-two solutions in all.

It will be noticed that all the solutions of one type are not always distinct. In general, from any solution seven others can be obtained at once. Of these eight solutions, four consist of the initial or fundamental solution and the three similar ones obtained by turning the board through one, two, or three right angles; the other four are the reflections of these in a mirror: but in any particular case it may happen that the reflections reproduce the originals, or that a rotation through one or two right angles makes no difference. Thus on boards of 4^2, 5^2, 6^2, 7^2, 8^2, 9^2, 10^2, 11^2, 12^2 cells there are respectively 1, 2, 1, 6, 12, 46, 92, 341, 1784 fundamental solutions; while altogether there are respectively 2, 10, 4, 40, 92, 352, 724, 2680, 14200 solutions.

The following collection of fundamental solutions may interest the reader. Each position on the board of the queens is indicated by a number, but as necessarily one queen is on each column, I can use a simpler notation than that explained on page 162. In this case the first digit represents the number of the cell occupied by the queen in the first column reckoned from one end of the column, the second digit the number in the second column, and so on. Thus on a board of 4^2 cells the solution 3142 means that one queen is on the 3rd square of the first column, one on the 1st square of the second column, one on the 4th square of the third column, and one on the 2nd square of the fourth column. If a fundamental solution gives rise to only four solutions, the number which indicates it is placed in curved brackets, ( ); if it gives rise to only two solutions, the number which indicates it is placed in square brackets, [ ]; the other fundamental solutions give rise to eight solutions each.

On a board of 4^2 cells there is 1 fundamental solution – namely, [3142].

On a board of 5^2 cells there are 2 fundamental solutions – namely, 14253, [25314]. It may be noted that the cyclic solutions 14253, 25314, 31425, 42531, 53142 give five superposable arrangements by which five white queens, five black queens, five red queens, five yellow queens, and five blue queens can be put simultaneously on the board so that no queen can be taken by any other queen of the same colour.

On a board of 6^2 cells there is 1 fundamental solution – namely, (246135). The four solutions are superposable. The puzzle for this case was sold in the streets of London for a penny, a small wooden board being ruled in the manner shown in the diagram and having holes drilled in it at the points marked by dots. The object is to put six pins into the holes so that no two are connected by a straight line.

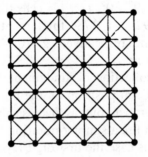

On a board of 7^2 cells there are 6 fundamental solutions: namely, 1357246, 3572461, (5724613), 4613572, 3162574, (2574136). It may be noted that the solution 1357246 gives by cyclic permutations seven superposable arrangements.

On a board of 8^2 cells, the fourth cell from some corner is always occupied. There are 12 fundamental solutions: namely, 41582736, 41586372, 42586137, 42736815, 42736851, 42751863, 42857136, 42861357, 46152837, (46827135), 47526138, 48157263. The arrangement in this order is due to J.M. Andreas. The 7th solution is the only one in which no three queens are in a straight line. It is impossible* to find eight superposable solutions; but we can in five typical ways pick out six solutions which can be superposed, and to some of these it is possible to add 2 sets of 7 queens, thus filling 62 out of the 64 cells with 6 sets of 8 queens and 2 sets of 7 queens, no one of which can take another of the same set. Here is such a solution: 16837425, 27368514, 35714286, 41586372, 52473861, 68241753, 73625140, 04152637. Similar superposition problems can be framed for boards of other sizes.

Norman Anning has pointed out that seven of the twelve fundamental solutions (nos. 1, 2, 5, 7, 8, 9, 11) can be found on a single infinite "satin pattern" by suitably choosing eight-by-eight squares. The basic pattern consists of two straight rows of four, consecutive members of each row being related by a knight's move.

On any board empirical solutions may be found with but little difficulty. Mr. Derrington has constructed the following table of solutions:

2.4.1.3	for a board of 4^2 cells
2.4.1.3.5	,, ,, 5^2 ,,
2.4.6.1.3.5	,, ,, 6^2 ,,
2.4.6.1.3.5.7	,, ,, 7^2 ,,
2.4.6.8.3.1.7.5	,, ,, 8^2 ,,
2.4.1.7.9.6.3.5.8	,, ,, 9^2 ,,
2.4.6.8.10.1.3.5.7.9	,, ,, 10^2 ,,
2.4.6.8.10.1.3.5.7.9.11	,, ,, 11^2 ,,
2.4.6.8.10.12.1.3.5.7.9.11	,, ,, 12^2 ,,
2.4.6.8.10.12.1.3.5.7.9.11.13	,, ,, 13^2 ,,
9.7.5.3.1.13.11.6.4.2.14.12.10.8	,, ,, 14^2 ,,
15.9.7.5.3.1.13.11.6.4.2.14.12.10.8	,, ,, 15^2 ,,
2.4.6.8.10.12.14.16.1.3.5.7.9.11.13.15	,, ,, 16^2 ,,
2.4.6.8.10.12.14.16.1.3.5.7.9.11.13.15.17	,, ,, 17^2 ,,
2.4.6.8.10.12.14.16.18.1.3.5.7.9.11.13.15.17	,, ,, 18^2 ,,
2.4.6.8.10.12.14.16.18.1.3.5.7.9.11.13.15.17.19	,, ,, 19^2 ,,
12.10.8.6.4.2.20.18.16.14.9.7.5.3.1.19.17.15.13.11	,, ,, 20^2 ,,
21.12.10.8.6.4.2.20.18.16.14.9.7.5.3.1.19.17.15.13.11	,, ,, 21^2 ,,

*See Thorold Gosset, *The Messenger of Mathematics*, Cambridge, July 1914, vol. XLIV, p. 48.

and so on. The rule is obvious except when n is of the form $6m+2$ or $6m+3$.

MAXIMUM PIECES PROBLEM

The Eight Queens Problem suggests the somewhat analogous question of finding the maximum number of kings – or more generally of pieces of one type – which can be put on a board so that no one can take any other, and the number of solutions possible in each case.

In the case of kings the number is 16; for instance, one solution is when they are put on the cells 11, 13, 15, 17, 31, 33, 35, 37, 51, 53, 55, 57, 71, 73, 75, 77. For queens, it is obvious that the problem is covered by the analysis already given, and the number is 8. For bishops the number is 14, the pieces being put on the boundary cells; for instance, one solution is when they are put on the cells 11, 12, 13, 14, 15, 16, 17, 81, 82, 83, 84, 85, 86, 87; there are 256 solutions. For knights the number is 32; for instance, they can be put on all the white or on all the black cells, and there are 2 fundamental solutions. For rooks it is obvious that the number is 8, and there are in all 8! solutions.

MINIMUM PIECES PROBLEM

Another problem of a somewhat similar character is the determination of the minimum number of kings – or more generally of pieces of one type – which can be put on a board so as to command or occupy all the cells.

For kings the number is 9; for instance, they can be put on the cells 11, 14, 17, 41, 44, 47, 71, 74, 77. For queens the number is 5; for instance, they can be put on the cells 18, 35, 41, 76, 82. For bishops the number is 8; for instance, they can be put on the cells 41, 42, 43, 44, 45, 46, 47, 48. For knights the number is 12; for instance, they can be put on the cells 26, 32, 33, 35, 36, 43, 56, 63, 64, 66, 67, and 73, constituting four triplets arranged symmetrically. For rooks the number is 8, and the solutions are obvious.

For queens the problem has also been discussed for a board

of n^2 cells where n has various values.* One queen can be placed so as to command all the cells when $n=2$ or 3, and there is only 1 fundamental solution. Two queens are required when $n=4$; and there are 3 fundamental solutions – namely, when they are placed on the cells 11 and 33, or on the cells 12 and 42, or on the cells 22, 23: these give 12 solutions in all. Three queens are required when $n=5$; and there are 37 fundamental solutions, giving 186 solutions in all. Three queens are also required when $n=6$, but there is only 1 fundamental solution – namely, when they are put on the cells 11, 35, and 53, giving 4 solutions in all. Four queens are required when $n=7$; one of the five fundamental solutions is when they are put on the cells 12, 26, 41, 55. When $n=8$ there are 638 fundamental solutions.

Jaenisch proposed also the problem of the determination of the minimum number of queens which can be placed on a board of n^2 cells so as to command all the unoccupied cells, subject to the restriction that no queen shall attack the cell occupied by any other queen. In this case three queens are required when $n=4$, for instance, they can be put on the cells 11, 23, 42; and there are 2 fundamental solutions, giving 16 solutions in all. Three queens are required when $n=5$, for instance, they can be put on the cells 11, 24, 43, or on the cells 11, 34, 53; and there are 2 fundamental solutions in all. Four queens are required when $n=6$, for instance, when they are put on the cells 13, 36, 41, 64; and there are 17 fundamental solutions. Four queens are required when $n=7$, and there is only 1 fundamental solution – namely, that already mentioned, when they are put on the cells 12, 26, 41, 55, which gives 8 solutions in all. Five queens are required when $n=8$, and there are no less than 91 fundamental solutions – for instance, one is when they are put on the cells 11, 23, 37, 62, 76.

I leave to any of my readers who may be interested in such questions the discussion of the corresponding problems for the

*C.F. de Jaenisch, *Applications de l'Analyse mathématique au Jeu des Échecs*, Leningrad, 1862, Appendix, pp. 244 *et seq.*; see also *L'Intermédiaire des Mathématiciens*, Paris, 1901, vol. VIII, p. 88.

other pieces,* and of the number of possible solutions in each case.

A problem of the same nature would be the determination of the minimum number of queens (or other pieces) which can be placed on a board so as to protect one another and command all the unoccupied cells. For queens the number is 5; for instance, they can be put on the cells 24, 34, 44, 54, and 84. For bishops the number is 10; for instance, they can be put on the cells 24, 25, 34, 35, 44, 45, 64, 65, 74, and 75. For knights the number is 14; for instance, they can be put on the cells 32, 33, 36, 37, 43, 44, 45, 46, 63, 64, 65, 66, 73, and 76: the solution is semi-symmetrical; two other arrangements are possible. For rooks the number is 8, and a solution is obvious. I leave to any who are interested in the subject the determination of the number of solutions in each case.

In connection with this class of problems, I may mention two other questions, to which Captain Turton first called my attention, of a somewhat analogous character.

The first of these is to place eight queens on a chess-board so as to command the fewest possible squares. Thus, if queens are placed on cells 21, 22, 62, 71, 73, 77, 82, 87, eleven cells on the board will not be in check; the same number can be obtained by other arrangements. Is it possible to place the eight queens so as to leave more than eleven cells out of check? I have never succeeded in doing so, nor in showing that it is impossible to do it.

The other problem is to place m queens (m being less than 5) on a chess-board so as to command as many cells as possible For instance, four queens can be placed in several ways on the board so as to command 58 cells besides those on which the queens stand, thus leaving only 2 cells which are not commanded; for instance, queens may be placed on the cells 35, 41,

*The problem for knights was discussed in *L'Intermédiaire des Mathématiciens*, Paris, 1896, vol. III, p. 58; 1897, vol. IV, pp. 15–17, 254; 1898, vol. V, pp. 87, 230–231.

76, and 82. Analogous problems with other pieces will suggest themselves.

There are endless similar questions in which combinations of pieces are involved. For instance, if queens are put on the cells 35, 41, 76, and 82, they command or occupy all but two cells, and these two cells may be commanded or occupied by a queen, a king, a rook, a bishop, or a pawn. If queens are put on the cells 22, 35, 43, and 54, they command or occupy all but three cells, and two of these three cells may be commanded by a knight which occupies the third of them.

RE-ENTRANT PATHS ON A CHESS-BOARD

Another problem connected with the chess-board consists in moving a piece in such a manner that it shall move successively on to every possible cell once, and only once.

Knight's re-entrant path.   I begin by discussing the classical problem of a knight's tour. The literature* on this subject is so extensive that I make no attempt to give a full account of the various methods for solving the problem, and I shall content myself by putting together a few notes on some of the solutions I have come across, particularly on those due to De Moivre, Euler, Vandermonde, Warnsdorff, and Roget.

On a board containing an even number of cells the path may or may not be re-entrant, but on a board containing an odd number of cells it cannot be re-entrant. For, if a knight begins on a white cell, its first move must take it to a black cell, the next to a white cell, and so on. Hence, if its path passes through all the cells, then on a board of an odd number of cells the last move must leave it on a cell of the same colour as that on which it started, and therefore these cells cannot be connected by one move.

---

*For a bibliography see A. van der Linde, *Geschichte und Literatur des Schachspiels*, Berlin, 1874, vol. II, pp. 101–111. On the problem and its history see a memoir by P. Volpicelli in *Atti della Reale Accademia dei Lincei*, Rome, 1872, vol. xxv, pp. 87–162; also *Applications de l'Analyse mathématique au Jeu des Echecs*, by C.F. de Jaenisch, 3 vols., Leningrad, 1862–3; and General Parmentier, *Association Française pour l'Avancement des Sciences*, 1891, 1892, 1894.

34	49	22	11	36	39	24	1
21	10	35	50	23	12	37	40
48	33	62	57	38	25	2	13
9	20	51	54	63	60	41	26
32	47	58	61	56	53	14	3
19	8	55	52	59	64	27	42
46	31	6	17	44	29	4	15
7	18	45	30	5	16	43	28

30	21	6	15	28	19
7	16	29	20	5	14
22	31	8	35	18	27
9	36	17	26	13	4
32	23	2	11	34	25
1	10	33	24	3	12

De Moivre's Solution   Euler's Thirty-six Cell Solution

The earliest solutions of which I have any knowledge are those given at the beginning of the eighteenth century by De Montmort and De Moivre.* They apply to the ordinary chess-board of 64 cells, and depend on dividing (mentally) the board into an inner square containing sixteen cells surrounded by an outer ring of cells two deep. If initially the knight is placed on a cell in the outer ring, it moves round that ring always in the same direction so as to fill it up completely – going into the inner square only when absolutely necessary. When the outer ring is filled up, the order of the moves required for filling the remaining cells presents but little difficulty. If initially the knight is placed on the inner square, the process must be reversed. The method can be applied to square and rectangular boards of all sizes. It is illustrated sufficiently by De Moivre's solution, which is given above, where the numbers indicate the order in which the cells are occupied successively. I place by its side a somewhat similar re-entrant solution, due to Euler, for a board of 36 cells. If a chess-board is used it is convenient to place a counter on each cell as the knight leaves it.

*They were sent by their authors to Brook Taylor who seems to have previously suggested the problem. I do not know where they were first published; they were quoted by Ozanam and Montucla; see Ozanam, 1803 edition, vol. I, p. 178; 1840 edition, p. 80.

The earliest serious attempt to deal with the subject by mathematical analysis was made by Euler* in 1759: it was due to a suggestion made by L. Bertrand of Geneva, who subsequently (in 1778) issued an account of it. This method is applicable to boards of any shape and size, but in general the solutions to which it leads are not symmetrical, and their mutual connection is not apparent.

Euler commenced by moving the knight at random over the board until it has no move open to it. With care this will leave only a few cells not traversed: denote them by $a, b, \ldots$. His method consists establishing certain rules by which these vacant cells can be interpolated into various parts of the circuit, and by which the circuit can be made re-entrant.

The following example, mentioned by Legendre as one of exceptional difficulty, illustrates the method. Suppose that we

55	58	29	40	27	44	19	22
60	39	56	43	30	21	26	45
57	54	59	28	41	18	23	20
38	51	42	31	8	25	46	17
53	32	37	a	47	16	9	24
50	3	52	33	36	7	12	15
1	34	5	48	b	14	c	10
4	49	2	35	6	11	d	13

22	25	50	39	52	35	60	57
27	40	23	36	49	58	53	34
24	21	26	51	38	61	56	59
41	28	37	48	3	54	33	62
20	47	42	13	32	63	4	55
29	16	19	46	43	2	7	10
18	45	14	31	12	9	64	5
15	30	17	44	1	6	11	8

Figure i   Figure ii

have formed the route given in figure i above, namely, 1, 2, 3, ... , 59, 60; and that there are four cells left untraversed, namely, a, b, c, d.

We begin by making the path 1 to 60 re-entrant. The cell 1 commands a cell p, where p is 32, 52, or 2. The cell 60 commands a cell q, where q is 29, 59, or 51. Then, if any of these values of

*Mémoires de Berlin for 1759, Berlin, 1766, pp. 310–337; or Commentationes Arithmeticae Collectae, Leningrad, 1849, vol. I, pp. 337–355.

p and q differ by unity, we can make the route re-entrant. This is the case here if $p = 52$, $q = 51$. Thus the cells 1, 2, 3, ..., 51; 60, 59, ..., 52 from a re-entrant route of 60 moves. Hence, if we replace the numbers 60, 59, ..., 52 by 52, 53, ..., 60, the steps will be numbered consecutively. I recommend the reader who wishes to follow the subsequent details of Euler's argument to construct this square on a piece of paper before proceeding further.

Next, we proceed to add the cells a, b, d to this route. In the new diagram of 60 cells formed as above the cell a commands the cells there numbered 51, 53, 41, 25, 7, 5, and 3. It is indifferent which of these we select: suppose we take 51. Then we must make 51 the last cell of the route of 60 cells, so that we can continue with a, b, d. Hence, if the reader will add 9 to every number on the diagram he has constructed, and then replace 61, 62, ..., 69 by 1, 2, ..., 9, he will have a route which starts from the cell occupied originally by 60, the 60th move is on to the cell occupied originally by 51, and the 61st, 62nd, 63rd moves will be on the cells a, b, d respectively.

It remains to introduce the cell c. Since c commands the cell now numbered 25, and 63 commands the cell now numbered 24, this can be effected in the same way as the first route was made re-entrant. In fact, the cells numbered 1, 2, ..., 24; 63, 62, ..., 25, c form a knight's path. Hence we must replace 63, 62, ..., 25 by the numbers 25, 26, ..., 63, and then we can fill up c with 64. We have now a route which covers the whole board.

Lastly, it remains to make this route re-entrant. First, we must get the cells 1 and 64 near one another. This can be effected thus. Take one of the cells commanded by 1, such as 28, then 28 commands 1 and 27. Hence the cells 64, 63, ..., 28; 1, 2, ..., 27 from a route; and this will be represented in the diagram if we replace the cells numbered 1, 2, ..., 27 by 27, 26, ..., 1.

The cell now occupied by 1 commands the cells 26, 38, 54, 12, 2, 14, 16, 28; and the cell occupied by 64 commands the cells 13, 43, 63, 55. The cells 13 and 14 are consecutive, and therefore the cells 64, 63, ..., 14; 1, 2, ..., 13 form a route. Hence we must replace the numbers 1, 2, ..., 13 by 13, 12, ..., 1, and we obtain a re-entrant route covering the whole board, which is represented in the second of the diagrams given on page 177. Euler showed how seven other re-entrant routes can be deduced from any given re-entrant route.

It is not difficult to apply the method so as to form a route which begins on one given cell and ends on any other given cell.

58	43	60	37	52	41	62	35
49	46	57	42	61	36	53	40
44	59	48	51	38	55	34	63
47	50	45	56	33	64	39	54
22	7	32	1	24	13	18	15
31	2	23	6	19	16	27	12
8	21	4	29	10	25	14	17
3	30	9	20	5	28	11	26

50	45	62	41	60	39	54	35
63	42	51	48	53	36	57	38
46	49	44	61	40	59	34	55
43	64	47	52	33	56	37	58
26	5	24	1	20	15	32	11
23	2	27	8	29	12	17	14
6	25	4	21	16	19	10	31
3	22	7	28	9	30	13	18

Euler's Half-board Solution   Roget's Half-board Solution

Euler next investigated how his method could be modified so as to allow of the imposition of additional restrictions.

An interesting example of this kind is where the first 32 moves are confined to one-half of the board. One solution of of this is delineated above. The order of the first 32 moves can be determined by Euler's method. It is obvious that, if to the number of each such move we add 32, we shall have a corresponding set of moves from 33 to 64 which would cover the other half of the board; but in general the cell numbered 33 will not be a knight's move from that numbered 32, nor will 64 be a knight's move from 1.

Euler, however, proceeded to show how the first 32 moves might be determined so that, if the half of the board containing the corresponding moves from 33 to 64 was twisted through two right angles, the two routes would become united and re-entrant. If x and y are the numbers of a cell reckoned from two consecutive sides of the board, we may call the cell whose distances are respectively x and y from the opposite sides a complementary cell. Thus the cells (x, y) and $(9-x, 9-y)$ are complementary, where x and y denote respectively the column and row occupied by the cell. Then in Euler's solution the numbers in complementary cells differ by 32: for instance, the cell $(3, 7)$ is complementary to the cell $(6, 2)$, the one is occupied by 57, the other by 25.

Roget's method, which is described later, can also be applied to give half-board solutions. The result is indicated above. The close of Euler's memoir is devoted to showing how the method could be applied to crosses and other rectangular figures. I may note in particular his elegant re-entrant symmetrical solution for a square of 100 cells.

The next attempt of any special interest is due to Vandermonde,* who reduced the problem to arithmetic. His idea was to cover the board by two or more independent routes taken at random, and then to connect the routes. He defined the position of a cell by a fraction x/y, whose numerator x is the number of the cell from one side of the board, and whose denominator y is its number from the adjacent side of the board; this is equivalent to saying that x and y are the co-ordinates of a cell. In a series of fractions denoting a knight's path, the differences between the numerators of two consecutive fractions can be only one or two, while the corresponding differences between their denominators must be two or one respectively. Also x and y cannot be less than 1 or greater than 8. The notation is convenient, but Vandermonde applied it merely to obtain a particular solution of the problem for a board of 64 cells: the method by which he effected this is analogous to that established by Euler, but it is applicable only to squares of an even order. The route that he arrives at is defined in his notation by the following fractions: 5/5, 4/3, 2/4, 4/5, 5/3, 7/4, 8/2, 6/1, 7/3, 8/1, 6/2, 8/3, 7/1, 5/2, 6/4, 8/5, 7/7, 5/8, 6/6, 5/4, 4/6, 2/5, 1/7, 3/8, 2/6, 1/8, 3/7, 1/6, 2/8, 4/7, 3/5, 1/4, 2/2, 4/1, 3/3, 1/2,

*L'Historie de l'Académie des Sciences for 1771, Paris, 1774, pp. 566–574.

3/1, 2/3, 1/1, 3/2, 1/3, 2/1, 4/2, 3/4, 1/5, 2/7, 4/8, 3/6, 4/4, 5/6, 7/5, 8/7, 6/8, 7/6, 8/8, 6/7, 8/6, 7/8, 5/7, 6/5, 8/4, 7/2, 5/1, 6/3.

The path is re-entrant, but unsymmetrical. Had he transferred the first three fractions to the end of this series, he would have obtained two symmetrical circuits of thirty-two moves joined unsymmetrically and might have been enabled to advance further in the problem. Vandermonde also considered the case of a route in a cube.

In 1773 Collini* proposed the exclusive use of symmetrical routes arranged without reference to the initial cell, but connected in such a manner as to permit of our starting from it. This is the foundation of the modern manner of attacking the problem. The method was re-invented in 1825 by Pratt,† and in 1840 by Roget, and has been subsequently employed by various writers. Neither Collini nor Pratt showed skill in using this method. The rule given by Roget is described later.

One of the most ingenious of the solutions of the knight's path is that given in 1823 by Warnsdorff.‡ His rule is that the knight must be always moved to one of the cells from which it will command the fewest squares not already traversed. The solution is not symmetrical and not re-entrant; moreover, it is difficult to trace practically. The rule has not been proved to be true, but no exception to it is known: apparently it applies also to all rectangular boards which can be covered completely by a knight. It is somewhat curious that in most cases a single false step, except in the last three or four moves, will not affect the result.

Warnsdorff added that when, by the rule, two or more cells are open to the knight, it may be moved to either or any of them indifferently. This is not so, and with great ingenuity two or three cases of failure have been constructed, but it would require exceptionally had luck to happen accidentally on such a route.

---

*Solution du Problème du Cavalier au Jeu des Échecs, Mannheim, 1773.
†Studies of Chess, sixth edition, London, 1825.
‡H.C. Warnsdorff, Des Rösselsprunges einfachste und allgemeinste Lösung, Schmalkalden, 1823; see also Jaenisch, vol. II, pp. 56–61, 273–289.

The above methods have been applied to boards of various shapes, especially to boards in the form of rectangles, crosses, and circles.*

All the more recent investigations impose additional re-strictions: such as to require that the route shall be re-entrant, or more generally that it shall begin and terminate on given cells.

The simplest solution with which I am acquainted is due to De Lavernède, but is more generally associated with the name of Roget, whose paper in 1840 attracted general notice to it.†
It divides the whole route into four circuits, which can be com-bined so as to enable us to begin on any cell and terminate on any other cell of a different colour. Hence, if we like to select this last cell at a knight's move from the initial cell, we obtain a re-entrant route. On the other hand, the rule is applicable only to square boards containing $(4n)^2$ cells: for example, it could not be used on the board of the French *jeu des dames*, which contains 100 cells.

Roget began by dividing the board of 64 cells into four quarters. Each quarter contains 16 cells, and these 16 cells can be arranged in 4 groups, each group consisting of 4 cells which form a closed knight's path. All the cells in each such path are denoted by the same letter *l*, *e*, *a*, or *p*, as the case may be. The path of 4 cells indicated by the consonants *l* and the path indicated by the consonants *p* are diamond-shaped: the paths indicated respectively by the vowels *e* and *a* are square-shaped, as may be seen by looking at one of the four quarters in figure (i) below.

Now, all the 16 cells on a complete chess-board which are marked with the same letter can be combined into one circuit, and wherever the circuit begins we can make it end on any other cell in the circuit, provided it is of a different colour from

---

*See, e.g., T. Ciccolini's work *Del Cavallo degli Scacchi*, Paris, 1836.

†J.E.T. de Lavernède, *Mémoires de l'Académie royale du Gard*, Nimes, 1839, pp. 151–179; P.M. Roget, *Philosophical Magazine*, April 1840, series 3, vol. xvi, pp. 305–309; see also *Quarterly Journal of Mathematics* for 1877, vol. xiv, pp. 354–359; and *Leisure Hour*, Sept. 13, 1873, pp. 587–590, and Dec. 20, 1873, pp. 813–815.

l	*e*	*a*	*p*	*l*	*e*	*a*	*p*
a	*p*	*l*	*e*	*a*	*p*	*l*	*e*
e	*l*	*p*	*a*	*e*	*l*	*p*	*a*
p	*a*	*e*	*l*	*p*	*a*	*e*	*l*
l	*e*	*a*	*p*	*l*	*e*	*a*	*p*
a	*p*	*l*	*e*	*a*	*p*	*l*	*e*
e	*l*	*p*	*a*	*e*	*l*	*p*	*a*
p	*a*	*e*	*l*	*p*	*a*	*e*	*l*

34	51	32	15	38	53	18	3
31	14	35	52	17	2	39	54
50	33	16	29	56	37	4	19
13	30	49	36	1	20	55	40
48	63	28	9	44	57	22	5
27	12	45	64	21	8	41	58
62	47	10	25	60	43	6	23
11	26	61	46	7	24	59	42

Roget's Solution (i)   Roget's Solution (ii)

the initial cell. If it is indifferent on what cell the circuit terminates, we may make the circuit re-entrant, and in this case we
can make the direction of motion round each group (of 4 cells)
the same. For example, all the cells marked *p* can be arranged
in the circuit indicated by the successive numbers 1 to 16 in
figure ii above. Similarly all the cells marked *a* can be combined
into the circuit indicated by the numbers 17 to 32; all the *l* cells
into the circuit 33 to 48; and all the *e* cells into the circuit 49
to 64. Each of the circuits indicated above is symmetrical and
re-entrant. The consonant and the vowel circuits are said to
be of opposite kinds.

The general problem will be solved if we can combine the
four circuits into a route which will start from any given cell,
and terminate on the 64th move on any other given cell of a
different colour. To effect this Roget gave the two following
rules.

First. If the initial cell and the final cell are denoted the one
by a consonant and the other by a vowel, take alternately circuits indicated by consonants and vowels, beginning with the
circuit of 16 cells indicated by the letter of the initial cell and
concluding with the circuit indicated by the letter of the final
cell.

Second. If the initial cell and the final cell are denoted both by consonants or both by vowels, first select a cell, Y, in the same circuit as the final cell, Z, and one move from it; next select a cell, X, belonging to one of the opposite circuits and one move from Y. This is always possible. Then, leaving out the cells Z and Y, it always will be possible, by the rule already given, to travel from the initial cell to the cell X in 62 moves, and thence to move to the final cell on the 64th move.

In both cases, however, it must be noticed that the cells in each of the first three circuits will have to be taken in such an order that the circuit does not terminate on a corner, and it may be desirable also that it should not terminate on any of the border cells. This will necessitate some caution. As far as is consistent with these restrictions, it is convenient to make these circuits re-entrant, and to take them and every group in them in the same direction of rotation.

As an example, suppose that we are to begin on the cell numbered 1 in figure (ii) above, which is one of those in a p circuit, and to terminate on the cell numbered 64, which is one of those in an e circuit. This falls under the first rule: hence first we take the 16 cells marked p, next the 16 cells marked a, then the 16 cells marked l, and lastly the 16 cells marked e. One way of effecting this is shown in the diagram. Since the cell 64 is a knight's move from the initial cell, the route is re-entrant. Also each of the four circuits in the diagram is symmetrical, re-entrant, and taken in the same direction, and the only point where there is any apparent breach in the uniformity of the movement is in the passage from the cell numbered 32 to that numbered 33.

A rule for re-entrant routes, similar to that of Roget, has been given by various subsequent writers, especially by De Polignac* and by Lacquière,† who have stated it at much greater length. Neither of these authors seems to have been aware of Roget's theorems. De Polignac, like Roget, illustrates

---

*Comptes Rendus, April 1861; and Bulletin de la Société Mathématique de France, 1881, vol. IX, pp. 17–24.

†Bulletin de la Société Mathématique de France, 1880, vol. VIII, pp. 82–102, 132–158.

the rule by assigning letters to the various squares in the way explained above, and asserts that a similar rule is applicable to all even squares.

Roget's method can also be applied to two half-boards, as indicated in the figure given above on page 179.

The method which Jaenisch gives as the most fundamental is not very different from that of Roget. It leads to eight forms, similar to that in the diagram printed below, in which the sum of the numbers in every column and every row is 260; but, although symmetrical, it is not, in my opinion, so easy to reproduce as that given by Roget. Other solutions, notably those by Moon and by Wenzelides, were given in former editions of this work. The two re-entrant routes printed below, each

63	22	15	40	1	42	59	18
14	39	64	21	60	17	2	43
37	62	23	16	41	4	19	58
24	13	38	61	20	57	44	3
11	36	25	52	29	46	5	56
26	51	12	33	8	55	30	45
35	10	49	28	53	32	47	6
50	27	34	9	48	7	54	31

15	20	17	36	13	64	61	34
18	37	14	21	60	35	12	63
25	16	19	44	5	62	33	56
38	45	26	59	22	55	4	11
27	24	39	6	43	10	57	54
40	49	46	23	58	3	32	9
47	28	51	42	7	30	53	2
50	41	48	29	52	1	8	31

Jaenisch's Solution   Two Half-board Solutions

covering 32 cells, and together covering the board, are remarkable as constituting a magic square.*

It is as yet impossible to say how many solutions of the problem exist. Legendre† mentioned the question, but Minding‡ was the earliest writer to attempt to answer it. More recent investigations have shown that on the one hand the

*See A. Rilly, *Le Problème du Cavalier des Échecs*, Troyes, 1905.
†*Théorie des Nombres*, Paris, 2nd edition, 1830, vol. II, p. 165.
‡*Cambridge and Dublin Mathematical Journal*, 1852, vol. VII, pp. 147–156; and *Crelle's Journal*, 1853, vol. XLIV, pp. 73–82.

number of possible routes is less* than the number of combinations of 168 things taken 63 at a time, and on the other hand is greater than 122,802512, since this latter number is the number of re-entrant paths of a particular type.†

Analogous problems.   Similar problems can be constructed in which it is required to determine routes by which a piece moving according to certain laws (e.g. a chess-piece such as a king, etc.) can travel from a given cell over a board so as to occupy successively all the cells, or certain specified cells, once, and only once, and terminate its route in a given cell. Euler's method can be applied to find routes of this kind: for instance, he applied it to find a re-entrant route by which a piece that moved two cells forward like a castle and then one cell like a bishop would occupy in succession all the black cells on the board.

King's re-entrant path. ‡   As an example here is a re-entrant tour of a king which moves successively to every cell of the board. I give it because the numbers indicating the cells successively occupied form a magic square. Of course, this also gives a solution of a re-entrant route of a queen covering the board.

61	62	63	64	1	2	3	4
60	11	58	57	8	7	54	5
12	59	10	9	56	55	6	53
13	14	15	16	49	50	51	52
20	19	18	17	48	47	46	45
21	38	23	24	41	42	27	44
37	22	39	40	25	26	43	28
36	35	34	33	32	31	30	29

King's Magic Tour on a Chess-board

*Jaenisch, vol. II, p. 268.
†Kraïtchik, pp. 360, 402.
‡Cf. I. Ghersi, *Matematica dilettevole e curiosa*, Milan, 1921, p. 320 (fig. 261).

Rook's re-entrant path.   There is no difficulty in constructing re-entrant tours for a rook which moves successively to every cell of the board. For instance, if the rook starts from the cell 11, it can move successively to the cells 18, 88, 81, 71, 77, 67, 61, 51, 57, 47, 41, 31, 37, 27, 21, and so back to 11: this is a symmetrical route. Of course, this also gives a solution of a re-entrant route for a king or a queen covering the board. If we start from any of the cells mentioned above, the rook takes sixteen moves. If we start from any cell in the middle of one of these moves, it will take seventeen moves to cover this route, but I believe that in most cases wherever the initial cell be chosen, sixteen moves will suffice, though in general the route will not be symmetrical. On a board of n^2 cells it is possible to find a route by which a rook can move successively from its initial cell to every other cell once, and only once. Moreover,* starting on any cell, its path can be made to terminate, if n be even, on any other cell of a different colour, and if n be odd, on any other cell of the same colour.

Bishop's re-entrant path.   As yet another instance, a bishop can traverse all the cells of one colour on the board in seventeen moves if the initial cell is properly chosen;† for instance, starting from the cell 11, it may move successively to the cells 55, 82, 71, 17, 28, 46, 13, 31, 86, 68, 57, 48, 15, 51, 84, 66, 88. One more move will bring it back to the initial cell. From the nature of the case, it must traverse some cells more than once.

MISCELLANEOUS PROBLEMS

We may construct numerous such problems concerning the determination of routes which cover the whole or part of the board subject to certain conditions. I append a few others which may tax the ingenuity of those not accustomed to such problems.

Routes on a chess-board.   One of the simplest is the determination of the path taken by a rook, placed in the cell 11,

---

*L'Intermédiaire des Mathématiciens, Paris, 1901, vol. VIII, pp. 153–154.
†H.E. Dudeney, The Tribune, Dec. 3, 1906; Amusements in Mathematics, p. 225.

which moves, one cell at a time, to the cell 88, so that in the course of its path it enters every cell once, and only once. This can be done, though I have seen good mathematicians puzzled to effect it. A hasty reader is apt to misunderstand the conditions of the problem.

Another simple problem of this kind is to move a queen from the cell 33 to the cell 66 in fifteen moves, entering every cell once and only once, and never crossing its own track or entering a cell more than once.*

A somewhat similar, but more difficult, question is the determination of the greatest distance which can be travelled by a queen starting from its own square in five consecutive moves, subject to the condition that it never crosses its own track or enters a cell more than once.† In calculating the distance, it may be assumed that the paths go through the centres of the cells. If the length of the side of a cell is 1 inch, the distance exceeds 33·97 inches.

Another familiar problem can be enunciated as follows. Construct a rectangular board of mn cells by ruling $m+1$ vertical lines and $n+1$ horizontal lines. It is required to know how many routes can be taken from the top left-hand corner to the bottom right-hand corner, the motion being along the ruled lines and its direction being always either vertically downwards or horizontally from left to right. The answer is the number of permutations of $m+n$ things, of which m are alike of one kind and n are alike of another kind: this is equal to $(m+n)!/m!n!$. Thus on a square board containing 16 cells (i.e. one-quarter of a chess-board), where $m=n=4$, there are 70 such routes; while on a common chess-board, where $m=n=8$, there are no less than 12870 such routes. A rook, moving according to the same law, can travel from the top left-hand cell to the bottom right-hand cell in $(m+n-2)!/(m-1)!(n-1)!$ ways. Similar theorems can be enunciated for a parallelepiped.

Another question of this kind is the determination of the number of closed routes though mn points arranged in m rows

*H.E. Dudeney, *The Tribune*, Oct. 3, 1906.
†*Ibid.*, Oct. 2, 1906.

and n columns, following the lines of the quadrilateral network, and passing once and only once through each point.*

Guarini's problem.   One of the oldest European problems
connected with the chess-board is the following, which was
propounded in 1512. It was quoted by Lucas in 1894, but I
believe has not been published otherwise than in his works and
the earlier editions of this book. On a board of nine cells, such
as that drawn below, the two white knights are placed on the

a	C	d
D		B
b	A	c

two top corner cells (a, d), and the two black knights on the
two bottom corner cells (b, c): the other cells are left vacant.
It is required to move the knights so that the white knights
shall occupy the cells b and c, while the black shall occupy the
cells a and d. The solution is obvious.

Queens' problems.   Another problem consists in placing
sixteen queens on a board so that no three are in a straight
line.† One solution is to place them on the cells 15, 16, 25, 26,
31, 32, 41, 42, 57, 58, 67, 68, 73, 74, 83, 84. It is, of course, assumed that each queen is placed on the middle of its cell.

LATIN SQUARES

A *Latin square* of order n consists of n distinct symbols, each
occurring n times, arranged in the form of a square matrix in
such a way that every row and every column is a permutation
of the n symbols. If the symbols are $0, 1, \ldots, n-1$, interpreted
as the residue-classes modulo n, a special kind of Latin square,
denoted by $(1, 1)_n$ (as in the first of the following examples), may
be constructed as an *addition table* (cf. page 108). Somewhat

*See C.F. Sainte-Marie in *L'Intermédiaire des Mathématiciens*, Paris, vol. XI,
March 1904, pp. 86–88.

†H.E. Dudeney, *The Tribune*, November 7. 1906; *Amusements in Mathematics*, p. 222.

2	0	1
1	2	0
0	1	2

$(1, 1)_3$

2	1	0
1	0	2
0	2	1

$(2, 1)_3$

1	2	3	0
2	3	0	1
3	0	1	2
0	1	2	3

$(1, 3)_4$

4	1	3	0	2
3	0	2	4	1
2	4	1	3	0
1	3	0	2	4
0	2	4	1	3

$(2, 1)_5$

more generally, in the Latin square $(\alpha, \beta)_n$, the bottom row contains the numbers 0, α, 2α, ... (reduced modulo n), the left column contains 0, β, 2β, ..., and every other cell contains the sum (mod n) of the numbers which thus determine its column and row.

Clearly, α and β must be prime to n. If also $\alpha + \beta$ and $\alpha - \beta$ are prime to n (as in the last of the above examples) we have a *diagonal* Latin square: the n symbols in each diagonal (as well as in each row and column) are all different.

Similarly, for any prime power $q = p^k$ and any two elements of $GF(p^k)$ satisfying $\alpha\beta \neq 0$, a Latin square $[\alpha, \beta]_q$ can be constructed as an addition table for the multiples of α and β in that Galois field. This is a diagonal Latin square if also $\alpha^2 \neq \beta^2$. If $q = 4$ or 9, we can interpret the elements of $GF(q)$ as numbers expressed in the scale of q, and then convert them to the denary scale. The resulting Latin square will be different from $(\alpha, \beta)_q$.

Eulerian squares. Two Latin squares of order n are said to be *orthogonal* if in their superposition all the n^2 ordered pairs of symbols occur exactly once. The resulting matrix, whose entries consist of ordered pairs of symbols, is called an *Eulerian square* or a "Graeco-Latin square." The latter name arose from the notion of superposing a square of Greek letters and a square of Latin letters, but it is more convenient to use the same n symbols in both positions. The first of the following examples may be denoted by $(12, 11)_3$, because the left digits form the Latin square $(1, 1)_3$ while the right digits form the orthogonal Latin square $(2, 1)_3$. Similarly, the second example is $(12, 21)_5$, while the third is a combination of $[1, 10]_{2^2}$ and $[1, 11]_{2^2}$ with the elements 10 and 11 replaced by 2 and 3 as if they were numbers in the binary scale. This Latin square of order 4 may be regarded as a solution for Ozanam's problem of the Magic Card Square*: to place the sixteen court cards

22	01	10
11	20	02
00	12	21

34	41	03	10	22
13	20	32	44	01
42	04	11	23	30
21	33	40	02	14
00	12	24	31	43

12	03	30	21
31	20	13	02
23	32	01	10
00	11	22	33

(taken out of a pack) in the form of a square so that no row, no column, and neither of the diagonals shall contain more than one card of each suit and one card of each rank.

More generally, two Latin squares $(\alpha', \beta')_n$ and $(\alpha, \beta)_n$ are orthogonal whenever the determinant $\alpha\beta' - \alpha'\beta$ is prime to n. Since also $\alpha, \beta, \alpha', \beta'$ must be prime to n, this construction is valid only when n is odd. The resulting Eulerian square $(\alpha'\alpha, \beta'\beta)_n$ is *diagonal* if $\alpha \pm \beta$ and $\alpha' \pm \beta'$ are prime to n, as in the above example $(12, 21)_5$.

Similarly, the "Galois square" $[\alpha', \beta']_q$ (whose order is a prime power q) is orthogonal to $[\alpha, \beta]_q$ whenever $\alpha\beta' - \alpha'\beta \neq 0$ in GF(q). In particular, the $q-1$ Latin squares $[1, \beta]_q$, where β runs over all the non-zero elements of GF(q), are mutually orthogonal. For instance, the two diagonal orthogonal Latin squares $[1, 10]_{2^2}$ and $[1, 11]_{2^2}$ (mentioned above) are both orthogonal to $[1, 1]_{2^2}$ (which is not diagonal).

However, there exist Latin squares of order 4 which have no orthogonal mate. One instance is $(1, 3)_4$ (see page 190). Such a square is said to be *inextensible*.

*Ozanam, 1723 edition, vol. IV, p. 434.

When Eulerian squares of orders m and n are given, we can easily construct one of order mn. Since the Latin squares $(1, 1)_n$ and $(1, 2)_n$ are orthogonal for every odd n, while $[1, 1]_q$ and $[1, 10]_q$ are orthogonal whenever $q = 2^k$ with $k > 1$, it follows that Eulerian squares exist for every odd order and for every doubly even order.

Euler's officers problem. Eulerian squares are so named because it was Euler who first considered the possibility of orthogonal Latin squares of singly-even order. Order 2 is obviously impossible. In 1782, Euler expressed the case of order 6 as follows. Is it possible to arrange 36 officers, each having one of six different ranks and belonging to one of six different regiments, in a square formation of six by six, so that each row and each file shall contain just one officer of each rank and just one from each regiment? He conjectured that the answer is No, and this was proved by Tarry 118 years later.* Unfortunately, Euler made the same conjecture for orders 10, 14, etc. For its refutation, see the frontispiece and chapter x.

Eulerian cubes. A *Latin cube* of order n may be defined as a cube formed by n distinct symbols, each occurring n^2 times, so that every row, every column, and every file is a permutation of the n symbols. In particular, a three-dimensional addition table modulo n provides a Latin cube $(\alpha, \beta, \gamma)_n$ whenever α, β, γ are all prime to n. Three such cubes can be superposed so as to form an Eulerian cube $(\alpha''\alpha'\alpha, \beta''\beta'\beta, \gamma''\gamma'\gamma)_n$ provided the determinant

$$\begin{vmatrix} \alpha & \beta & \gamma \\ \alpha' & \beta' & \gamma' \\ \alpha'' & \beta'' & \gamma'' \end{vmatrix}$$

is prime to n. We shall make use of such a cube on page 220.

*L. Euler, *Verhandelingen Zeeuwsch Genootschap der Wetenschappen*, 1782, vol. IX, pp. 85–239, or *Commentationes Arithmeticae*, Leningrad, 1849, vol. II, pp. 202–361; G. Tarry, *Comptes Rendus de l'Association Française pour l'Avancement de Science naturel*, 1900, vol. I, pp. 122–123; 1901, vol. II, pp. 170–203. See also R.A. Fisher and F. Yates, *Proceedings of the Cambridge Philosophical Society*, 1934, vol. XXX, pp. 429–507, and Albert Sade, *Annals of Mathematical Statistics*, 1951, vol. XXII, pp. 306–307.

CHAPTER VII

MAGIC SQUARES

A *Magic Square* consists of a number of integers arranged in the form of a square, so that the sum of the numbers in every row, in every column, and in each diagonal is the same. If the integers are the consecutive numbers from 1 to n^2, the square is said to be of the nth order, and it is easily seen that in this case the sum of the numbers in every row, column, and diagonal is equal to $\frac{1}{2}n(n^2+1)$. Unless otherwise stated, I confine my account to such magic squares – that is, to squares formed with consecutive integers from 1 upwards. The same rules cover similar problems with n^2 numbers in arithmetical progression.

Thus the first 16 integers, arranged in either of the forms in figures i and ii below, represent magic squares of the fourth

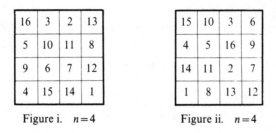

16	3	2	13
5	10	11	8
9	6	7	12
4	15	14	1

Figure i. $n=4$

15	10	3	6
4	5	16	9
14	11	2	7
1	8	13	12

Figure ii. $n=4$

order, the sum of the numbers in each row, column, and diagonal being 34. Similarly, figure iii on page 195 and figure xvi on page 206 represent magic squares of the fifth order; figure vi on page 197 represents a magic square of the sixth order; figure xiv on page 201 and figure xvii on page 206 represent magic squares of the seventh order; figure xiii, xx, xxviii, xxxix represent magic squares of the eighth order;

figures xxii, xxix represent magic squares of the ninth order; and figure ix represents a magic square of the tenth order.

The formation of these squares is an old amusement, and in times when mystical ideas were associated with particular numbers it was natural that such arrangements should be studied. Magic squares were constructed in China before the Christian era: their introduction into Europe appears to have been due to Moschopulus, who lived at Constantinople in the early part of the fifteenth century. The famous Cornelius Agrippa (1486–1535) constructed magic squares of the orders 3, 4, 5, 6, 7, 8, 9, which were associated by him with the seven astrological "planets": namely, Saturn, Jupiter, Mars, the Sun, Venus, Mercury, and the Moon. A magic square engraved on a silver plate was sometimes prescribed as a charm against the plague, and one – namely, that represented in figure i on the preceding page – appears in the picture of Melancholy, engraved in 1514 by Albert Dürer, the numbers in the middle cells of the bottom row giving the date of the work. The mathematical theory of the construction of these squares was taken up in France in the seventeenth century, and later has been a favourite subject with writers in many countries.*

It is convenient to use the following terms. The spaces or small squares occupied by the numbers are called *cells*. It is customary to call the rows first, second, etc., reckoning from the top, and the columns first, second, etc., reckoning from the left. The hth and $(n+1-h)$th rows (or columns) are said to be *complementary*. The kth cell in the hth row is said to be *skewly related* to the $(n+1-k)$th cell in the $(n+1-h)$th row. Skewly related cells are situated symmetrically to the centre of the square.

Magic squares of any order higher than two can be constructed at sight. The rule to be used varies according as the order n is odd, that is, of the form $2m+1$; or singly-even,

---

*For a sketch of the history of the subject and its bibliography see S. Günther's *Geschichte der mathematischen Wissenschaften*, Leipzig, 1876, chapter IV; W. Ahrens, *Mathematische Unterhaltungen und Spiele*, Leipzig, 1901, chapter XII; and W.S. Andrews, *Magic Squares and Cubes*, Chicago, 1917.

that is, of the form $2(2m+1)$; or doubly-even, that is, of the form $4m$. In each case, I now give the simplest rule with which I am acquainted.

MAGIC SQUARES OF AN ODD ORDER

A magic square of the nth order, where $n=2m+1$, can be constructed by the following rule due to De la Loubère.* First, the number 1 is placed in the middle cell of the top row. The successive numbers are then placed in their natural order in a diagonal line which slopes upwards to the right, except that (i) when the top row is reached, the next number is written in the bottom row as if it came immediately above the top row; (ii) when the right-hand column is reached, the next

17	24	1	8	15
23	5	7	14	16
4	6	13	20	22
10	12	19	21	3
11	18	25	2	9

Figure iii. $n=5$

31	43	00	12	24
42	04	11	23	30
03	10	22	34	41
14	21	33	40	02
20	32	44	01	13

Figure iv. $n=5$

number is written in the left-hand column, as if it immediately succeeded the right-hand column; and (iii) when a cell which has been filled up already or when the top right-hand square is reached, the path of the series drops to the row vertically below it and then continues to mount again. Probably a glance at the diagram in figure iii, showing the construction by this rule of a square of the fifth order, will make the rule clear.

The reason why such a square is magic can be best explained by taking a particular case, for instance $n=5$, and expressing all the numbers in the scale of notation whose radix is 5 (or n, if the magic square is of order n). To simplify the discussion,

*S. De la Loubère, *Du Royaume de Siam* (Eng. Trans.), London, 1693, vol. ii, pp. 227–247. De la Loubère was the envoy of Louis xiv to Siam in 1687–1688, and there learned this method.

let us at the same time diminish each number by unity; this clearly makes no difference to the magic properties. The resulting square (figure iv) may be regarded as an Eulerian square.* Since every row and every column contains one of each of the 5 possible unit digits and one of each of the radix digits, the magic property of the rows and columns is automatically assured; so is that of the leading diagonal. The other diagonal suffers from a repetition of the radix digit 2 (in general, m), but it still has the correct sum. Moreover, every number from 0 to $n^2 - 1$ occurs once and only once.

The reader can easily apply this rule to construct a magic square of the third order. Such a square, called the *lo-shu*,† is attributed to the Chinese Emperor Yu, about 2200 B.C., and is still used as a charm by various Oriental peoples. Even in the West it is generally to be seen on the deck of any large passenger ship, for scoring in such games as "shuffleboard."

MAGIC SQUARES OF A SINGLY-EVEN ORDER

A magic square of the nth order, where $n = 2(2m+1)$, can be constructed by the following rule due to Ralph Strachey.‡ Divide the square into four equal quarters A, B, C, D. Construct in A, by De la Loubère's method, a magic square with

A	C
B	D

the numbers 1 to u^2 where $u = n/2$. Construct by the same rule, in B, C, D, similar magic squares with the numbers $u^2 + 1$ to $2u^2$, $2u^2 + 1$ to $3u^2$, and $3u^2 + 1$ to $4u^2$. Clearly the resulting composite square is magic in columns. In the middle row of A

*See page 190. The Eulerian square that underlies De la Loubère's rule is derivable from $(12, 11)_n$ by reversing the order of the rows, and cyclically permuting the columns.

†D.E. Smith, *History of Mathematics*, Boston, 1925, vol. II, p. 591.

‡Communicated to me in a letter, August 1918.

8	1	6	26	19	24
3	5	7	21	23	25
4	9	2	22	27	20
35	28	33	17	10	15
30	32	34	12	14	16
31	36	29	13	18	11

Figure v. Initial Quarter-Squares

35	1	6	26	19	24
3	32	7	21	23	25
31	9	2	22	27	20
8	28	33	17	10	15
30	5	34	12	14	16
4	36	29	13	18	11

Figure vi. Final Square, $n = 6$

27+8	0+1	0+6	18+8	18+1	18+6
0+3	27+5	0+7	18+3	18+5	18+7
27+4	0+9	0+2	18+4	18+9	18+2
0+8	27+1	27+6	9+8	9+1	9+6
27+3	0+5	27+7	9+3	9+5	9+7
0+4	27+9	27+2	9+4	9+9	9+2

Figure vii. Final Square, $n = 6$

take the m cells next but one to the left-hand side, in each of the other rows take the m cells nearest to the left-hand side, and interchange the numbers in these cells with the numbers in the corresponding cells in D. Next interchange the numbers in the cells in each of the $m-1$ columns next to the right-hand side of C with the numbers in the corresponding cells in B. Of course, the resulting square remains magic in columns. It will also now be magic in rows and diagonals, since the construction is equivalent to writing in each of the quarters A, B, C, D equal magic squares of the order u made with the numbers 1 to u^2, and then superposing on them a magic square of the nth order made with the four radix numbers 0, u^2, $2u^2$, $3u^2$, each repeated u^2 times. The component squares being magic, the square resulting from their superposition must be magic, and they are so formed that their superposition ensures that every number from 1 to n^2 appears once and only once in the resulting square.

<u>17</u>	<u>24</u>	1	8	15	67	74	51	58	<u>65</u>
<u>32</u>	<u>5</u>	7	14	16	73	55	57	64	<u>66</u>
4	<u>6</u>	<u>13</u>	20	22	54	56	63	70	<u>72</u>
<u>10</u>	<u>12</u>	19	21	3	60	62	69	71	<u>53</u>
<u>11</u>	<u>18</u>	25	2	9	61	68	75	52	<u>59</u>
92	99	76	83	90	42	49	26	33	40
98	80	82	89	91	48	30	23	39	41
79	81	88	95	97	29	31	38	45	47
85	87	94	96	78	35	37	44	46	28
86	93	100	77	84	36	43	50	27	34

Figure viii. Initial Quarter-Squares, $n = 10$

92	99	1	8	15	67	74	51	58	40
98	80	7	14	16	73	55	57	64	41
4	81	88	20	22	54	56	63	70	47
85	87	19	21	3	60	62	69	71	28
86	93	25,	2	9	61	68	75	52	34
17	24	76	83	90	42	49	26	33	65
23	5	82	89	91	48	30	32	39	66
79	6	13	95	97	29	31	38	45	72
10	12	94	96	78	35	37	44	46	53
11	18	100	77	84	36	43	50	27	59

Figure ix. Final Square, $n = 10$

Figures v and vi show the application of the rule to the construction of a magic square of the sixth order. In figure v, those numbers in the cells in the initial quarter-square A which are to be interchanged vertically with the numbers in the corresponding cells in D are underlined; figure vi represents the final square obtained; and figure vii shows how these interchanges serve to bring the radix numbers to a position which makes the square magic.

Since this construction is novel, I add figures to show the application of the rule to the formation of a square of the tenth order: in figure viii those numbers in the quarter-squares A and C which have to be interchanged vertically with the numbers in the corresponding cells in D and B are underlined, while figure ix represents the final magic square of the tenth order thus obtained.

MAGIC SQUARES OF A DOUBLY-EVEN ORDER

A magic square of the fourth order (figure xi), scarcely different from Dürer's (figure i), can be constructed by writing the num-

Figure x

16	2	3	13
5	11	10	8
9	7	6	12
4	14	15	1

Figure xi. $n = 4$

Figure xii

64	2	3	61	60	6	7	57
9	55	54	12	13	51	50	16
17	47	46	20	21	43	42	24
40	26	27	37	36	30	31	33
32	34	35	29	28	38	39	25
41	23	22	44	45	19	18	48
49	15	14	52	53	11	10	56
8	58	59	5	4	62	63	1

Figure xiii. $n = 8$

bers from 1 to 16 in their natural order in rows of four, and
then replacing the numbers in the *diagonal* cells by their complements. The same rule* applies to a magic square of any
doubly-even order, if we change the numbers in those cells
which are crossed by the diagonals of every component block
of 4^2 cells. Figures xii and xiii show the case when $n = 8$.

BORDERED SQUARES

One other general method, due to Frénicle, of constructing
magic squares of any order should be mentioned. By this
method, to form a magic square of the nth order we first construct one of the $(n-2)$th order, add to every number in it an
integer, and then surround it with a border of the remaining
numbers in such a way as to make the resulting square magic.
In this manner from the magic square of the 3rd order we can
build up successively squares of the orders 5, 7, 9, etc. – that is,
every odd magic square. Similarly from a magic square of the
4th order we can build up successively all higher even magic
squares.

The method of construction will be clear if I explain how
the square in figure xiv, where $n = 7$, is built up. First the inner
magic square of the $(n-2)$th order is formed by any rule we
like to choose: the sum of the numbers in each line being
$(n-2)\{(n-2)^2+1\}/2$. To every number in it, $2n-2$ is added:
thus the sum of the numbers in each row, column, and diagonal
is now $(n-2)\{n^2+1\}/2$. The numbers not used are 1, 2, ...,
$2n-2$, and their complements, $n^2, n^2-1, ..., n^2-2n+3$. These
reserved numbers are placed in the $4(n-1)$ border cells so that
complementary numbers occur at the end of each row, column,
and diagonal of the inner square: this makes the sum of the
numbers in each of these latter lines equal to $n(n^2+1)/2$. It only
remains to make the sum of the numbers in each of the border
lines also have this value: such an arrangement is easily made
by trial and error. With a little patience a magic square of any
order can be thus built up, border upon border, and, of course,

*Kraïtchik, p. 176. R.V. Heath has devised a similar rule for constructing
a Magic Cube.

it will have the property that, if each border is suc̲
stripped off, the remaining square will still be magic. This ıs
a method of construction much favoured by self-taught mathematicians.

Definite rules for arranging the numbers in the border cells
have been indicated,* though usually not in a precise form.
Recently, Mr. J. Travers† has proposed a simple rule for constructing an odd square, where $n = 2m + 1$. Rather than describe

46	1	2	3	42	41	40
45	35	13	14	32	31	5
44	34	28	21	26	16	6
7	17	23	25	27	33	43
12	20	24	29	22	30	38
11	19	37	36	18	15	39
10	49	48	47	8	9	4

Figure xiv. A Bordered Square, $n = 7$

	1	2	3			
		13	14			5
			21		16	6
7	17	23	25			
12	20	24		22		
11	19			18	15	
10				8	9	4

Figure xv

it in words, I illustrate its application to the above example
(which is in fact a multiple bordered square). Figure xv indicates the proper positions for the numbers $1, 2, \ldots, m; m + 1;$
$m + 2, \ldots, 2m; 2m + 1; 2m + 2, \ldots, 3m; 3m + 1, \ldots, 4m$. The
complementary numbers are easily inserted afterwards. It may
interest my readers to see if they can evolve a similar simple
rule for the formation of bordered even squares.‡

NUMBER OF SQUARES OF A GIVEN ORDER

One unsolved problem in the theory is the determination of

*For instance, see *Japanese Mathematics* by D.E. Smith and Y. Mikami,
Chicago, 1914, pp. 116–120: in this work, in the diagram on p. 120, the numbers 8 and 29 should be interchanged.

†*Education Outlook*, 1936. For another good rule, see R.V. Heath, *Scripta
Mathematica*, 1936, vol. IV, p. 67.

‡A somewhat more complicated rule for even squares has been given by
J. Travers, *Engineering Gazette*, Aug. 13, 1938, p.6.

the number of magic squares of the fifth (or any higher) order. There is, in effect, only one magic square of the third order, though by reflections and rotations it can be presented in 8 forms. There are 880 magic squares of the fourth order, but by reflections and rotations these can be presented in 7040 forms. The problem of the number of magic squares of the fifth order is incompletely solved. From the square given in figure iv (page 195) formed by De la Loubère's method, we can get 720 distinct squares; for we can permute the unit digits 0, 1, 2, 3, 4 in 5! ways, and the radix digits 0, 1, 3, 4 in 4! ways. We thus obtain 2880 magic squares of the fifth order, though only 720 of them are distinct. Bachet gave a somewhat similar construction,* in which he began by placing 1 in the cell immediately above the middle cell; his method gives another 720 distinct magic squares of the fifth order. There are, however, numerous other rules for constructing odd magic squares, and De la Hire showed that by methods known in his day, and apart from mere reflections and rotations, there were 57600 magic squares of the fifth order which could be formed by the methods he enumerated; taking account of other methods, it is now known that the total number of magic squares of the fifth order considerably exceeds thirteen million.

SYMMETRICAL AND PANDIAGONAL SQUARES

With the exception of determining the number of squares of a given order, we may fairly say that the theory of the construction of magic squares, as defined above, has been sufficiently worked out. Accordingly, attention has of late been chiefly directed to the construction of squares which, in addition to being magic, satisfy other conditions. It has been suggested that we might impose on the construction of a square of order n the condition that the sum of any two numbers in cells skewly related to one another shall be constant and equal to $n^2 + 1$. Such squares are called *symmetrical* (or *associated*). Dürer's square (figure i) is symmetrical; so are all De la Loubère's squares (such as figure iii), and all squares of doubly-even order

*See Kraïtchik, p. 128.

that are constructed by the rule described above (such as figures xi and xiii). There are no symmetrical magic squares of singly-even order.*

One of the earliest additional conditions to be suggested was that the square should be magic along the broken diagonals as well as along the two ordinary diagonals.† Such squares are called *pandiagonal*. They are also known as nasik, perfect, and diabolic squares.

A magic pandiagonal square of the fourth order‡ was inscribed at Khajuraho, India, as long ago as the eleventh or twelfth century. A slightly different 'square of the same kind is represented in figure ii on page 193. In it the sum of the numbers in each row, column, and in the two diagonals is 34, as also is the sum of the numbers in the six broken diagonals formed by the numbers 15, 9, 2, 8, the numbers 10, 4, 7, 13, the numbers 3, 5, 14, 12, the numbers 6, 4, 11, 13, the numbers 3, 9, 14, 8, and the numbers 10, 16, 7, 1.

It follows from the definition that if a pandiagonal square be cut into two pieces along a line between any two rows or any two columns, and the two pieces be interchanged, the new square so formed will be also pandiagonally magic. Hence it is obvious that by one vertical and one horizontal transposition of this kind any number can be made to occupy any specified cell.

I have already remarked that there is essentially only one magic square of the third order; this is not pandiagonal, so the order of a pandiagonal square must exceed three. Moreover, there are no pandiagonal squares of singly-even order.§

---

*C. Planck, *The Monist*, Chicago, 1919, vol. xxix, p. 308.

†Squares of this type were mentioned by P. De la Hire, J. Sauveur, and Euler. Attention was again called to them by A.H. Frost in the *Quarterly Journal of Mathematics*, London, 1878, vol. xv, pp. 34–49, and subsequently their properties have been discussed by several writers. Besides Frost's papers I have made considerable use of a paper by E. McClintock in the *American Journal of Mathematics*, vol. xix, 1897, pp. 99–120.

‡D.E. Smith, *History of Mathematics*, Boston, 1925, vol. ii, p. 594. For the general theory of such squares, see J.B. Rosser and R.J. Walker, *Bulletin of the American Mathematical Society*, 1938, vol. xliv, pp. 416–420.

§C. Planck, *loc. cit.*

Pandiagonal squares of odd order, not divisible by 3, can be constructed by a rule somewhat analogous to De la Loubère's; and new rules have been devised to cover the cases where the order is divisible by 3.

Of the 880 magic squares of the fourth order, 48 are pandiagonal. Rosser and Walker have shown that there are exactly 3600 pandiagonal squares of the fifth order, more than 38 million of the seventh order, and more than $6\frac{1}{2}$ billion* of the eighth order.

Generalization of De la Loubère's rule. It is convenient to name the n^2 cells by pairs of co-ordinates in such a way that the cells of the bottom row (from left to right) are $(0, 0)$, $(1, 0)$, ..., $(n-1, 0)$, while those of the left-hand column (from bottom to top) are $(0, 0)$, $(0, 1)$, ..., $(0, n-1)$. It follows that the cell (x, y) lies in the $(n-y)$th row and in the $(x+1)$th column. The conventions (i) and (ii) on page 195 are equivalent to reducing the co-ordinates modulo n, so that the cells $(n+x, y)$ and $(x, n+y)$ are indentified with (x, y).

In De la Loubère's method, the step from 1 to 2 increases both co-ordinates by unity – i.e. it is made by the "vector" $(1, 1)$. Similarly, the "cross-step" from n to $n+1$ is made by the vector $(0, -1)$. This method may evidently be generalized by taking other vectors: say (a, b) for the step from ξ to $\xi+1$ ($\xi = 1, 2, ..., n-1$), and $(a+a', b+b')$ for the cross-step from $\xi'n$ to $\xi'n+1$ ($\xi' = 1, 2, ..., n-1$). Consequently the step from 1 to $n+1$ is $(n-1)(a, b)+(a+a', b+b') \equiv (a', b')$ (mod n); the step from 1 to $\xi'n+1$ is $(a'\xi', b'\xi')$; and the step from 1 to $\xi'n+\xi+1$ is $(a\xi+a'\xi', b\xi+b'\xi')$.

Let (i, j) be the cell occupied by the number 1. The position of any other number s can be deduced by expressing $s-1$ as $\xi'\xi$ in the scale of n, that is, by expressing s in the form $\xi'n+\xi+1$, where ξ' and ξ are integers less than n (and positive or zero).

Applying the vector $(a\xi+a'\xi', b\xi+b'\xi')$, the required position is found to be $(i+a\xi+a'\xi', j+b\xi+b'\xi')$. Any other number, say $s+X'n+X$, must have a different position; therefore the

*For American readers, $6\frac{1}{2}$ *trillion.*

congruences

$$aX + a'X' \equiv 0, \quad bX + b'X' \equiv 0 \quad (\text{mod } n)$$

must imply $X \equiv X' \equiv 0$. The condition for this is that $ab' - a'b$ shall be prime to n. (In De la Loubère's case $ab' - a'b = -1$.)

If also a, b, a', b' are prime to n (and therefore not zero), the square will be magic as far as rows and columns are concerned. For, the numbers in a column are given by the n solutions of a congruence of the form $a\xi + a'\xi' \equiv c$, and these involve all the possible values, $0, 1, \ldots, n-1$, for both ξ and ξ' (in some combination); similarly for the numbers in a row. Since a, b, a', b', and $ab' - a'b$ are all prime to n, n must be odd. For, if n were even, a, b, a', b' would have to be odd, and then $ab' - a'b$ would be even. Accordingly we may write $n = 2m + 1$.

Finally, let us choose i and j so as to put the middle number (for which $\xi = \xi' = m$) into the middle cell (m, m). This requires

$$i \equiv (1 - a - a')m, \quad j \equiv (1 - b - b')m.$$

(In De la Loubère's case, $i = m$, $j = 2m$; hence the position of 1 in the middle of the top row.) It now follows that the square is symmetrical – i.e. that the numbers in the skewly related cells (x, y) and $(2m - x, 2m - y)$ are complementary. This ensures the magic property for the diagonals.

It is convenient to denote this magic square by the symbol $\begin{pmatrix} a & a' \\ b & b' \end{pmatrix}_n$. Thus De la Loubère's square is $\begin{pmatrix} 1 & -1 \\ 1 & -2 \end{pmatrix}_n$, and Bachet's is $\begin{pmatrix} 1 & -1 \\ 1 & 1 \end{pmatrix}_n$. What we have proved is that such a square can be constructed whenever a, b, a', b', and $ab' - a'b$ are all prime to n.

If also $a + b$, $a - b$, $a' + b'$, $a' - b'$ are prime to n, the square will be *pandiagonal*. For, the numbers in a generalized diagonal are given by the n solutions of a congruence of the form $(a \pm b)\xi + (a' \pm b')\xi' \equiv c$. It is impossible to satisfy all these conditions when n is a multiple of 3; in such cases, therefore, the method has to be modified.

When the conditions are all satisfied, the values of i and j

(Pandiagonal Squares)

7	20	3	11	24
13	21	9	17	5
19	2	15	23	6
25	8	16	4	12
1	14	22	10	18

Figure xvi. $n = 5$

26	21	9	4	48	36	31
44	39	34	22	17	12	7
20	8	3	47	42	30	25
38	33	28	16	11	6	43
14	2	46	41	29	24	19
32	27	15	10	5	49	37
1	45	40	35	23	18	13

Figure xvii. $n = 7$

obtained above render the square symmetrical as well as pan-diagonal; but, of course, the square will still be pandiagonal if the position of the number 1 is chosen arbitrarily. For instance, n being prime to 6, the conditions are all satisfied by $\begin{pmatrix} 1 & -1 \\ 2 & -3 \end{pmatrix}_n$. Thus, to form a pandiagonal square of the fifth order (figure xvi) we may put 1 in any cell; proceed by four successive steps, like a knight's move, of one cell to the right, and two cells up, writing consecutively numbers 2, 3, 4, 5 in each cell, until we come to a cell already occupied; then take one step, like a rook's move, one cell down; and so on until the square is filled. Again, in the square $\begin{pmatrix} 1 & 1 \\ 2 & -3 \end{pmatrix}_n$ the successive numbers can be written in with a knight's move at *every* step, as in figure xvii; but, since $ab' - a'b = -5$, this formation is impossible when n is 5 or a multiple of 5.

Arnoux's method.* If the numbers in figures xvi and xvii are all diminished by 1 and expressed in the scale of $n (= 5$ or $7)$, the results are immediately recognizable as the Eulerian squares $(23, 44)_5$ and $(62, 43)_7$. Conversely, it is obvious that any diagonal Eulerian square, composed of the digit-pairs 00, 01, ..., $(n-1)(n-1)$, becomes a magic square when its entries are interpreted as numbers in the scale of n.

*G. Arnoux, *Arithmétique graphique. Les espaces arithmétiques hypermagiques*, Paris, 1894, p. 51. Cf. Kraïtchik, pp. 130–146.

Moreover, any Eulerian square of the form $(\alpha'\alpha, \beta'\beta)_n$ (even if not *diagonally* Eulerian) becomes a magic square when its rows and columns are cyclically permuted in such a way as to put the middle number mm into the middle cell (m, m); for, this permutation makes the square symmetrical – i.e. it puts complementary numbers into the cells $(m+x, m+y)$ and $(m-x, m-y)$. In fact, the number in the cell $(m+x, m+y)$ has digits congruent to $m+\alpha'x+\beta'y, m+\alpha x+\beta y \pmod{n}$. (In order to make the numbers run from 1 to n^2, instead of 0 to n^2-1, we merely have to add 1 throughout.)

On the other hand, the above generalization of De la Loubère's rule puts the number $(m+\xi')n+(m+\xi)+1$ into the cell $(m+a\xi+a'\xi', m+b\xi+b'\xi')$. The two methods lead to the same square if the congruences

$$x \equiv a\xi+a'\xi', \quad k \equiv b\xi+b'\xi', \quad \xi \equiv \alpha x+\beta y, \quad \xi' \equiv \alpha'x+\beta'y \pmod{n}$$

are consistent, i.e. if

$$a\alpha+a'\alpha' \equiv b\beta+b'\beta' \equiv 1 \quad \text{and} \quad a\beta+a'\beta' \equiv b\alpha+b'\alpha' \equiv 0.$$

Under these circumstances, $\begin{pmatrix} a & a' \\ b & b' \end{pmatrix}$ and $\begin{pmatrix} \alpha & \beta \\ \alpha' & \beta' \end{pmatrix}$ are called "inverse matrices," since, on multiplying them together by the usual rule for determinants, the result is $\begin{pmatrix} 1 & 0 \\ 0 & 1 \end{pmatrix} \pmod{n}$. In fact, the two methods correspond to the two aspects of a transformation of co-ordinates.* Clearly, the statement that a, b, a', b', and $ab'-a'b$ are prime to n is equivalent to the statement that $\alpha, \beta, \alpha', \beta'$, and $\alpha\beta'-\alpha'\beta$ are prime to n.

When n is prime to 6, we can choose $\alpha, \beta, \alpha', \beta'$ so as to make $\alpha \pm \beta, \alpha' \pm \beta'$ (as well as $\alpha, \beta, \alpha', \beta', \alpha\beta'-\alpha'\beta$) all prime to n. Then the square $(\alpha'\alpha, \beta'\beta)_n$ is not merely diagonal, but *pandiagonal*. (There is no need to permute the rows or columns, unless the extra quality of symmetry is desired.) The simplest example, from this point of view, is $(12, 21)_n$. (See page 191.)

Margossian's method.† I next describe an extension of

Arnoux's method which enables us to construct a pandiagonal square whose order is a multiple of 4 or an odd multiple of 3 (but not 3 itself, or course). Figure xviii shows the square

23	31	03	11
02	10	22	30
21	33	01	13
00	12	20	32

Figure xviii

32	21	02	11
03	10	33	20
31	22	01	12
00	13	30	23

Figure xix

$(12, 21)_4$ (which is not properly Eulerian, since 2 divides 4). Though not magic as it stands, this square becomes magic, in fact pandiagonal, when every digit 2 is replaced by 3, and vice versa, as in figure xix. Figure ii shows the same square in the ordinary notation.

More generally, α being any even* number, the square $(1\alpha, \alpha1)_{2\alpha}$ becomes pandiagonally magic when the digits $\alpha, \alpha + 1, \ldots, 2\alpha - 1$ are replaced by $2\alpha - 1, 2\alpha - 2, \ldots, \alpha$. The unit digits in a row consist of two particular digits, each repeated α times; Margossian's substitution makes these two digits have the

74	63	54	43	04	13	24	33
05	12	25	32	75	62	55	42
76	61	56	41	06	11	26	31
07	10	27	30	77	60	57	40
73	64	53	44	03	14	23	34
02	15	22	35	72	65	52	45
71	66	51	46	01	16	21	36
00	17	20	37	70	67	50	47

Figure xx.   A Pandiagonal Square, $n = 8$ (in the scale of 8)

*This would fail if α were odd, since then the numbers in the square would not be all distinct. In fact, $\alpha\beta' - \alpha'\beta$ $(= \alpha^2 - 1)$ would not be prime to n $(= 2\alpha)$.

35	48	52	65	78	82	05	18	22
04	17	21	34	47	51	64	77	81
63	76	80	03	16	20	33	46	50
32	45	58	62	75	88	02	15	28
01	14	27	31	44	57	61	74	87
60	73	86	00	13	26	30	43	56
38	42	55	68	72	85	08	12	25
07	11	24	37	41	54	67	71	84
66	70	83	06	10	23	36	40	53

Figure xxi

uniform sum $2\alpha - 1$; similarly for the radix digits in a column. A square formed by this rule (with $\alpha = 4$) is shown in figure xx.

Similarly, a square of the form $(13, 31)_{3m}$, where m is odd, becomes pandiagonally magic when the digits, $0, 1, \ldots, 3m - 1$ are so permuted that, if they are written in their new order in m rows of three, the sum of the columns is uniform. For this purpose, $\frac{1}{2}(m - 1)$ of the m triads $0, 1, 2; 3, 4, 5; \ldots$ can be left unchanged if two of the rest are cyclically permuted (opposite

35	47	50	85	67	70	15	27	00
14	26	02	34	46	52	84	66	72
83	68	71	13	28	01	33	48	51
30	45	57	80	65	77	10	25	07
12	24	06	32	44	56	82	64	76
81	63	78	11	23	08	31	43	58
37	40	55	87	60	75	17	20	05
16	22	04	36	42	54	86	62	74
88	61	73	18	21	03	38	41	53

Figure xxii.   A Pandiagonal Symmetrical Square, $n = 9$
(in the scale of 9)

ways) and the remaining $\frac{1}{2}(m-3)$ are reversed. Thus, for a square of the fifteenth order, the digits 0, 1, 2, 3, 4, 5 can be left unchanged if the rest are replaced by 7, 8, 6; 11, 9, 10; 14, 13, 12. In the special case when $m=3$, i.e. for a square of the ninth order, we may cyclically permute the rows and columns of $(13, 31)_9$ so as to put 44 into the middle cell (figure xxi) and then replace the digits 0, 1, 2, 3, 4, 5, 6, 7, 8 by 1, 2, 0, 3, 4, 5, 8, 6, 7; the resulting square (figure xxii) is not merely pandiagonal* but also symmetrical.

MAGIC SQUARES OF NON-CONSECUTIVE NUMBERS

Although it is impossible to make a pandiagonal or symmetrical magic square of singly-even order, using *consecutive* numbers, C. Planck† has devised a method for making such squares with numbers that are *almost* consecutive. For a square of singly-even order n, he uses the numbers from 1 to n^2+3, omitting the "middle number" $\frac{1}{2}n^2+2$ and any two other even numbers whose sum is n^2+4. In particular, he might simply omit the multiples of $\frac{1}{4}n^2+1$. The sum of the numbers in a row, column, or diagonal is $\frac{1}{2}n(n^2+4)$. Moreover, for the pandiagonal square, the sum of the numbers in any component square of $(\frac{1}{2}n)^2$ cells is $\frac{1}{8}n^2(n^2+4)$.

28	1	26	36	8	21
3	35	7	27	23	25
34	24	22	2	29	9
4	32	19	12	39	14
13	17	15	37	5	33
38	11	31	6	16	18

Figure xxiii
A Pandiagonal Square, $n=6$

28	1	26	21	8	36
3	35	7	25	23	27
34	24	22	9	29	2
38	11	31	18	16	6
13	17	15	33	5	37
4	32	19	14	39	12

Figure xxiv
A Symmetrical Square, $n=6$

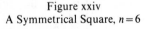

*Arnoux, *Arithmétique graphique*, pp. 152–154.
†*The Monist* (Chicago), 1919, vol. XXIX, pp. 307–316.

Figures xxiii and xxiv show the case when $n=6$, omitting the multiples of 10.

Another problem involving non-consecutive numbers is that of constructing a magic square of *primes*. The first of the following examples is due to H.E. Dudeney, the second (figure xxvi) to E. Bergholt and C.D. Shuldham. Such squares, of orders 5, 6, ..., 12, have been constructed by H.A. Sayles and J.N. Muncey.* (Notice that all these gentlemen were unaware that the number 1 is not properly included in the list of primes.) Muncey's square is remarkable in that it uses 1, 3, 5, 7, 11, ..., 827 with no intermediate primes omitted.

67	1	43
13	37	61
31	73	7

Figure xxv

3	71	5	23
53	11	37	1
17	13	41	31
29	7	19	47

Figure xxvi

Figure xxvii shows Bergholt's general form for any magic square of the fourth order† (pandiagonal if $a=b=d-c=\frac{1}{2}(A-B-C+D)$, symmetrical if $a+c=d=b-c$ and $A+C=B+D$; therefore never both pandiagonal and symmetrical, since then we should have $A-a=B$). Figure xxvi is obtained by taking $A=13$, $B=11$, $C=37$, $D=41$, $a=10$, $b=18$, $c=24$, $d=-2$.

$A-a$	$C+a+c$	$B+b-c$	$D-b$
$D+a-d$	B	C	$A-a+d$
$C-b+d$	A	D	$B+b-d$
$B+b$	$D-a-c$	$A-b+c$	$C+a$

Figure xxvii

DOUBLY-MAGIC SQUARES

For certain values of n (not less than 8) it is possible to construct a magic square of the nth order in such a way that if the number in each cell is replaced by its square the resulting square shall also be magic.*

17	50	43	04	32	75	66	21
31	76	65	22	14	53	40	07
00	47	54	13	25	62	71	36
26	61	72	35	03	44	57	10
45	02	11	56	60	27	34	73
63	24	37	70	46	01	12	55
52	15	06	41	77	30	23	64
74	33	20	67	51	16	05	42

Figure xxviii. A Doubly-Magic Pandiagonal Square, $n = 8$
(in the scale of 8)

Here are two examples. Figure xxviii represents a pandiagonal magic square of the eighth order, due to M.H. Schots.† The numbers are expressed in the scale of 8 in order to exhibit the underlying Eulerian square. After adding 1 throughout, the sum of the numbers in each line is equal to 260, and the sum of their squares is equal to 11180. Figure xxix represents R.V. Heath's doubly-magic square of the ninth order, written in the scale of 9.

Trebly-magic squares. The construction of squares which shall be magic for the original numbers, for their squares, and

---

*See M. Coccoz in *L'Illustration,* May 29, 1897. The subject has been studied by Messrs. G. Tarry, B. Portier, M. Coccoz, A. Rilly, E. Barbette, and W.S. Andrews. More than 200 such squares have been given by Rilly in his *Études sur les Triangles et les Carrés Magiques aux deux premiers degrés.* Troyes, 1901.

†*Bulletin de la classe des Sciences de l'Académie royale de Belgique,* 1931, pp. 339–361. Cf. *Sphinx,* 1931, p. 137. It should be observed that this is *not* a pandiagonal *Eulerian* square, since the broken diagonals contain repeated digits (although they have the proper sum).

76	82	64	15	27	00	41	53	38
11	23	08	46	52	34	75	87	60
45	57	30	71	83	68	16	22	04
62	74	86	07	10	25	33	48	51
03	18	21	32	44	56	67	70	85
37	40	55	63	78	81	02	14	26
84	66	72	20	05	17	58	31	43
28	01	13	54	36	42	80	65	77
50	35	47	88	61	73	24	06	12

Figure xxix.   A Doubly-Magic Symmetrical Square, $n = 9$
(in the scale of 9)

for their cubes has also been studied. Such a square of order 64 was found by Cazalas.* Captain William H. Benson, of Carlisle, Pennsylvania, has discovered one of order 32.

OTHER MAGIC PROBLEMS

Other problems, closely related to magic squares, will suggest themselves; the following will serve as specimens.

Magic domino squares. An ordinary set of dominoes, ranging from double zero to double six, contains 28 dominoes. Each domino is a rectangle formed by fixing two small square blocks together side by side: of these 56 blocks, eight are blank, on each of eight of them is one pip, on each of another eight of them are two pips, and so on. It is required to arrange the dominoes so that the 56 blocks form a square of 7 by 7 bordered by one line of 7 blank squares, and so that the sum of the pips in each row, column, and diagonal of the square is equal to 24. A solution† is given in figure xxx.

If we select certain dominoes out of the set and reject the

*J.J.A.M.E. Cazalas, *Carrés Magiques au degré n*, Hermann, Paris, 1934, p. 114.

†See *L'Illustration*, July 10, 1897; *Scientific American*, December 1969, vol. 221, no. 6, pp. 122–127.

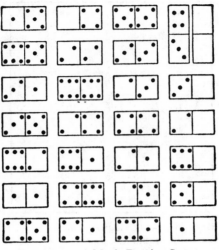

Figure xxx.   A Magic Domino Square

Figure xxxi.   Magic Domino Squares

others we can use them to make various magic puzzles. As instance, I give in figure xxxi two examples of magic squares of this kind due respectively to Escott and Dudeney.

Cubic and octahedral dice. The faces of a cube can obviously be numbered from 1 to 6 in such a way that the pairs of opposite faces sum to 7. A cube so marked is called a *die*. Dice are used in pairs, but no attention seems to have been paid to specifying whether a pair should be identical or enantiomorphous.

Another kind of die might be made by numbering the faces of a polyhedron from 1 to F in such a way that the faces around a vertex give a constant sum. If there are m faces around each vertex, the constant sum must be equal to $\frac{1}{2}m(F+1)$; hence either m is even or F is odd. Every regular polyhedron has an even number of faces, so the only one which could be made into this kind of die is the octahedron (for which $m=4$ and $F=8$). The faces of the octahedron can be numbered (so that those around each vertex sum to 18) in three essentially distinct ways, each of which occurs in two enantiomorphous forms. Pairs of opposite faces have a constant difference, 1 or 2 or 4.

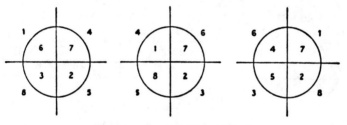

Figure xxxii.   Octahedral Dice*

Interlocked hexagons.   The problem of numbering the faces of the Rotating Ring of n tetrahedra (page 154), so that the faces around each vertex sum to $3(4n+1)$, appears to be insoluble. But R.V. Heath has elegantly solved the corresponding problem for the network of $36n$ triangles obtained by dividing each face of the ring into nine. Such a network of triangles may be regarded as a symmetrical "map" on a torus (page 237), even when $n<6$. (When $n\geq6$, the map can be drawn on the rotating ring in nine distinct ways.) Figure xxxiii shows the simplest case ($n = 2$), but a similar method† applies in every case. The six triangles around any vertex sum

*The third of these is due to J.M. Andreas.
†The method is precisely the same for all singly-even values of n. When n is odd or doubly-even, the ends have to be joined as in figure xxxiv. In the latter case, the regular arrangement of numbers has to be broken after $9n$ and $27n$.

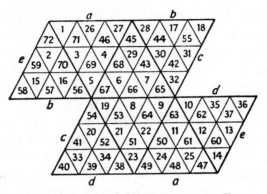

Figure xxxiii.   Interlocked Hexagons on a Torus

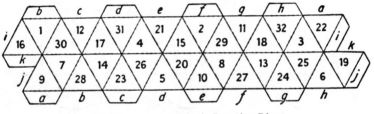

Figure xxxiv.   A Magic Rotating Ring

to $3(36n+1)$; so do the six triangles adjacent to these (which form a kind of star). Heath has succeeded in covering many other surfaces with such interlocked hexagons.

Figure xxxiv shows Heath's special arrangement of the numbers from 1 to 32 on the ring of eight tetrahedra, which is magic in a quite different sense. The four faces of each tetrahedron sum to 66; "corresponding" faces, one from each tetrahedron, sum to 132 (for instance, $9+7+17+31+10+8+18+32=132$), and so do eight sets of eight faces which wind helically around the ring (for instance, $1+12+31+21+2+11+32+22=132$).

MAGIC CUBES

A *Magic Cube* of the nth order consists of the consecutive numbers from 1 to n^3, arranged in the form of a cube, so that the sum of the numbers in every row, every column, every file, and in each of the four diagonals (or "diameters") is the same namely, $\frac{1}{2}n(n^3+1)$. This sum occurs in $3n^2+4$ ways. I do not know of any rule for constructing magic cubes of singly-even order. But such cubes of any odd or doubly-even order can be constructed by a natural extension of the methods already used for squares.

As in the two-dimensional case (page 204), we proceed from cell to cell by steps, only now the vectors that make the steps are three-dimensional. In fact, we define the step (a, b, c) as taking us a cells "east," b cells "north," and c cells "up." As before, all the movements are taken cyclically, so that the numbers a, b, c may at any stage be reduced modulo n. We can make the same step $n-1$ times before coming to an occupied cell. Then we make one cross-step $(a+a',\ b+b',\ c+c')$, follow this by $n-1$ more steps (a, b, c), and so on. There will be no difficulty until we try to write the number n^2+1. At that stage the cross-step has to be altered, say to $(a+a'+a'',\ b+b'+b'',\ c+c'+c'')$, and this new kind of cross-step has to be used again after each multiple of n^2 has been reached. If a'', b'', c'' are suitably chosen, the whole cube can now be filled up. In fact, the position of the number $\xi''n^2+\xi'n+\xi+1$ is derived from that of the number 1 by applying ξ steps (a, b, c), ξ' steps (a', b', c'), and ξ'' steps (a'', b'', c''). Thus, if 1 is in the cell whose co-ordinates are (i, j, k), then $\xi''n^2+\xi'n+\xi+1$ is in that whose co-ordinates are

$$(i+a\xi+a'\xi'+a''\xi'',\ j+b\xi+b'\xi'+b''\xi'',\ k+c\xi+c'\xi'+c''\xi'').$$

In order that different numbers may always occupy different positions, the congruences

$$aX+a'X'+a''X''\equiv0,$$
$$bX+b'X'+b''X''\equiv0,$$
$$cX+c'X'+c''X''\equiv0 \quad (\text{mod } n)$$

must imply $X\equiv X'\equiv X''\equiv0$. The condition for this is that the determinant

$$\begin{vmatrix} a & a' & a'' \\ b & b' & b'' \\ c & c' & c'' \end{vmatrix}$$

shall be prime to n. The magic property of the rows, columns, and files is secured by making all the two-rowed minors of the determinant likewise prime to n. (This is impossible if n is even.) The magic property of the diagonals is secured by adjusting i, j, k so as to place the middle number in the middle cell; the cube will then be *symmetrical*.

(*Bottom layer*)(*Middle layer*) (*Top layer*)

4	12	26	20	7	15	18	23	1
11	25	6	9	14	19	22	3	17
27	5	10	13	21	8	2	16	24

Figure xxxv.   A Magic Cube, $n = 3$

It is convenient to denote this magic cube by the symbol

$$\begin{pmatrix} a & a' & a'' \\ b & b' & b'' \\ c & c' & c'' \end{pmatrix}_n$$

What we have proved is that such a cube can be constructed whenever the determinant and all its first minors are prime to n. Clearly, these conditions are satisfied, for any odd value of n, if

$$a = a' = b = b'' = c' = c'' = 1, \quad a'' = b' = c = 0,$$

and $i = j = k = \frac{1}{2}(n+1)$. In practice it is easiest to insert first the numbers $1, n^2 + 1, 2n^2 + 1, \ldots$, proceeding by steps $(0, 1, 1)$; then to insert the rest of the numbers $1, n+1, 2n+1, \ldots$, proceeding by steps $(1, 0, 1)$; and finally to fill in the remaining numbers, proceeding by steps $(1, 1, 0)$. Figure xxxv shows the three "horizontal" layers of the magic cube

$$\begin{pmatrix} 1 & 1 & 0 \\ 1 & 0 & 1 \\ 0 & 1 & 1 \end{pmatrix}_3$$

The reader may be interested to try the same rule with $n = 5$. The resulting cube is *pandiagonal*, having the magic property not merely in its four main diagonals, but in all the broken diagonals as well. In such a cube the magic sum occurs $(3+4)n^2$ times, instead of only $3n^2+4$.

It can be shown that the cube

$$\begin{pmatrix} a & b & 0 \\ b & 0 & a \\ 0 & a & b \end{pmatrix}_n$$

is pandiagonal whenever a, b, $a+b$, and $a^2 \pm ab \pm b^2$ (with all four distributions of sign) are prime to n. Thus $a=b=1$ gives a solution whenever n is not divisible by 3 (nor by 2); $a=2$, $b=1$ gives another solution whenever n is divisible by none of 2, 3, 5, 7.

If a, b, $a \pm b$, and a^2+b^2 are prime to n, the cube, without necessarily being pandiagonal itself, has pandiagonal squares for its n layers in each of the three principal directions, so that the magic sum occurs $(3+6)n^2$ times. For instance, $a=2$, $b=1$ gives a solution whenever n is prime to 30, and in particular when n is 7 or any larger prime. This cube will itself be pandiagonal if n is prime to 210, and in particular when $n=11$ or any larger prime. The magic sum will then occur $(3+4+6)n^2$ times. Rosser and Walker have shown that a magic cube can be pandiagonal in this stricter sense when $n=8$ or any multiple of 8, and when $n=9$ or any larger odd number, but in no other cases.

Figure xxxvi shows the same cube as figure xxxv, but with every number diminished by 1 and expressed in the scale of 3. Apart from cyclic permutations of the layers, this is just the Eulerian cube $(122, 212, 221)_3$; for, the row, column, and file involving 000 contain the elements ± 122, ± 212, ± 221 of $GF(3^3)$, and the whole cube constitutes an addition table based on this row, column, and file.* Since

$$\begin{pmatrix} 1 & 1 & 0 \\ 1 & 0 & 1 \\ 0 & 1 & 1 \end{pmatrix} \quad \text{and} \quad \begin{pmatrix} 2 & 2 & 1 \\ 2 & 1 & 2 \\ 1 & 2 & 2 \end{pmatrix}$$

are inverse matrices (mod 3), the two methods of construction again correspond to the two aspects of a transformation of co-ordinates.

*Arnoux, *Arithmétique graphique*, p. 63.

010	102	221	201	020	112	122	211	000
101	220	012	022	111	200	210	002	121
222	011	100	110	202	021	001	120	212

Figure xxxvi.   An Eulerian Cube, $n = 3$

232	310	032	110	013	131	213	331	230	312	030	112	011	133	211	333
020	102	220	302	201	323	001	123	022	100	222	300	203	321	003	121
212	330	012	130	033	111	233	311	210	332	010	132	031	113	231	313
000	122	200	322	221	303	021	103	002	120	202	320	223	301	023	101

Figure xxxvii. $n = 4$

60	37	12	21,	7	26	55	42	57	40	9	24	6	27	54	43
13	20	61	36	50	47	2	31	16	17	64	33	51	46	3	30
56	41	8	25	11	22	59	38	53	44	5	28	10	23	58	39
1	32	49	48	62	35	14	19	4	29	52	45	63	34	15	18

Figure xxxviii.   A Pandiagonal Magic Cube, $n = 4$

(First layer)   (Second layer)

1	8	61	60	48	41	20	21
62	59	2	7	19	22	47	42
52	53	16	9	29	28	33	40
15	10	51	54	34	39	30	27
32	25	36	37	49	56	13	12
35	38	31	26	14	11	50	55
45	44	17	24	4	5	64	57
18	23	46	43	63	58	3	6

(Fourth layer)   (Third layer)

Figure xxxix.   A Magic Square which is also a Pandiagonal Cube

Figure xxxvii shows the cube $(122, 212, 221)_4$ (which is not properly Eulerian, since 2 divides 4). This becomes magic, in fact pandiagonal, when every digit 2 is replaced by 3, and vice versa. The result, in the ordinary notation, is shown in figure xxxviii. In fact, Margossian's substitution (page 208) gives a pandiagonal cube $(1\alpha\alpha, \alpha1\alpha, \alpha\alpha1)_{2\alpha}$ for every even value of α.

Figure xxxix, by R.V. Heath, is remarkable as being both a magic square of the eighth order and a pandiagonal cube of the fourth order. As a magic square, it has the interesting property that *alternate* numbers in any row or column, or in either diagonal, sum to 130. As a pandiagonal cube, it excels figure xxxviii in that the four horizontal layers (which are the four quarters of the eighth-order square) are themselves magic squares.

CHAPTER VIII

MAP-COLOURING PROBLEMS

This chapter and the next are concerned with the branch of mathematics known as Topology or Analysis Situs, which differs from Geometry in having no connection with the idea of straightness, flatness, or measurement. Here every oval is equivalent to a circle, every spheroid to a sphere; in fact, no distinction is made between any two figures derivable from one another by the kind of transformations that are familiar as crumpling and stretching, without tearing or joining. (One's thoughts turn naturally to indiarubber.) But topology does distinguish between a solid sphere and a hollow sphere, and between either of these and a *torus* or anchor-ring. It also distinguishes between a simple closed curve and a knotted curve, provided these are definitely understood to lie in the same three-dimensional space; in four-dimensional space, any such knot could be untied without breaking the circuit.

THE FOUR-COLOUR CONJECTURE

I shall first mention a famous conjecture which appears to be simple, and seems to be true, but has never yet been proved, although more complicated theorems of a similar type are fairly easy to prove. The conjecture is that *not more than four colours are necessary in order to colour a map of a country (divided into districts) in such a way that no two contiguous districts shall be of the same colour.* By contiguous districts are meant districts having a common *line* as part of their boundaries: districts which touch only at points are not contiguous in this sense. The map is drawn on a simply-connected surface, such as a plane or a sphere. The number of districts is finite, and no district consists of two or more disconnected pieces.

The map may or may not fill up the whole surface. (If the map is plane, and fills up the plane, one district at least must have an infinite area, since the number is finite.) Of course, some maps can be coloured with fewer than four colours; thus a chess-board requires only two, a hexagonal tessellation three.

In 1853, Francis Guthrie,* a graduate student at University College, London, was drawing a map of England. He noticed

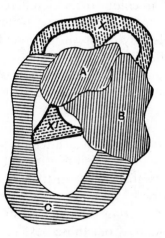

that four colours suffice for distinguishing the counties. Recognizing the possibility of a general theorem, he discussed the matter with his brother Frederick, who asked Professor De Morgan for an explanation. De Morgan was sufficiently interested to mention it in a letter to Sir William Hamilton, but then the conjectured theorem was forgotten till 1878, when Cayley† stated that he could not obtain a rigorous proof of it.

Probably the following argument, though not a formal demonstration, will satisfy the reader that the result is true. Let A, B, C be three contiguous districts, and let X be any other district contiguous with all of them. Then X must lie either wholly outside the external boundary of the area ABC or

*See *Proceedings of the Royal Society of Edinburgh*, July 19, 1880, vol. x, p. 728.

†*Proceedings of the London Mathematical Society*, 1878, vol. ix, p. 148, and *Proceedings of the Royal Geographical Society*, London, 1879, N.S., vol. i, pp. 259–261, where some of the difficulties are indicated.

wholly inside the internal boundary, that is, it must occupy a position either like X or like X'. In either case there is no possible way of drawing another area Y which shall be contiguous with A, B, C, and X. In other words, it is possible to draw on a plane four areas which are contiguous, but it is not possible to draw five such areas. If A, B, C are not contiguous, each with the other, or if X is not contiguous with A, B, and C, it is not necessary to colour them all differently, and thus the most favourable case is that already treated. Moreover any of the above areas may diminish to a point, and finally disappear without affecting the argument.

That we may require at least four colours is obvious from the above diagram, since in that case the areas A, B, C, and X would have to be coloured differently.

A proof of the proposition involves difficulties of a high order, which as yet have baffled all attempts to surmount them. This is partly due to the fact that if, using only four colours, we build up our map, district by district, and assign definite colours to the districts as we insert them, we can always contrive the addition of two or three fresh districts which cannot be coloured differently from those next to them, and which accordingly upset our scheme of colouring. But by starting afresh, it would seem that we can always rearrange the colours so as to allow of the addition of such extra districts.

The argument by which the truth of the proposition was formerly supposed to be demonstrated was given by A.B. Kempe* in 1879, but there is a flaw in it.

In 1880, Tait published a solution† depending on the theory of graphs. A *graph* (in this sense) is simply an arrangement of

---

*He sent his first demonstration across the Atlantic to the *American Journal of Mathematics*, 1879, vol. II, pp. 193–200; but subsequently he communicated it in simplified forms to the London Mathematical Society, *Proceedings*, 1879, vol. X, pp. 229–231, and to *Nature*, Feb. 26, 1880, vol. XXI, pp. 399–400. The flaw in the argument was pointed out by P.J. Heawood, *Quarterly Journal of Mathematics*, 1890, vol. XXIV, p. 337.

†*Proceedings of the Royal Society of Edinburgh*, July 19, 1880, vol. X, pp. 728–729; *Philosophical Magazine*, January 1884, series 5, vol. XVII, p. 41. See also Coxeter, *Journal of Recreational Mathematics*, 1961, vol. II, pp. 3–12, or *Leonardo*, 1971, vol. IV, pp. 273–277.

points, called *vertices*, certain pairs of which are joined by line-segments, called *edges* (not necessarily straight). By imagining the graph to be constructed in three-dimensional space, we can ensure that two edges never "cross" but only meet in a common vertex. An example is the set of vertices and edges of a polyhedron. Graphs occur also in electrical engineering and in the structural formulae of chemistry. The latter application makes it natural to use the term *valency* for the number of edges at a vertex, and so to call a graph *trivalent* if every vertex belongs to just three edges. In this special case the numbers of vertices and edges are evidently $2n$ and $3n$ for some n.

We shall restrict consideration to *connected* graphs, such that any vertex can be reached from any other by a path along a chain of consecutively adjacent edges. An edge is called an *isthmus* if its removal would make the graph fall apart into two disconnected pieces. A graph is said to be *planar* if it can be drawn on a plane or on a sphere (without extraneous intersections).

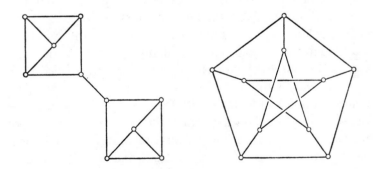

Tait believed that every trivalent graph could have its edges coloured with three colours in such a way that the three edges at any vertex all have different colours. It may be true that such a three-colouring is possible whenever the graph is planar and has no isthmus. Without these limitations it is false, as we can see by a glance at the accompanying figures, one with an isthmus and the other non-planar. In each case there are

10 vertices and 15 edges, but neither graph can be three-coloured.

Petersen* explained the difficulty as follows. He proved that any trivalent graph can be coloured with two colours, say red and green, so that each vertex belongs to one red and two green edges. The green edges form certain loops (polygons). What Tait apparently assumed is that the two-colouring could always be arranged so that each polygon would have an even number of sides, enabling him to colour these sides alternately blue and yellow, instead of all green.

Let us assume that the improved form of Tait's conjecture can in fact be justified, i.e., that every planar trivalent graph without an isthmus can be three-coloured. His argument that four colours will suffice for a map is divided into two parts and is as follows.

First, suppose that no point of the map is on the boundaries of more than three districts. Then the boundary lines may be regarded as the edges of a trivalent graph; also, by Tait's theorem, the boundary lines can be marked with three colours β, γ, δ, so that no two like colours meet at a point of junction. Suppose this done. Now take four colours, A, B, C, D, wherewith to colour the map. Paint one district with the colour A; paint the district adjoining A and divided from it by the line β with the colour B; the district adjoining A and divided from it by the line γ with the colour C; the district adjoining A and divided from it by the line δ with the colour D. Proceed in this way so that a line β always separates the colours A and B, or the colours C and D; a line γ always separates A and C, or D and B; and a line δ always separates A and D, or B and C. It is easy to see that, if we come to a district bounded by districts already coloured, the rule for crossing each of its boundaries will give the same colour; this also follows from the fact that, if we regard β, γ, δ as indicating certain operations, then an

*See Julius Petersen (1839–1910), *L'Intermédiaire des Mathématiciens*, vol. v, 1898, pp. 225–227; and vol. vi, 1899, pp. 36–38. Also *Acta Mathematica*, Stockholm, vol. xv, 1891, pp. 193–220. The simplest non-planar graph that *can* be given a Tait colouring has six vertices, say 1, 2, 3, 4, 5, 6, and nine edges, each joining an odd vertex to an even vertex.

operation like δ may be represented as equivalent to the effect of the two other operations β and γ performed in succession in either order. Thus for such a map the problem is solved.

In the second case, suppose that at some point four or more boundary lines meet, so that the valency of the graph is greater than 3. At any such point introduce a small district as indicated below: this will reduce the problem to the first case. The small district thus introduced may be coloured by the previous rule; but after the rest of the map is coloured this district will have served its purpose, it may be then made to contract without limit to a mere point and will disappear leaving the boundaries as they were at first.

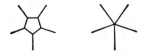

For a further discussion of the problem* it is desirable to develop the notion of reducing the map to a "standard" map, of simpler type than the original, and such that, if the reduced map can be coloured with four colours, so can the original. (*A fortiori*, if the reduced map can be coloured with five or more colours, so can the original.)

We suppose the map drawn on a sphere; if it does not already cover the whole sphere, we regard "the rest of the world" as one more district. Secondly, we reduce to vertices of valency three, not as above (since that method increases the number of districts), but by observing that a higher vertex must involve a pair of non-contiguous districts, which can be given the same

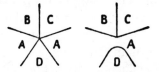

colour; we then open out the vertex, and merge these two into one district. Thirdly, we get rid of districts having one, two,

*Heawood, *Quarterly Journal of Mathematics*, 1890, vol. XXIV, p. 333. This simplified account of Heawood's paper is due to L.A. Pars.

or three sides. (We merely have to remove one side, and merge
the district into an adjacent district.) Fourthly, we get rid of

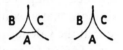

four-sided districts. (Of the four districts that surround such a
district, at least one pair are non-contiguous; this pair can be
merged with the four-sided district. If the map so formed can
be coloured with four colours, so can the original; for, when
we restore the four-sided district, only three colours at most

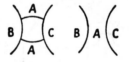

surround it, and we have a colour to spare for it.) Fifthly, we
get rid of ring-shaped districts, so that each district is bounded
by a single continuous line, and no district encloses one or
more others. (A ring-shaped district may be broken by a

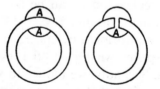

corridor joining one district inside to one outside, and these
two districts may be merged with the corridor.)

If we started with a fairly simple sort of map, the chances are
that we have reduced the map away, so that we are left with a
single district covering the sphere, and the theorem is proved
for that particular map. At worst we shall be left with a
"standard" map, in which no district has fewer than five sides.

Such a map may be regarded as a polyhedron, having F
faces (the districts), E edges (boundary lines separating pairs
of contiguous districts), and V vertices. A district having n
neighbours may reasonably be called an n-gon; thus the F

districts consist of n-gon for various of n. Since each vertex belongs to three districts, and each edge to two, we have

$$3V = 2E = \Sigma n,$$

where the Σ indicates summation over the F districts. By Euler's Formula $F - E + V = 2$ (which we shall prove on pages 232–233),

$$\Sigma(6-n) = 6F - \Sigma n = 6F - 3\Sigma n + 2\Sigma n = 6(F - E + V) = 12.$$

Thus $n < 6$ in at least one case: there is at least one pentagon.

We can now prove by induction that a standard map, and therefore any map, can be coloured with *six* colours. Consider a district having five sides, and merge it with one of the contiguous districts by removing one side. If the new map can be coloured with six colours, so can the original; for, when we restore the five-sided district, only five colours are adjacent to it, and we have a colour to spare for it. But we can reduce, step by step, and the six-colour theorem is proved. (The argument is rather subtler than it looks at first sight. For, when we remove the side, the new map will not necessarily be "standard," and we may have to reduce it by one of the above processes before we can apply the same argument again. I shall refer to this later as "the crude induction argument.")

By a subtler argument, likewise due to Heawood,* we can show that *five* colours are always sufficient. Consider a five-sided district, P. Of the districts Q, R, S, T, U which touch it, there must be at least one non-contiguous pair, say Q and S. Merge Q, P, S into a single district P' by deleting two sides of P. If the resulting map can be coloured with five colours, so can the original; for the districts P', R, T, U use up at most four colours, and there is a colour left for P when we return to the original map. Again we reduce, step by step, until the five-colour theorem is proved.

But this is as far as we can go. The gap between the five colours that are always sufficient and the four that are generally necessary has never been bridged, save for small values of F.

*Ibid., p. 337.

Since $6-n$ is zero or negative whenever $n > 5$, the above equation

$$\Sigma(6-n) = 12$$

tells us that every standard map contains (at least) twelve pentagons. The simplest standard map is the dodecahedron. We can colour this with four colours. It follows that every map having not more than 12 districts can be coloured with four colours. In other words, any map that requires five colours must have at least 13 districts, including at least 12 pentagons.

These numbers have been improved by various investigators.[*] In 1976, Jean Mayer showed that a map requiring five colours must have at least 96 districts. In that same year, K. Appel and W. Haken caused a sensational stir by announcing a computer-generated "proof" of the four-colour theorem; however, many mathematicians still feel that more confirmation is needed.

In his second paper[†] Heawood showed how the problem may be reduced to purely number-theoretical considerations. Consider again Tait's graph of $3n$ edges joining pairs of $2n$ vertices (page 226). Through each vertex pass three edges, marked β, γ, δ. We may distribute the vertices into two classes – say "positive" and "negative" – according as the incident edges β, γ, δ occur in a counter-clockwise or clockwise sense. Then every district has a number of positive vertices, and a number of negative vertices, which differ by a multiple of 3. For, in the succession of sides of a district, taken clockwise, each positive vertex takes us one step onwards, and each negative vertex one step backwards, in the cyclic order β, γ, δ; and when we get round to the starting-point, the difference between the numbers of onward and backward steps must be a multiple of 3. Conversely, given such a distribution of the vertices into

[*] *Transactions of the American Mathematical Society,* 1922, vol. xliv, pp. 225–236; *Annals of Mathematics,* 1927, series 2, vol. xxviii, pp. 1–15; *Bulletin of the American Mathematical Society,* 1936, vol. xlii, p. 491; *Historia Mathematica,* 1976, vol. iii, p. 60.

[†]*Quarterly Journal of Mathematics,* 1897, vol. xxix, pp. 277–278.

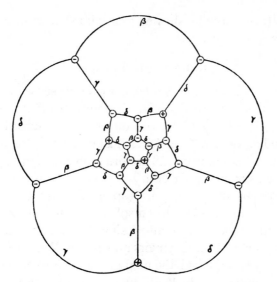

two classes, the edges can be consistently marked β, γ, δ, and then Tait's argument shows that the given map can be coloured with four colours.

The description of the vertices as positive or negative is equivalent to associating with them $2n$ variables x_1, x_2, ..., x_{2n}, having the values ± 1. The colouring problem thus reduces to the solution of a system of $n+2$ $(=F)$ congruences of the form

$$x_a + x_b + \ldots \equiv 0 \ (\text{mod } 3),$$

without any zeros. Each of the $2n$ $(=V)$ variables occurs in just three of the congruences. The question now is whether every such system is soluble.

Considering "extended" congruences of the form

$$x_a + x_b + \ldots \equiv \rho \ (\text{mod } 3),$$

where ρ may take any of the values 0, 1, 2, Heawood found* that, as the number n increases, the number of "failures" (i.e. of insoluble systems of congruences) diminishes rapidly.

*Proceedings of the London Mathematical Society, 1932, series 2, vol. XXXIII, pp. 253–286.

But in a later paper he showed that failures (for the extended congruences) always occur.

UNBOUNDED SURFACES

A surface is said to be *orientable* if a positive sense of rotation can be assigned consistently at all points of it.* A surface in ordinary space is orientable or non-orientable according as it is two-sided or one-sided. For instance, the "paradromic rings" described on page 128 are orientable or non-orientable according as m is even or odd.

In order to investigate the topology of an unbounded† surface (such as the surface of a sphere or a torus), we cover it with a map, i.e. we divide it into F simply connected "districts" by means of E arcs, joining (pairs of) V points. When the surface is crumpled or stretched, the map is supposed to go with it, so that the numbers F, E, V are unaltered. I proceed to prove that the number $F - E + V$ is a property of the surface itself, and not merely of a particular map, i.e., that if a second map on the same surface has F' districts, E' lines, and V' points, then $F' - E' + V' = F - E + V$. For this purpose let us superpose the two maps,‡ allowing the lines of each to break up the districts of the other, so as to make a third map, having (say) f districts, e lines, and v points. The v vertices of this third map consist of the $V + V'$ vertices of the other two,§ together with the points where their lines cross.

Let us modify the first map by admitting these crossing-points as vertices, and consequently breaking up the lines on which they lie. Since E and V are equally increased, the value of $F - E + V$ is unchanged. The remaining lines and vertices

---

*To be more precise, the positive sense of rotation may be defined at any point by a "directrix," that is, a small circle with an arrow-head marked on its circumference. A surface is *non-orientable* if there can be found on it a closed path such that the directrix is reversed when moved around this path.

†I.e. without any periphery (or "rim," or "edge"). For an interesting *bounded* surface, see Steinhaus, *Mathematical Snapshots*, New York, 1938, pp. 117, 233, 237.

‡This method of proof was suggested to me by Professor J.W. Alexander.

§This is no loss of generality in supposing these $V + V'$ points to be distinct, since a slight distortion would shift any vertex of either map.

of the third map may now be added in the form of successive "chains," each joining two given vertices and dividing a given district into two parts. Such a chain consists of (say) n new lines meeting consecutively at $n-1$ new vertices ($n \geq 1$). Its insertion therefore increases F by 1, E by n, and V by $n-1$, leaving $F - E + V$ unchanged. Continuing in this way until the third map is complete, we thus find that $f - e + v = F - E + V$. Similarly, $f - e + v = F' - E' + V'$. Hence $F' - E' + V' = F - E + V$, as required.

The invariant $\chi = F - E + V$ is called the *characteristic** of the surface. We can find it for any given surface by investigating a conveniently simple map on that surface. For instance, in the case of the sphere we can use the tetrahedron, for which $F = V = 4$ and $E = 6$, with the conclusion that $\chi = 2$. We have thus proved Euler's Formula

$$F - E + V = 2$$

which we used on pages 132 and 229.

From any given surface, we can derive a topologically different surface by adding a "handle," which may be thought of as a bent prism connecting two separate n-gons of a map on the given surface. Such a prism has $2n$ vertices, all of which belong to the original map, $3n$ edges, of which all but n belong to the original map, and $n + 2$ faces, of which 2 (the bases) are the faces of contact, which are supposed to be removed. Hence the operation of adding the handle leaves V unchanged, increases E by n, and increases F by $n-2$; altogether, it diminishes the characteristic by 2.

The most general (unbounded) orientable surface may be regarded as a sphere with p handles. The number p is called the *genus* of the surface. Thus the sphere is of genus zero; the torus, of genus one. The above argument shows that the characteristic is equal to $2 - 2p$. Thus every orientable surface has an even characteristic.

*It "characterizes" the surface in the following sense. Two unbounded surfaces, both orientable or both non-orientable, which have the same characteristic, are homeomorphic (or topologically equivalent). Some authors prefer to change the sign, i.e. to define the characteristic as $-V + E - F$.

The characteristic of a non-orientable surface may be either even or odd (but cannot be greater than 1). In ordinary space there is no unbounded non-orientable surface that does not cut itself.

DUAL MAPS

Given any map (covering an unbounded surface), we can define a *dual* map (covering the same surface), each of whose vertices lies within a corresponding district of the first map, while each of its lines crosses a corresponding line of the first map. (For an example, see page 106.) Clearly, the vertices of the first map lie in separate districts of the second, and the relationship between the two maps is symmetrical. The dualizing process replaces the numbers F, E, V by V, E, F, respectively. The dual of a "standard" map has triangular districts throughout. Reciprocal polyhedra may be regarded as a special case of dual maps.

The great dodecahedron, $\{5, \frac{5}{2}\}$ (page 145), is a map of twelve pentagons on a surface of genus 4, since $F - E + V = 12 - 30 + 12 = 2 - 8$. This map being self-dual (or rather, dual to an identical map), the polyhedra $\{5, \frac{5}{2}\}$ and $\{\frac{5}{2}, 5\}$ are topologically equivalent (or "homeomorphic"). In the same sense, the polyhedra $\{3, \frac{5}{2}\}$ and $\{\frac{5}{2}, 3\}$ are equivalent to the ordinary icosahedron and dodecahedron, respectively.

The rotating ring of tetrahedra (page 154) provides a map of $4n$ triangles on a torus. The genus is 1, since $F - E + V = 4n - 6n + 2n = 0$.

MAPS ON VARIOUS SURFACES

The simplest non-orientable surface is the projective plane, which can be regarded as a sphere with antipodes identified, or as a disc with diametrically opposite points identified. The simplest map on this surface is obtained by drawing a diameter so as to divide the disc into two districts (digons). Since $F = E = 2$ and $V = 1$, the characteristic is $\chi = 1$. The following more interesting map (with $F = 6$, $E = 15$, $V = 10$) can be derived from the dodecahedron by identifying antipodes.

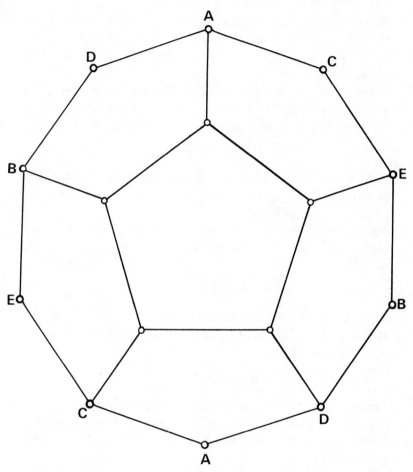

For a standard map on any unbounded surface, the argument used on page 229 shows that

$$\Sigma(6-n) = 6\chi.$$

This is positive when $\chi = 1$, just as it was when $\chi = 2$. Thus the "crude induction argument," which proves that every map on a sphere can be coloured with 6 colours, suffices to prove that *every map on the projective plane can be coloured with 6 colours.* The above "hemi-dodecahedron," in which every two of the six faces are adjacent, provides an example to show that

6 colours are *necessary*; and, as we have just seen, 6 colours are always *sufficient*. There is thus no gap waiting to be filled: the colouring problem for the projective plane is completely solved.

This is the simplest case ($\chi = 1$) of Heawood's "Map-Colour Theorem" (1890): *To colour any map on an unbounded surface of characteristic $\chi < 2$ requires at most $[N]$ colours, where*

$$N = \tfrac{1}{2}(7 + \sqrt{49 - 24\chi})$$

and $[N]$ means the greatest integer not exceeding N. (Actually he wrote k for $-\chi$.) His subtle and condensed argument may be paraphrased as follows.

It is not necessary to "standarize" the map; all we require is that each vertex should belong to at least three districts, so that

$$3V \leqq 2E = \Sigma n.$$

Since $\chi = F - E + V$, we have $E \leqq 3(E - V) = 3(F - \chi)$. Notice also that N is the positive root of the quadratic equation $N^2 - 7N + 6\chi = 0$ or

$$6(1 - \chi/N) = N - 1.$$

We are now ready to prove the following lemma: *Every map on a surface of characteristic $\chi < 2$ includes at least one district having fewer than $[N]$ neighbours*; that is, at least one n-gon with $n < [N]$.

Since this is obvious when $F \leqq [N]$, we may restrict consideration to a map having more than $[N]$ districts, so that $F > N$. Since the case when $\chi = 1$ was disposed of on page 235, we may now assume that $\chi \leqq 0$. It follows that

$$\Sigma n = 2E \leqq 6(F - \chi) = 6(1 - \chi/F)F \leqq 6(1 - \chi/N)F$$
$$= (N - 1)F < [N]F,$$

whereas, if it were true that every $n \geqq [N]$, we would have $\Sigma n \geqq [N]F$. Thus the lemma has been proved.

For a proof of the Map-Colour Theorem itself (asserting that, if $\chi < 2$, $[N]$ colours always suffice), we make the inductive assumption that it holds for every map of $F - 1$ districts, and then we consider a given F-district map, paying special attention to one district having fewer than $[N]$ neighbours. Modifying the given map by making this district shrink to a single point, we obtain a new map which, having only $F - 1$ districts, can be coloured with $[N]$ colours. Let this be done, and let the same colouring be applied to the original map, that is, to all its districts except the special one. Then, even if the neighbours all have different colours there are at most $[N] - 1$ of them, and we still have an $[N]$th colour left for the special district itself. Since this argument can be applied when $F = [N] + 1$, then with $[F] = [N] + 2$, and so on, the $[N]$-colour theorem has now been established for all values of F.

For instance, on the torus, which is the orientable surface with $\chi = 0$, seven colours suffice, and it is easy to draw a seven-colour map consisting of seven mutually contiguous hexagons.

On page 128 we considered an amusing surface called a *Möbius strip*. This is a bounded surface, its boundary being a single circuit, that is, a topological circle, just like the boundary of a disc. Theoretically (though not in ordinary space!) we can derive an unbounded surface by fitting a disc to this boundary. The unbounded surface so obtained is easily recognized to be a projective plane,* for which $\chi = 1$. Another possibility is to fit two Möbius strips together along their common boundary; the unbounded surface so obtained is the *Klein bottle,†* which is the non-orientable surface with $\chi = 0$. As this surface has the same characteristic as the torus, Heawood's theorem tells us that its maps require at most seven colours. In this single case we can go farther: it was proved by P. Franklin‡ that *every map on the Klein bottle can be coloured with six colours.* This is the only case in which Heawood's number $[N]$ is not both necessary and sufficient. With a prodigious amount

---

*See, for instance, Coxeter, *Introduction to Geometry*, p. 383.

†D. Hilbert and S. Cohn-Vossen, *Geometry and the Imagination*, New York, 1952.

‡*Journal of Mathematics and Physics*, 1934, vol. XIII, pp. 363–369. (A particularly interesting paper.)

of labour, Gerhard Ringel and J.W.T. Youngs* succeeded in proving that *every unbounded surface except the Klein bottle carries a map of* $[N]$ *mutually contiguous districts.*

PITS, PEAKS, AND PASSES

Returning to the sphere (for which $\chi = 2$ and $N = 4$), let us briefly consider an application of Euler's Formula to the geographical problem† of relating the numbers of pits, peaks, and passes on a dry planet. A *peak* (or summit) is a point of locally maximum altitude, and a *pit* (or bottom) is a point of locally minimum altitude. Each point of either kind is surrounded by a nested system of contour lines (of constant altitude). A *pass* is a point where a contour line crosses itself. Bisecting the angles formed by the crossing of the contour line, we find four directions, in each of which we can proceed along a line of (greatest) slope, either upward to a peak along a *watershed* or downward to a pit along a *watercourse*. Clearly, the peaks and the pairs of opposite watersheds may be regarded as the vertices and edges of a map whose districts (the "dales") each surround a pit. Hence, if there are F pits, E passes, and V peaks,

$$F - E + V = 2.$$

Similarly, the pits and the pairs of opposite watercourses are the vertices and edges of the dual map, whose districts (the "hills") each surround a peak.

COLOURING THE ICOSAHEDRON

The faces of a tetrahedron can be map-coloured with four given colours in two enantiomorphous ways, and those of a dodecahedron‡ in four ways, consisting of two enantiomorphous pairs. The faces of an octahedron or of a cube can be coloured

---

Proceedings of the National Academy of Sciences, U.S.A., 1968, vol. LX, pp. 438–445.

†A. Cayley and J. Clerk Maxwell, *Philosophical Magazine*, 1859, series 4, vol. XVIII, pp. 264–268, and 1870, series 4, vol. XL, pp. 421–427. See also D.A. Moran, *American Mathematical Monthly*, 1970, vol. LXXVII, p. 1096.

‡ See the elegant self-unfolding model at the end of Steinhaus's *Mathematical Snapshots*.

with two or three colours, respectively, in just one way. It is therefore interesting to observe that the faces of an icosahedron can be coloured with three given colours in as many as 144 ways. This number seems to have been first obtained by J.M. Andreas.

It is easily found to be impossible to colour the icosahedron in such a way that no face is surrounded by three that are all coloured alike. In fact, there are always just two such faces. Let the three colours be white, black, and grey, and suppose that one black face is surrounded by three white faces. Between each pair of these there occur two further faces, which must be black and grey (since they are contiguous). Beyond these, again, occur two more faces, which may be grey and black, white and black, grey and white, or both white. Let these four possibilities be denoted by a, b, c, d, respectively, with or without a dash (') according as the first-mentioned "black and grey" occur clockwise or counterclockwise when we go round the original black face. The colouring (of $1+3+6+6$ of the twenty faces) may now be represented by a symbol consisting of three of the letters a, b, c, d (repetitions allowed), with a certain distribution of dashes. The colouring is unaltered by cyclic permutation of the three letters, and it is natural to let the same symbol describe any new colouring that may result from permuting the colours. Thus, in the above definitions for a, b, c, d, *black* merely means the colour of the special face, *white* the colour of the three surrounding faces, and *grey* the remaining colour.

Given any colouring, we can derive another by making the special face grey instead of black. This is equivalent to leaving it black, and interchanging black and grey everywhere else, which involves changing b into c, c into b, and adding or removing dashes. (Thus a^2b becomes a'^2c'.)

In the twelve cases depicted above, the colouring of the first sixteen faces unequivocally determines that of the remaining four. (The icosahedron has been projected radially on its circum-sphere, and then stereographically on a plane. One vertex has been projected to infinity. In each case, the special black face is the one just above the centre of the drawing.)

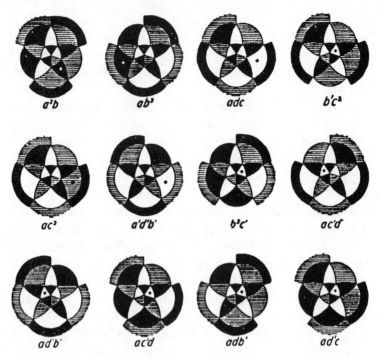

a²b   ab³   adc   b'c²

ac²   a'd'b'   b²c'   ac'd

ad'b'   ac'd   adb'   ad'c

I have already remarked that there is always a second face surrounded by three that are all coloured alike. (In the drawings, this face is distinguished by a spot.) After a suitable permutation of the colours, this will be a black face surrounded by white faces, and then an alternative symbol can be assigned, to describe the colouring from this second point of view. This change of aspect leaves a^2b, ab^2, adc, and $b'c^2$ unaltered, but interchanges ac^2 and $a'd'b'$, b^2c' and $ac'd'$, $ad'b'$ and $ac'd$, adb' and $ad'c$. Since neither of the two special faces takes precedence over the other, it is desirable to combine the two symbols, and to describe the colouring types as $(a^2b)^2$, $(ab^2)^2$, $(adc)^2$, $(b'c^2)^2$, $(ac^2)(a'd'b')$, $(b^2c')(ac'd')$, $(ad'b')(ac'd)$, $(adb')(ad'c)$.

Each of these eight colouring types gives rise to another by reflection in a mirror. The new symbol is derived by inserting or removing dashes, and reversing the cyclic order of the letters in each triad. Thus $(b^2c')(ac'd')$ leads to the enantiomorph $(b'^2c)(a'dc)$.

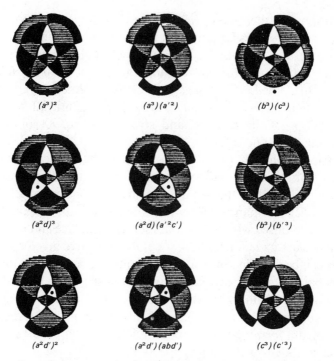

$(a^3)^2$ $(a^3)(a'^2)$ $(b^3)(c^3)$

$(a^2d)^3$ $(a^2d)(a'^2c')$ $(b^3)(b'^3)$

$(a^2d')^2$ $(a^2d')(abd')$ $(c^3)(c'^3)$

In the nine cases depicted above, the colouring of the
first 16 faces does not determine that of the remaining four;
but the ambiguity is removed by the use of double symbols.

In order to show that these and their enantiomorphs cover
all possibilities, we merely have to make a list of all cyclic
triads of letters which do not include any of the following
consecutive pairs: aa', $a'a$, ab', ba', $a'c$, $c'a$, bb', $b'b$, cc', $c'c$, bc,
$c'b'$, $b'c'$, cb, $b'd$, $d'b$, $b'd'$, db, cd, $d'c'$, cd', dc', d^2, dd', $d'd$, d'^2. For,
each of these pairs leads to two adjacent faces having the same
colour. For the same reason, the triads acb' and $a'bc'$ must
also be ruled out. The remaining triads are just those which
have been considered above.

The actual enumeration now proceeds as follows. Each of
the 14 colourings $(a^3)^2$, $(a'^3)^2$, $(a^2b)^2$, $(a'^2b')^2$, $(ab^2)^2$, $(a'b'^2)^2$,
$(a^2d)^2$, $(a'^2d')^2$, $(adc)^2$, $(a'c'd')^2$, $(a^2d')^2$, $(a'^2d)^2$, $(b'c^2)^2$, $(bc'^2)^2$ is
unchanged by transposing a certain pair of colours; therefore
these (by cyclic permutation of the colours) give rise to 42

solutions. In the case of $(b^3)(b'^3)$ or of $(c^3)(c'^3)$, transposition of a pair of colours is equivalent to reflection in a mirror; these give rise to 12 solutions. Finally, the 15 colourings $(a^3)(a'^3)$, $(b^3)(c^3)$, $(b'^3)(c'^3)$, $(a^2d)(a'^2c')$, $(a'^2d')(a^2c)$, $(ac^2)(a'd'b')$, $(a'c'^2)$ (abd), $(a^2d')(abd')$, $(a'^2d)(a'db')$, $(b^2c')(ac'd')$, $(b'^2c)(a'dc)$, $(ad'b')$ $(ac'd)$, $(a'bd)(a'd'c)$, $(adb')(ad'c)$, $(a'bd')(a'c'd)$ admit all six permutations of the colours, and so give rise to 90 solutions, making the grand total $42 + 12 + 90 = 144$.

The six solutions $(a^3)(a'^3)$ are special, in that reflection leaves them entirely unaltered. We may call these six *reflexible* solutions, and describe the rest as 69 enantiomorphous pairs.

L.B. Tuckerman remarks that the faces of an icosahedron can be coloured with five colours so that each face and its three neighbours have four different colours. For five given colours this can be done in four ways, consisting of two enantiomorphous pairs. One pair can be derived from the other by making any odd permutation of the five colours (e.g. by interchanging two colours). The faces coloured alike belong to five regular tetrahedra forming the compound described on page 135. (Cf. Coxeter, *Regular Polytopes*, pp. 50, 106.)

CHAPTER IX

UNICURSAL PROBLEMS

I propose to consider in this chapter some problems which arise out of the theory of unicursal curves. I shall commence with *Euler's Problem and Theorems*, and shall apply the results briefly to the theories of *Mazes* and *Geometrical Trees*. The reciprocal unicursal problem of the *Hamilton Game* will be discussed in the latter half of the chapter.

EULER'S PROBLEM

Euler's problem has its origin in a memoir* presented by him in 1736 to the St. Petersburg Academy, in which he solved a question then under discussion, as to whether it was possible from any point in the town of Königsberg to take a walk in such a way as to cross every bridge in it once and only once and return to the starting point.

The town is built near the mouth of the river Pregel, which there takes the form indicated below and includes the island of Kneiphof. In the eighteenth century there were seven bridges in the positions shown in the diagram, and it is easily seen that with such an arrangement the problem is insoluble. (Since then, two more bridges have been built). Euler, however, did not confine himself to the case of Königsberg, but discussed the general problem of any number of islands connected in any way by bridges. It is evident that the question will not be affected if we suppose the islands to diminish to points and the bridges to lengthen out. In this way we ultimately obtain

*"Solutio problematis ad Geometriam situs pertinentis." *Commentarii Academiae Scientiarum Petropolitanae* for 1736, Leningrad, 1741, vol. VIII, pp. 128–140. This has been translated into French by M. Ch. Henry; see Lucas, vol. I, part 2, pp. 21–33.

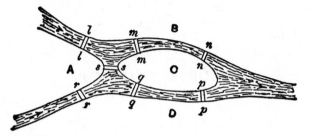

a geometrical figure or network. In the Königsberg problem this figure is of the shape indicated below, the areas being represented by the points *A*, *B*, *C*, *D*, and the bridges being represented by the lines *l*, *m*, *n*, *p*, *q*, *r*, *s*.

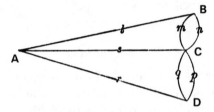

Euler's problem consists therefore in finding whether a given geometrical figure can be described by a point moving so as to traverse every line in it once and only once. A more general question is to determine how many strokes are necessary to describe such a figure so that no line is traversed twice: this is covered by the rules hereafter given. The figure may be either in three or in two dimensions, and it may be represented by lines, straight, curved, or tortuous, joining a number of given points, or a model may be constructed by taking a number of rods or pieces of string furnished at each end with a hook so as to allow of any number of them being connected together at one point.

The theory of such figures is included as a particular case in the propositions proved by Listing in his *Topologie*.* I

*Göttinger Studien, 1847, part x. See also Tait on "Listing's *Topologie*," *Philosophical Magazine*, London, January 1884, series 5, vol. XVII, pp. 30–46; and *Collected Scientific Papers*, Cambridge, vol. II, 1900, pp. 85–98. The problem was discussed by J.C. Wilson in his *Traversing of Geometrical Figures*, Oxford, 1905.

shall, however, adopt here the methods of Euler, and I shall begin by giving some definitions, as it will enable me to put the argument in a more concise form.

A *node* (or isle) is a point to or from which lines are drawn. A *branch* (or bridge or path) is a line connecting two consecutive nodes. An *end* (or hook) is the point at each termination of a branch. The *order* of a node is the number of branches which meet at it. A node to which only one branch is drawn is a *free* node or a free end. A node at which an even number of branches meet is an *even* node: evidently the presence of a node of the second order is immaterial. A node at which an odd number of branches meet is an *odd* node. A figure is closed if it has no free end: such a figure is often called a closed network.

A *route* consists of a number of branches taken in consecutive order and so that no branch is traversed twice. A *re-entrant* route terminates at a point from which it started. A figure is described *unicursally* when the whole of it is traversed in one route.

The following are Euler's results. (i) In any network the number of odd nodes is even. (ii) A figure which has no odd node can be described unicursally, in a re-entrant route, by a moving point which starts from any point on it. (iii) A figure which has two and only two odd nodes can be described unicursally by a moving point which starts from one of the odd nodes and finishes at the other. (iv) A figure which has more than two odd nodes cannot be described completely in one route; to which Listing added the corollary that a figure which has $2n$ odd nodes, and no more, can be described completely in n separate routes. I now proceed to prove these theorems.

First. *The number of odd nodes in any network is even.*

Suppose the number of branches to be b. Therefore the number of hooks is $2b$. Let k_n be the number of nodes of the nth order. Since a node of the nth order is one at which n branches meet, there are n hooks there.

$$\therefore \quad k_1 + 2k_2 + 3k_3 + 4k_4 + \ldots + nk_n + \ldots = 2b.$$

Hence $\quad k_1 + 3k_3 + 5k_5 + \ldots \quad$ is even.

$$\therefore \quad k_1 + k_3 + k_5 + \ldots \quad \text{is even.}$$

Second. *A figure which has no odd node can be described unicursally in a re-entrant route.*

Since the route is to be re-entrant, it will make no difference where it commences. Suppose that we start from a node A. Every time our route takes us through a node we use up one hook in entering it and one in leaving it. There are no odd nodes, therefore the number of hooks at every node is even: hence, if we reach any node except A, we shall always find a hook which will take us into a branch previously untraversed. Hence the route will take us finally to the node A from which we started. If there are more than two hooks at A, we can continue the route over one of the branches from A previously untraversed, but in the same way as before we shall finally come back to A.

It remains to show that we can arrange our route so as to make it cover all the branches. Suppose each branch of the network to be represented by a string with a hook at each end, and that at each node all the hooks there are fastened together. The number of hooks at each node is even, and if they are unfastened, they can be re-coupled together in pairs, the arrangement of the pairs being immaterial. The whole network will then form one or more closed curves, since now each node consists merely of two ends hooked together.

If this random coupling gives us one single curve, then the proposition is proved; for, starting at any point, we shall go along every branch and come back to the initial point. But if this random coupling produces anywhere an isolated loop, L, then where it touches some other loop, M, say at the node P, unfasten the four hooks there (viz., two of the loop L and two of the loop M) and re-couple them in any other order: then the loop L will become a part of the loop M. In this way, by altering the couplings, we can transform gradually all the separate loops into parts of only one loop.

For example, take the case of three isles, *A*, *B*, *C*, each connected with both the others by two bridges. The most unfavourable way of re-coupling the ends at *A*, *B*, *C* would be to make *ABA*, *ACA*, and *BCB* separate loops. The loops *ABA* and *ACA* are separate and touch at *A*; hence we should

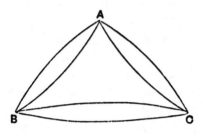

re-couple the hooks at *A* so as to combine *ABA* and *ACA* into one loop *ABACA*. Similarly, by re-arranging the couplings of the four hooks at *B*, we can combine the loop *BCB* with *ABACA*, and thus make only one loop.

I infer from Euler's language that he had attempted to solve the problem of giving a practical rule which would enable one to describe such a figure unicursally without knowledge of its form, but that in this he was unsuccessful. He, however, added that any geometrical figure can be described completely in a single route provided each part of it is described twice and only twice, for, if we suppose that every branch is duplicated, there will be no odd nodes and the figure is unicursal. In this case any figure can be described completely without knowing its form: rules to effect this are given below.

Third. *A figure which has two and only two odd nodes can be described unicursally by a point which starts from one of the odd nodes and finishes at the other odd node.*

This at once reduces to the second theorem. Let *A* and *Z* be the two odd nodes. Consider the new figure derived by adding an extra branch from *A* to *Z*. In this new figure, 'all the nodes are even, including *A* and *Z*; hence, by Euler's second proposition, it can be described unicursally, and, if the route begins at Z, it will end at Z. We may suppose ZA to be the first branch of the route. The effect of removing this branch, and so restoring the original figure, is just to make the route begin at A; but it still ends at Z.

Fourth. *A figure having 2n odd nodes, and no more, can be described completely in n separate routes.*

If any route starts at an odd node, and if it is continued until it reaches a node where no fresh path is open to it, this latter node must be an odd one. For every time we enter an even node there is necessarily a way out of it; and similarly every time we go through an odd node we use up one hook in entering and one hook in leaving, but whenever we reach it as the end of our route, we use only one hook. If this route is suppressed there will remain a figure with $2n - 2$ odd nodes. Hence n such routes will leave one or more networks with only even nodes. But each of these must have some node common to one of the routes already taken, and therefore can be described as a part of that route. Hence the complete passage will require n, and not more than n, routes. It follows, as stated by Euler, that, if there are more than two odd nodes, the figure cannot be traversed completely in one route.

The Königsberg bridges lead to a network with four odd nodes; hence, by Euler's fourth proposition, it cannot be described unicursally in a single journey, though it can be traversed completely in two separate routes.

The first and second diagrams figured below contain only even nodes, and therefore each of them can be described unicursally. The first of these is a pentagram $\{\frac{5}{2}\}$ (see page 144); the second is the so-called sign-manual of Mohammed, said to have been originally traced in the sand by the point of his scimitar without taking it off the ground or retracing any part of the figure – which, as it contains only even nodes, is possible. The third diagram is taken from Tait's article: it contains only two odd nodes, and can therefore be described unicursally if we start from one of them and finish at the other.

The polygon $\{\frac{5}{2}\}$ was used by the Pythagoreans as a sign, known as the triple triangle or pentagram star, by which they could recognize one another. It was considered symbolical of health, and probably the angles were denoted by the letters of the word ὑγίεια, the diphthong εἰ being replaced by a θ. Iambli-

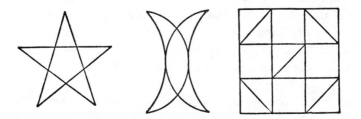

chus, who is our authority for this, tells us that a certain Pythagorean, when travelling, fell ill at a roadside inn where he had put up for the night; he was poor and sick, but the landlord, who was a kind-hearted fellow, nursed him carefully and spared no trouble or expense to relieve his pains. However, in spite of all efforts, the student got worse. Feeling that he was dying, and unable to make the landlord any pecuniary recompense, he asked for a board, on which he inscribed the pentagram star; this he gave to his host, begging him to hang it up outside so that all passers-by might see it, and assuring him that the result would recompense him for his charity. The scholar died and was honourably buried, and the board was duly exposed. After a considerable time had elapsed, a traveller one day riding by saw the sacred symbol; dismounting, he entered the inn, and after hearing the story, handsomely remunerated the landlord. Such is the anecdote, which, if not true, is at least well found.

As another example of a unicursal diagram, I may mention the geometrical figure formed by taking a $(2n+1)$-gon and joining every angular point to every other angular point. The edges of an octahedron also form a unicursal figure. On the other hand, a chess-board, divided as usual by straight lines into 64 cells, has 28 odd nodes: hence it would require 14 separate pen-strokes to trace out all the boundaries without going over any more than once. Again, the diagram on page 170 has 20 odd nodes, and therefore would require 10 separate pen-strokes to trace it out.

I turn next to discuss in how many ways we can describe a unicursal figure, all of whose nodes are even.*

Let us consider first how the problem is affected by a path which starts from a node A of order $2n$ and returns to it, forming a closed loop L. If this loop were suppressed, we should have a figure with all its nodes even, the node A being now of the order $2(n-1)$. Suppose the original figure can be described in N ways, and the reduced figure in N' ways. Then each of these N' routes passes $(n-1)$ times through A, and in any of these passages we could describe the loop L in either sense as a part of the path. Hence $N = 2(n-1)N'$.

Similarly if the node A on the original figure is of the order $2(n+l)$, and there are l independent closed loops which start from and return to A, we shall have

$$N = 2^l n(n+1)(n+2)\ldots(n+l-1)N',$$

where N' is the number of routes by which the figure obtained by suppressing these l loops can be described.

By the use of these results, we may reduce any unicursal figure to one in which there are no closed loops of the kind described above. Let us suppose that in this reduced figure there are k nodes. We can suppress one of these nodes, say A, provided we replace the figure by two or more separate figures each of which has not more than $k-1$ nodes. For suppose that the node A is of the order $2n$. Then the $2n$ paths which meet at A may be coupled in n pairs in $1.3.5\ldots(2n-1)$ ways and each pair will constitute either a path through A, or (in the special case where both members of the pair abut on another node B) a loop from A. This path or loop will form a portion of the route through A in which this pair of paths is concerned. Hence the number of ways of describing the

*See G. Tarry, *Association Française pour l'Avancement des Sciences*, 1886, pp. 49–53.

original figure is equal to the sum of the number of ways of describing $1.3.5\ldots (2n-1)$ separate simpler figures.

It will be seen that the process consists in successively suppressing node after node. Applying this process continually we finally reduce the figure to a number of figures without loops and in each of which there are only two nodes. If in one of these figures these nodes are each of the order $2n$, it is easily seen that it can be described in $2.(2n-1)!$ ways.

We know that a figure with only two odd nodes, A and B, is unicursal if we start at A (or B) and finish at B (or A). Hence the number of ways in which it can be described unicursally will be the same as the number required to describe the figure obtained from it by joining A and B. For if we start at A, it is obvious that at the B end of each of the routes which cover the figure we can proceed along BA to the node A whence we started.

This theory has been applied by Monsieur Tarry* to determine the number of ways in which a set of dominoes, running up to even numbers, can be arranged. This example will serve to illustrate the general method.

The 28 dominoes of an ordinary set are marked

$$
\begin{array}{llllllll}
6\text{--}6, & 6\text{--}5, & 6\text{--}4, & 6\text{--}3, & 6\text{--}2, & 6\text{--}1, & 6\text{--}0, \\
5\text{--}5, & 5\text{--}4, & 5\text{--}3, & 5\text{--}2, & 5\text{--}1, & 5\text{--}0, & 4\text{--}4, \\
4\text{--}3, & 4\text{--}2, & 4\text{--}1, & 4\text{--}0, & 3\text{--}3, & 3\text{--}2, & 3\text{--}1, \\
3\text{--}0, & 2\text{--}2, & 2\text{--}1, & 2\text{--}0, & 1\text{--}1, & 1\text{--}0, & 0\text{--}0.
\end{array}
$$

Dominoes are used in various games, in most, if not all, of which the pieces are played so as to make a line such that consecutive squares of adjacent dominoes are marked alike. Thus if 6–3 is on the table, the only domino which can be placed next to the 6 end is 6–6, 6–5, 6–4, 6–2, 6–1, or 6–0. Similarly the domino 3–5, 3–4, 3–3, 3–2, 3–1, or 3–0 can be placed next to the 3 end. Assuming that the doubles are played in due course, it is easy to see that such a set of dominoes will form a

*See the second edition of the French translation of this work, Paris, 1908, vol. ii, pp. 253–263; see also Lucas, vol. iv, pp. 145–150.

closed circuit.* We want to determine the number of ways in which such a line or circuit can be formed.

Let us begin by considering the case of a set of 15 dominoes marked up to double-four. Of these 15 pieces, 5 are doubles. The remaining 10 dominoes may be represented by the sides and diagonals of a regular pentagon 01, 02, etc. The intersections of the diagonals do not enter into the representation, and accordingly are to be neglected. Omitting these from our consideration, the figure formed by the sides and diagonals of the pentagon has five even nodes, and therefore is unicursal. Any unicursal route (e.g. 0–1, 1–3, 3–0, 0–2, 2–3, 3–4, 4–1, 1–2, 2–4, 4–0) gives one way of arranging these 10 dominoes. Suppose there are a such routes. In any such route we may put

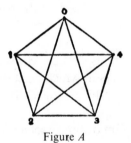

Figure A

each of the five doubles in either of two positions (e.g. in the route given above the double-two can be put between 0–2 and 2–3 or between 1–2 and 2–4). Hence the total number of unicursal arrangements of the 15 dominoes is $2^5 a$. If we arrange the dominoes in a straight line, then as we may begin with any of the 15 dominoes, the total number of arrangements is $15 . 2^5 . a$.

We have next to find the number of unicursal routes of the pentagon delineated above in figure A. At the node 0 there are four paths which may be coupled in three pairs. If 0 1 and 0 2 are coupled, as also 0 3 and 0 4, we get figure B. If 0 1 and 0 3 are coupled, as also 0 2 and 0 4, we get figure C. If 0 1 and 0 4 are coupled, as also 0 2 and 0 3, we get figure D. Let us denote the number of ways of describing figure B by b, of describing

---

*Hence if we remove one domino, say 5–4, we know that the line formed by the rest of the dominoes must end on one side in a 5 and on the other in a 4.

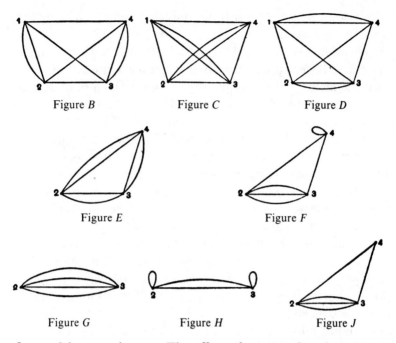

Figure B   Figure C   Figure D

Figure E   Figure F

Figure G   Figure H   Figure J

figure C by c, and so on. The effect of suppressing the node 0 in the pentagon A is to give us three quadrangles, B, C, D. And, in the above notation, we have $a = b + c + d$.

Take any one of these quadrangles, for instance D. We can suppress the node 1 in it by coupling the four paths which meet there in pairs. If we couple 1 2 with the upper of the paths 1 4, as also 1 3 with the lower of the paths 1 4, we get the figure E. If we couple 1 2 with the lower of the paths 1 4, as also 1 3 with the upper of the paths 1 4, we again get the figure E. If we couple 1 2 and 1 3, as also the two paths 1 4, we get the figure F. Then, as above, $d = 2e + f$. Similarly $b = 2e + f$, and $c = 2e + f$. Hence $a = b + c + d = 6e + 3f$.

We proceed to consider each of the reduced figures E and F. First take E, and in it let us suppress the node 4. For simplicity of description, denote the two paths 4 2 by β and β', and the two paths 4 3 by γ and γ'. Then we can couple β and γ, as also β' and γ', or we can couple β and γ', as also β' and γ: each of these couplings gives the figure G. Or we can couple β and β', as also γ and γ': this gives the figure H. Thus $e = 2g + h$.

Each of the figures G and H has only two nodes. Hence by the formulae given above, we have $g = 2.3.2 = 12$, and $h = 2.2.2 = 8$. Therefore $e = 2g + h = 32$. Next take the figure F. This has a loop at 4. If we suppress this loop we get the figure J, and $f = 2j$. But the figure J, if we couple the two lines which meet at 4, is equivalent to the figure G. Thus $f = 2j = 2g = 24$. Introducing these results we have $a = 6e + 3f = 192 + 72 = 264$, and therefore $N = 15.2^5 . a = 126720$. This gives the number of possible arrangements in line of a set of 15 dominoes. In this solution we have treated an arrangement from right to left as distinct from one which goes from left to right: if these are treated as identical we must divide the result by 2. The number of arrangements in a closed ring is $2^5 a$, that is 8448.

We have seen that this number of unicursal routes for a pentagon and its diagonals is 264. Similarly the number for a heptagon is $h = 129,976320$. Hence the number of possible arrangements in line of the usual set of 28 dominoes, marked up to double-six, is $28.3^7 . h$, which is equal to 7,959229,931520. The number of unicursal routes covering a polygon of nine sides is $n = 2^{17} . 3^{11} . 5^2 . 7 . 11 . 40787$. Hence the number of possible arrangements in line of a set of 45 dominoes marked up to double-eight is $45.4^9 . n$.*

MAZES

Everyone has read of the labyrinth of Minos in Crete and of Rosamund's Bower. A few modern mazes exist here and there – notably one, a very poor specimen of its kind, at Hampton Court – and in one of these, or at any rate on a drawing of one, most people have at some time threaded their way to the interior. I proceed now to consider the manner in which any such construction may be completely traversed even by one who is ignorant of its plan.

The theory of the description of mazes is included in Euler's theorems given above. The paths in the maze are what previously we have termed branches, and the places where two

---

*These numerical conclusions have also been obtained by algebraical methods: see M. Reiss, *Annali di Matematica*, Milan, 1871, vol. v, pp. 63–120.

or more paths meet are nodes. The entrance to the maze, the end of a blind alley, and the centre of the maze are free ends and therefore odd nodes.

If the only odd nodes are the entrance to the maze and the centre of it – which will necessitate the absence of all blind alleys – the maze can be described unicursally. This follows from Euler's third proposition. Again, no matter how many odd nodes there may be in a maze, we can always find a route which will take us from the entrance to the centre without retracing our steps, though such a route will take us through only a part of the maze. But in neither of the cases mentioned in this paragraph can the route be determined without a plan of the maze.

A plan is not necessary, however, if we make use of Euler's suggestion, and suppose that every path in the maze is duplicated. In this case we can give definite rules for the complete description of the whole of any maze, even if we are entirely ignorant of its plan. Of course, to walk twice over every path in a labyrinth is not the shortest way of arriving at the centre, but, if it is performed correctly, the whole maze is traversed, the arrival at the centre at some point in the course of the route is certain, and it is impossible to lose one's way.

I need hardly explain why the complete description of such a duplicated maze is possible, for now every node is even, and hence, by Euler's second proposition, if we begin at the entrance we can traverse the whole maze; in so doing we shall at some point arrive at the centre, and finally shall emerge at the point from which we started. This description will require us to go over every path in the maze twice, and as a matter of fact the fact the two passages along any path will be always made in opposite directions.

If a maze is traced on paper, the way to the centre is generally obvious, but in an actual labyrinth it is not so easy to find the correct route unless the plan is known. In order to make sure of describing a maze without knowing its plan, it is necessary to have some means of marking the paths which we traverse and the direction in which we have traversed them – for example, by drawing an arrow at the entrance and end of every path traversed, or better perhaps by marking the wall on the right-hand side, in which case a path may not be entered when there is a mark on each side of it.

Of the various practical rules for threading a maze those enunciated by G. Tarry* seem to be the simplest. These I proceed to explain. After traversing a path PQ to a node Q, proceed (if you can) along any other path QR not previously traversed in the direction away from Q. If all the paths from Q, except QP, have been traversed in that direction, or if Q is the closed end of a blind alley, then return to P along the path QP.

When these rules have been followed consistently, you will have returned to your starting point and you will have traversed every path just once in each direction. (Notice that the maze is not necessarily two-dimensional. The same rules will enable you to escape from the Catacombs, provided you have a light to show you your own footprints in the dust.)

Few if any mazes of the type I have been considering (namely, a series of interlacing paths through which some route can be obtained leading to a space or building at the centre of the maze) existed in classical or medieval times. One class of what the ancients called mazes or labyrinths seems to have comprised any complicated buildings with numerous vaults and passages.† Such a building might be termed a labyrinth, but it is not what is now usually understood by the word. The above rules would enable anyone to traverse the whole of any structure of this kind. I do not know if there are any accounts or descriptions of Rosamund's Bower other than those by Drayton, Bromton, and Knyghton: in the opinion of some,

---

*Nouvelles Annales de Mathématiques, 1895, series 3, vol. XIV, pp. 187–190. See also Dénes König, Theorie der endlichen und unendlichen Graphen, New York, 1950, pp. 41–43.

†For instance, see the descriptions of the labyrinth at Lake Moeris given by Herodotus, bk. ii, c. 148; Strabo, bk. xvii, c. 1, art. 37; Diodorus, bk. i, cc. 61, 66; and Pliny, Hist. Nat., bk. xxxvi, c. 13, arts. 84–89. On these and other references see A. Wiedemann, Herodots zweites Buch, Leipzig, 1890, p. 522 et seq. See also Virgil, Aeneid, bk. v, c. v, 588; Ovid, Met., bk. viii, c. 5, 159; Strabo, bk. viii, c. 6.

these imply that the bower was merely a house, the passages in which were confusing and ill-arranged.

Another class of ancient mazes consisted of a tortuous path confined to a small area of ground and leading to a tree or shrine in the centre.* This is a maze in which there is no chance of taking a wrong turning; but, as the whole area can be occupied by the windings of one path, the distance to be traversed from the entrance to the centre may be considerable, even though the piece of ground covered by the maze is but small.

The traditional form of the labyrinth constructed for the Minotaur is a specimen of this class. It was delineated on the reverses of the coins of Cnossus, specimens of which are not uncommon; one form of it is indicated in the accompanying diagram (figure i). The design really is the same as that drawn in figure ii, as can be easily seen by bending round a circle the rectangular figure there given.

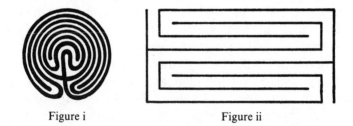

Figure i   Figure ii

Mr. Inwards has suggested† that this design on the coins of Cnossus may be a survival from that on a token given by the priests as a clue to the right path in the labyrinth there. Taking the circular form of the design shown above he supposed each circular wall to be replaced by two equidistant walls separated by a path, and thus obtained a maze to which the original design would serve as the key. The route thus indicated may be at once obtained by noticing that when a node

*On ancient and medieval labyrinths – particularly of this kind – see an article by Mr. E. Trollope in *The Archaeological Journal*, 1858, vol. xv, pp. 216–235, from which much of the historical information given above is derived.

†*Knowledge*, London, October 1892.

is reached (i.e. a point where there is a choice of paths), the path to be taken is that which is next but one to that by which the node was approached. This maze may be also threaded by the simple rule of always following the wall on the right-hand side, or always that on the left-hand side. The labyrinth may be somewhat improved by erecting a few additional barriers, without affecting the applicability of the above rules, but it cannot be made really difficult. This makes a pretty toy, but though the conjecture on which it is founded is ingenious, it has no historical justification. Another suggestion is that the curved line on the reverse of the coins indicated the form of the rope held by those taking part in some rhythmic dance; while others consider that the form was gradually evolved from the widely prevalent swastika.

Copies of the maze of Cnossus were frequently engraved on Greek and Roman gems; similar but more elaborate designs are found in numerous Roman mosaic pavements.* A copy of the Cretan labyrinth was embroidered on many of the state robes of the later Emperors, and, apparently thence, was copied on to the walls and floor of various churches.† At a later time in Italy and in France these mural and pavement decorations were developed into scrolls of great complexity, but consisting, as far as I know, always of a single line. Some of the best specimens now extant are on the walls of the cathedrals at Lucca, Aix in Provence, and Poitiers; and on the floors of the churches of Santa Maria in Trastevere at Rome, San Vitale at Ravenna,

Maze at Hampton Court

*See, e.g., Breton's *Pompeia*, p. 303.
†Ozanam, *Graphia aureae urbis Romae*, pp. 92, 178.

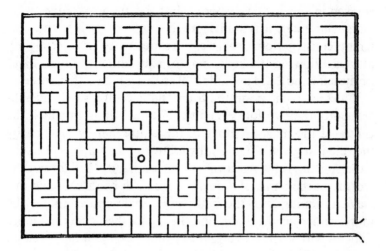

Notre Dame at St. Omer, and the cathedral at Chartres. It is possible that they were used to represent the journey through life as a kind of pilgrim's progress.

In England these mazes were usually, perhaps always, cut in the turf adjacent to some religious house or hermitage; and there are some slight reasons for thinking that, when traversed as a religious exercise, a *pater* or *ave* had to be repeated at every turning. After the Renaissance, such labyrinths were frequently termed Troy-Towns or Julian's Bowers. Some of the best specimens, which are still extant, or were so until recently, are those at Rockliff Marshes, Cumberland; Asenby, Yorkshire; Alkborough, Lincolnshire; Wing, Rutlandshire; Boughton-Green, Northamptonshire; Comberton, Cambridgeshire; Saffron Walden, Essex; and Chilcombe, near Winchester.

The modern maze seems to have been introduced – probably from Italy – during the Renaissance, and many of the palaces and large houses built in England during the Tudor and the Stuart periods had labyrinths attached to them. Those adjoining the royal palaces at Southwark, Greenwich, and Hampton Court were well known from their vicinity to the capital. The last of these was designed by London and Wise in 1690, for William III, who had a fancy for such conceits: a plan of it is given in various guide-books. For the majority of the sight-seers who enter, it is sufficiently elaborate; but it is an indifferent construction, for it can be described completely by always following the hedge on one side (either the right hand or the left hand), and no node is of an order higher than three.

Unless at some point the route to the centre forks and subsequently the two forks reunite, forming a loop in which the centre of the maze is situated, the centre can be reached by the rule just given – namely, by following the wall on one side, either on the right hand or on the left hand. No labyrinth is worthy of the name of a puzzle which can be threaded in this way. Assuming that the path forks as described above, the more numerous the nodes and the higher their order the more difficult will be the maze, and the difficulty might be increased considerably by using bridges and tunnels so as to construct a labyrinth in three dimensions. In an ordinary garden and on a small piece of ground, often of an inconvenient shape, it is not easy to make a maze which fulfils these conditions. On page 259 is a plan of one which I put up in my own garden on a plot of ground which would not allow of more than 36 by 23 paths, but it will be noticed that none of the nodes are of a high order.

TREES

Euler's original investigations were confined to a closed network. In the problem of the maze it was assumed that there might be any number of blind alleys in it, the ends of which formed free nodes. We may now progress one step further, and suppose that the network or closed part of the figure diminishes to a point. This last arrangement is known as a *tree*.

We can illustrate the possible form of these trees by rods, having a hook at each end. Starting with one such rod, we can attach at either end one or more similar rods. Again, on any free hook we can attach one or more similar rods, and so on. Every free hook, and also every point where two or more rods meet, are what hitherto we have called nodes. The rods are what hitherto we have termed branches or paths.

The theory of trees – which already plays a somewhat important part in certain branches of modern analysis, and possibly may contain the key to certain chemical and biological theories – originated in a memoir by Cayley,[*] written in 1856. The discussion of the theory has been analytical rather than geometrical. I content myself with noting the following results.

As steps towards finding t_n, the number of trees with n nodes (so that $t_1 = t_2 = t_3 = 1$, $t_4 = 2$, ...), Cayley observed that the number of trees with n labelled[†] nodes is n^{n-2}. He also considered the number T_n of *rooted* trees with n nodes, that is, trees having one special node called the root. (This concept arose from the enumeration of monosubstituted hydrocarbons.) Let

$$t(x) = \sum_{n=0}^{\infty} t_n x^n \quad \text{and} \quad T(x) = \sum_{n=0}^{\infty} T_n x^n$$

be the generating functions (or counting series) for trees and rooted trees, so that

$$t(x) = x + x^2 + \ x^3 + 2x^4 + 3x^5 + \ 6x^6 + 11x^7 + \ 23x^8 + \ldots,$$
$$T(x) = x + x^2 + 2x^3 + 4x^4 + 9x^5 + 20x^6 + 48x^7 + 115x^8 + \ldots .$$

Implicitly due to Cayley, and explicitly to Pólya,[‡] is the functional equation

$$T(x) = x \exp \sum_{r=1}^{\infty} \frac{1}{r} T(x^r).$$

R.E. Otter[§] discovered the neatest possible formula for the number of trees in terms of the number of rooted trees:

$$t(x) = T(x) - \tfrac{1}{2}\{T^2(x) - T(x^2)\}.$$

---

[*]*Philosophical Magazine*, March 1857, series 4, vol. XIII, pp. 172–176; or *Collected Works*, Cambridge, 1890, vol. III, no. 203, pp. 242–246: see also the paper on double partitions, *Philosophical Magazine*, November 1860, series 4, vol. XX, pp. 337–341.

[†]J.W. Moon, *Counting Labelled Trees* (Twelfth Biennial Seminar of the Canadian Mathematical Congress), W. Clowes, London, 1970.

[‡]*Acta Mathematica*, 1937, vol. LXVIII, pp. 145–254.

[§]*Annals of Mathematics*, 1948, vol. XLIX, pp. 583–599. See also F. Harary, *Graph Theory*, Addison-Wesley, Reading, Massachusetts, 1969, pp. 187–190, 233–234. His Appendix 3 gives the diagrams of all trees with $n \leqq 10$ nodes.

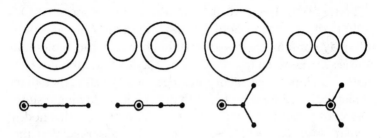

These results have an application to another topological problem: to find the number of ways in which n circles (or spheres) can be internal or external to one another. For instance, two circles may be outside one another, or one inside the other. The n circles can be put into correspondence with the nodes, other than the root, of a tree having n branches. Circles not contained in others correspond to nodes directly joined to the root. Thereafter, every branch indicates that one circle is inside another. This correspondence shows that the required number of ways is T_{n+1}. In the above diagrams (for the case when $n = 3$), the root of each tree is ringed.

THE HAMILTONIAN GAME

I turn next to consider some problems where it is desired to find a route which will pass once and only once through each node of a given geometrical figure. This is the reciprocal of the problem treated in the first part of this chapter, and is a far more difficult question.

The Hamiltonian Game consists in the determination of a route along the edges of a regular dodecahedron which will

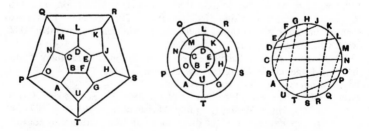

pass once and only once through every angular point. Sir William Hamilton,* who invented this game – if game is the right term for it – denoted the twenty angular points on the solid by letters which stand for various towns. The thirty edges constitute the only possible paths. The inconvenience of using a solid is considerable, and the dodecahedron may be represented conveniently in perspective by a flat board marked as shown in the first of the above diagrams. The second and third diagrams will answer our purpose equally well and are easier to draw.

The first problem is to go "all round the world," that is, starting from any town, to go to every other town once and only once and to return to the initial town; the order of the n towns to be first visited being assigned, where n is not greater than five.

Hamilton's rule for effecting this was given at the meeting of the British Association at Dublin in 1857. In our diagrams one solution is the route $ABCDEFGHJKLMNOPQR$ STU. There are 30 solutions in all, all equivalent under the symmetry of the dodecahedron.

A similar game may be played with other solids or with other diagrams on the plane, or with diagrams on other surfaces. Hamiltonian circuits are of great interest in connection with the Four-Colour Theorem discussed in chapter VIII. Let such a circuit be drawn in some map in the plane, going along edges of the map and passing exactly once through each vertex. The circuit separates the plane into two regions, the inside and the outside. Any edge of the map crossing the inside separates it into two smaller regions, any edge traversing one of these separates it into two more, and so on. Thus if the inside is traversed by exactly r edges, the number of districts of the map inside the circuit is $r + 1$. Moreover these districts can be coloured in two colours so that no two with a common edge have the same colour. We colour one district red, the ones next it

*See *Quarterly Journal of Mathematics*, London, 1862, vol. v, p. 305; or *Philosophical Magazine*, January 1884, series 5, vol. XVII, p. 42; also Lucas, vol. II, part vii.

green, the ones next those red, and so on. No contradiction can arise because each possible common edge separates the inside of the circuit. A similar argument applies to the outside of the circuit, in which one district is of infinite area. Hence any map on the plane having a Hamiltonian circuit satisfies the Four-Colour Theorem.

P.G. Tait conjectured that any standard map, in the sense of page 227, in which no two districts have more than one common edge, does indeed have a Hamiltonian circuit. A counter-example to this conjecture was found in 1946.* A most elegant disproof of the conjecture was found more recently by È. Ya. Grinbergs.†

His argument runs as follows. Consider, in any map, the inside of a Hamiltonian circuit, with its r diagonals and $r + 1$ districts. Let the number of districts with j edges be f_j. Then the number N of edges in the circuit itself can be computed as

$$N = \sum_{j=2}^{\infty} j f_j - 2r = \sum_{j=2}^{\infty} (j-2) f_j + 2.$$

But let the number of districts outside the circuit having j edges be f_j'. Replacing the inside of the circuit by the outside in the above argument, we find that the formula for N remains valid when f_j is replaced by f_j'. So by subtraction we have

$$\sum_{j=2}^{\infty} (j-2)(f_j - f_j') = 0.$$

Suppose we succeed in constructing a map in which, with a single exception, each face has a number of edges which leaves the remainder 2 on division by 3. Then the above equation is false for the map, and therefore the map has no Hamiltonian circuit. Such maps can in fact be found. Figure iii is an example constructed by Grinbergs. This map has no ring of fewer than 5 districts. Apart from the outside enneagon its districts are pentagons and octagons.

Using the same method (and the further observation that,

*W.T. Tutte, "On Hamiltonian Circuits," *Journal of the London Mathematical Society*, 1946, vol. XXI, pp. 98–101.

†*Latvian Mathematical Yearbook* (Izdat. Zinatne, Riga, 1968), vol. IV, pp. 51–58.

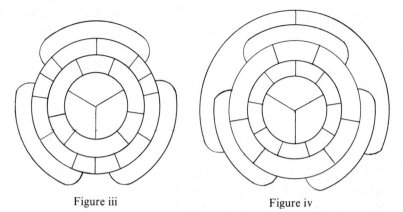

Figure iii   Figure iv

when three hexagons share a vertex, any Hamiltonian circuit must separate one of these hexagons from the other two); Grinbergs has constructed the somewhat simpler map shown in figure iv.

A theorem of C.A.B. Smith* states that in a network with three edges at each vertex the number of Hamiltonian circuits through a specified edge is even (possibly zero). To prove this we first define a *Tait cycle* as a set of circuits going along edges of the map, each with an even number of edges, such that just one of the circuits passes through each vertex. A *Tait colouring* is a colouring of the edges in three colours so that no two edges of the same colour meet at a vertex (see pp. 225–226). Two Tait colourings are counted the same if they differ only by a permutation of the three colours. It is easily seen that the following propositions hold:

(i) In a Tait colouring each pair of colours defines a Tait cycle.

(ii) A Tait cycle of m circuits is derivable in this way from exactly 2^{m-1} distinct Tait colourings.

(iii) Of the Tait cycles derived from a given Tait colouring exactly two go through each edge.

(iv) A Hamiltonian circuit is equivalent to a Tait cycle with $m = 1$.

*Smith's proof is still unpublished. The proof in the text is by W.T. Tutte, "On Hamiltonian Circuits." (See the first footnote on page 264.)

We now investigate the number of Tait cycles through a given edge A, counting a cycle of m circuits 2^{m-1} times. By (ii) and (iii) this number is even, being twice the number of Tait colourings. But 2^{m-1} is even whenever m is not 1, and is 1 if $m=1$. It follows that the number of Hamiltonian circuits through A is even.

It is a consequence of Smith's Theorem that if the network under consideration has one Hamiltonian circuit, then it has at least three.

Hamilton's problem has been extended to certain cases where the number of towns is infinite (the required route having no beginning and no end). For a route along the edges of the tessellation of squares $\{4, 4\}$, see König's book on

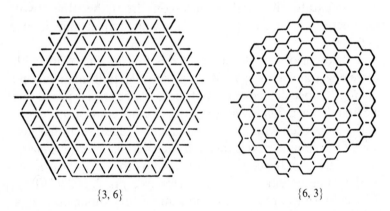

$\{3, 6\}$ $\{6, 3\}$

graphs.* \ similar treatment of the other regular tessellations ($\{3, 6\}$ and $\{6, 3\}$) is shown above.

The problem of the knight's path on a chess-board is somewhat similar in character to the Hamiltonian game. This I have already discussed in chapter vi.

DRAGON DESIGNS

Another kind of endless polygonal path in a plane can be constructed by folding a long strip of paper.† Beginning with

*Dénes König, *Theorie der endlichen und unendlichen Graphen*, Leipzig, 1936, p. 32.
†*Scientific American*, 1967, March, pp. 124–125; April, pp. 118–120; and

a horizontal strip, we fold the right half upward completely over the left; then fold the resulting doubled strip upward in the same way, so that the crease at the right comes over the left end; and continue as often as possible. (Seven times is the practical limit, but theoretically we can continue forever.) When the paper is opened up again, it displays an interesting sequence of creases. Denoting creases which turn upward by *U* and those which turn downward by *D*, we find that the sequence begins as follows:

$$U\,U D U\,U D D U\,U U D D U D D U\,U U D U\,U D D D U\,U D D U D D U\,U$$

where the first, second, fourth, eighth, . . . creases, here marked by dots, are termed "pivots." It is easy to see that the crease *k* steps beyond a pivot has opposite sense to the crease *k* steps before that pivot; this, with the fact that each pivot is a *U* (because the strip is always folded upward), provides a rule for writing down as many terms of the sequence as we like. There are alternative, equivalent rules, of which I mention one. If $n = 2^k m$ with *m* odd, then the *n*th crease is *U* or *D* according as $m \equiv 1$ or $m \equiv 3 \pmod 4$.

Next fold the paper at each crease, in the direction of that crease, to an angle of 90°. The resulting shape is known as the "dragon design" because of its pleasing reptilian shape; it is an infinite polygonal path (see figure v) along one quarter of the edges of the tessellation {4, 4}. In fact, if the design is repeated by quarter-turns about its initial point, so as to form four dragons issuing from this same point, tail to tail, these four designs will not cross one another; between them, if continued indefinitely, they will trace every edge of the tessellation exactly once.

If we join the starting point to the second vertex of the dragon design, the second vertex to the fourth, and so on, the result is a replica of the original design, expanded in the ratio $\sqrt{2}$ and

July, p. 115. For proofs of many of the properties cited here, and numerous figures, see C. Davis and D. Knuth, *Journal of Recreational Mathematics*, 1970, vol. 3, pp. 66–81, 133–149.

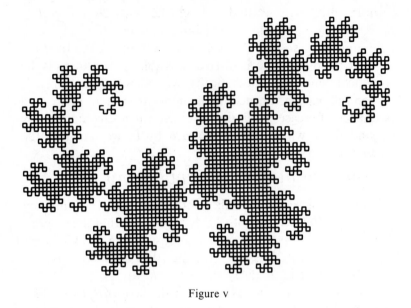

Figure v

rotated clockwise by 45°. (This statement has a counterpart bearing on the above infinite sequence of U's and D's: if we omit from the sequence every odd term, we get the original sequence back.) From this we see that the pivots lie on an equiangular spiral about the starting point.

Reversing the above similarity, which "rarefies" the dragon design by dropping out some of its vertices, we obtain a similarity which "compresses" it by inserting new ones. Namely, let all the vertices of a given dragon design serve as the even-numbered vertices of a dragon design with edge length $1/\sqrt{2}$ as big. Continuing this compression indefinitely, we get a sequence of smaller and smaller dragon designs, whose limit is a continuous *space-filling* curve.* Four of these limiting curves issuing from a single point include between them every point in the plane.

Many variants of the original paper-folding procedure have been examined. Perhaps the simplest is that in which, rather than folding upward each time, we fold the strip alternately

*H. Steinhaus, *Mathematical Snapshots*, New York, 1969, pp. 102–103.

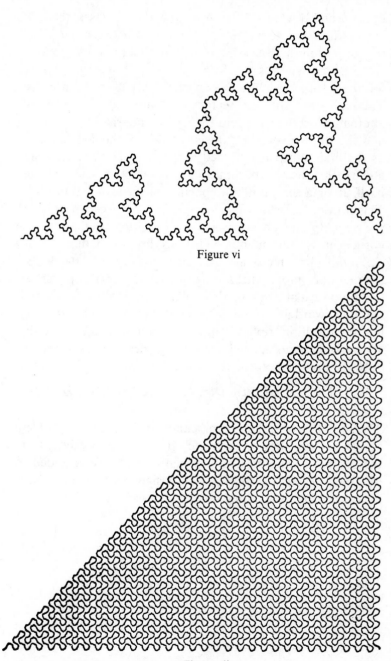

Figure vi

Figure vii

upward and downward. In the sequence of creases, then, the *pivots* will alternate between U's and D's; between the pivots, the rule is still that the kth crease beyond a pivot has opposite sense to the kth crease before. Figure vi shows the result of folding a strip so creased to an angle of 108° at each crease. The path stays within a sector of opening 36°, the pivots lie alternately on the right and the left boundaries of this sector, and the third pivot is τ^2 times as far from the starting-point as the first. When it is folded a little farther to reduce each angle to 90°, the path becomes a unicursal route, never crossing itself, along a certain infinite portion of the tessellation $\{4, 4\}$. This portion is just a sector of opening 45°, except that along the right boundary every even edge is omitted. (In figure vii the corners have been rounded off to show how the path proceeds.)

Similar phenomena are obtained by folding in thirds. Beginning once more with a long horizontal strip of paper, we fold it upward at a point a third of its length from its right end, and downward at a point an equal distance from its left; then fold the resulting tripled strip in the same way; and continue. When the paper is opened up, the sequence of creases, reading outward from the centre, is in part

$$UDDUDDUUDUDDUDD \mid UUDUUDUDDUUDUUD$$

The pivots have similar significance to those in the problem above, but now they occur at each $\frac{1}{2}(3^n + 1)$st crease away from the centre in each direction. When each crease is folded to 60°, there results the "ter-dragon design": a unicursal route, never crossing itself, along the whole tessellation $\{3, 6\}$.

CHAPTER X

COMBINATORIAL DESIGNS

In this chapter we return to the kind of problems that arose at the end of chapter VI. In the nineteenth century such problems were only of recreational interest, but more recently they have proved useful to statisticians. One important type of combinatorial design became available in 1856, when von Staudt discovered the possibility of a geometry involving only finitely many points. His work was forgotten for forty years, but the idea has become immensely fruitful.

A projective plane. A lady wishes to invite her seven friends for a series of dinner parties. Her table provides room for exactly three other persons. In addition, the lady wants each pair of friends to meet exactly once at her table. The question arises, how she should arrange her invitations over the days. Let us denote the friends by their names a, b, c, d, e, f, g, and the days on which the dinners take place by 1, 2, 3, ...

It is no restriction of generality if we assume that on day 1 the friends a, b, c are invited; otherwise we would rename the friends. However, concerning the invitations for day 2 the lady

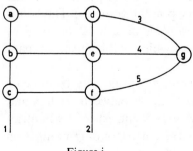

Figure i

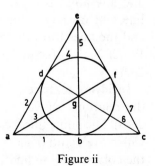

Figure ii

faces a dilemma. Should she invite any one friend among the set $\{a, b, c\}$ again, or should she invite a whole fresh team $\{d, e, f\}$, say? We shall show that she would run into trouble if she were to proceed with $\{d, e, f\}$ for day 2. Indeed, for the next days she then has to combine g with one out of $\{a, b, c\}$ and one out of $\{d, e, f\}$. She can only do so on days 3, 4, 5, but then her possibilities are exhausted, as can be seen from figure i. For day 6 there is no companion to join the table with a and e, say, who has met neither of them. So, the required arrangement is impossible in this way.

The conclusion is that on day 2 the other possibility should have been chosen. In fact, we have proved that on any two distinct days there should be one common guest. Once we know this, it is easy to complete a scheme for the invitations and to observe that seven days are needed. For instance, the following columns yield a solution:

$$
\begin{array}{cccccccc}
\text{Days} & 1 & 2 & 3 & 4 & 5 & 6 & 7 \\
& a & a & a & b & b & c & c \\
\text{Guests} & b & d & f & d & e & d & e \\
& c & e & g & f & g & g & f
\end{array}
$$

This scheme may be arranged in a more geometric way as shown in figure ii, by representing the friends by points and the days by lines. The seven points and the seven lines of the figure are related as follows. Any line contains three points, and any point is on three lines. Any pair of points is on one line, and any pair of lines has one point in common. The points and the lines form the *projective plane* of order 2, denoted by PG(2, 2), the precise definition of which will be given on page 283.

Incidence matrices. There is an algebraic way to indicate the relations between points and lines (between friends and days). The point-line *incidence matrix N* consists of 7×7 entries which are 1 and 0 according as the corresponding point is on the corresponding line or not:

$$
\begin{array}{c c}
 & \begin{array}{c} 1\ 2\ 3\ 4\ 5\ 6\ 7 \end{array} \\
\begin{array}{c} a \\ b \\ c \\ d \\ e \\ f \\ g \end{array} &
\left[\begin{array}{c c c c c c c}
1 & 1 & 1 & 0 & 0 & 0 & 0 \\
1 & 0 & 0 & 1 & 1 & 0 & 0 \\
1 & 0 & 0 & 0 & 0 & 1 & 1 \\
0 & 1 & 0 & 1 & 0 & 1 & 0 \\
0 & 1 & 0 & 0 & 1 & 0 & 1 \\
0 & 0 & 1 & 1 & 0 & 0 & 1 \\
0 & 0 & 1 & 0 & 1 & 1 & 0
\end{array}\right] = N.
\end{array}
$$

From the matrix N the properties of the projective plane can be read off. Any column contains three 1's, and any row contains three 1's. Any pair of rows has one 1 in common, and so does any pair of columns. These properties are maintained if the points and the lines are permuted; for instance, if N is changed into

$$
\begin{array}{c c}
 & \begin{array}{c} 7\ 2\ 3\ 5\ 1\ 6\ 4 \end{array} \\
\begin{array}{c} a \\ g \\ b \\ c \\ d \\ f \\ e \end{array} &
\left[\begin{array}{c c c c c c c}
0 & 1 & 1 & 0 & 1 & 0 & 0 \\
0 & 0 & 1 & 1 & 0 & 1 & 0 \\
0 & 0 & 0 & 1 & 1 & 0 & 1 \\
1 & 0 & 0 & 0 & 1 & 1 & 0 \\
0 & 1 & 0 & 0 & 0 & 1 & 1 \\
1 & 0 & 1 & 0 & 0 & 0 & 1 \\
1 & 1 & 0 & 1 & 0 & 0 & 0
\end{array}\right] = N'.
\end{array}
$$

Note that, whereas the matrix N was symmetric, the permuted matrix N' is antisymmetric. The matrix N' is a *circulant*, because of the relation between any pair of consecutive rows. We denote it as follows:

$$N' = \text{circ}(0\quad 1\quad 1\quad 0\quad 1\quad 0\quad 0).$$

An Hadamard matrix. The lady herself was present at each of the seven dinners. Suppose there is a farewell party on day 8 for all eight persons involved. We complete the incidence matrix N by a row and a column of ones to the 8×8 matrix F as follows:

$$F = \begin{bmatrix} 1 & 1 & 1 & 1 & 1 & 1 & 1 & 1 \\ 1 & 1 & 1 & 1 & 0 & 0 & 0 & 0 \\ 1 & 1 & 0 & 0 & 1 & 1 & 0 & 0 \\ 1 & 1 & 0 & 0 & 0 & 0 & 1 & 1 \\ 1 & 0 & 1 & 0 & 1 & 0 & 1 & 0 \\ 1 & 0 & 1 & 0 & 0 & 1 & 0 & 1 \\ 1 & 0 & 0 & 1 & 1 & 0 & 0 & 1 \\ 1 & 0 & 0 & 1 & 0 & 1 & 1 & 0 \end{bmatrix}, \quad H = \begin{bmatrix} - & - & - & - & - & - & - & - \\ - & - & - & - & + & + & + & + \\ - & - & + & + & - & - & + & + \\ - & - & + & + & + & + & - & - \\ - & + & - & + & - & + & - & + \\ - & + & - & + & + & - & + & - \\ - & + & + & - & - & + & + & - \\ - & + & + & - & + & - & - & + \end{bmatrix}.$$

Let J denote the matrix (of order 8), all of whose elements are 1. Let I denote the unit matrix (of order 8), which has 1's on the main diagonal and 0 elsewhere. The matrix $H = J - 2F$ has the property*:

$$HH^T = 8I.$$

This says that the *inner product* of any two distinct rows (= the sum of the products of the corresponding elements of any two distinct rows) is zero, and that the inner product of any row with itself equals the size of the matrix. Matrices with elements $+1$ and -1, which have this property, are called *Hadamard matrices*, after Jacques Hadamard. On page 108, where they were called anallagmatic pavements, a simple construction for many of them was described.

An error-correcting code. Combining the above 8×8 matrix F with its complement $J - F$, we obtain the following 8×16 matrix:

$$G^T = \begin{array}{c} \\ \begin{array}{cccccccccccccccc} 1 & 2 & 3 & 4 & 5 & 6 & 7 & 8 & 9 & 10 & 11 & 12 & 13 & 14 & 15 & 16 \end{array} \\ \begin{bmatrix} 1 & 1 & 1 & 1 & 1 & 1 & 1 & 1 & 0 & 0 & 0 & 0 & 0 & 0 & 0 & 0 \\ 1 & 1 & 1 & 1 & 0 & 0 & 0 & 0 & 0 & 0 & 0 & 0 & 1 & 1 & 1 & 1 \\ 1 & 1 & 0 & 0 & 1 & 1 & 0 & 0 & 0 & 0 & 1 & 1 & 0 & 0 & 1 & 1 \\ 1 & 1 & 0 & 0 & 0 & 0 & 1 & 1 & 0 & 0 & 1 & 1 & 1 & 1 & 0 & 0 \\ 1 & 0 & 1 & 0 & 1 & 0 & 1 & 0 & 0 & 1 & 0 & 1 & 0 & 1 & 0 & 1 \\ 1 & 0 & 1 & 0 & 0 & 1 & 0 & 1 & 0 & 1 & 0 & 1 & 1 & 0 & 1 & 0 \\ 1 & 0 & 0 & 1 & 1 & 0 & 0 & 1 & 0 & 1 & 1 & 0 & 0 & 1 & 1 & 0 \\ 1 & 0 & 0 & 1 & 0 & 1 & 1 & 0 & 0 & 1 & 1 & 0 & 1 & 0 & 0 & 1 \end{bmatrix} \end{array}$$

*The transposed matrix A^T of any square matrix A is obtained by reflecting A with respect to its main diagonal. If H is symmetric, obviously $H^T = H$.

We consider the 16 columns, each consisting of 8 co-ordinates. The number of co-ordinates in which any two columns differ is called their *distance*. We observe that any pair of columns has distance 4 or 8. The property that all distances are at least 4 turns the set of 16 columns into a useful *code* as follows.

Let each of the 16 columns be used as a binary code word for the communication of messages. Because of imperfections in the transmission channel it may happen that a message, which is correctly sent off, is received with one or more errors. If, for instance, the column (1 1 0 1 1 0 0 1) is received, and if no more than one error is made, then, thanks to the distances between the code words, the receiver can draw the conclusion that the 7th column was sent off. The receiver is able to correct any word which contains one error, and to detect (not to correct) the presence of two errors. Therefore, the present code is called *1-error-correcting* and *2-error-detecting*.

The present code is a *linear (8, 4) code*. This means the following. We add columns modulo 2, as on page 37, that is, using

$$0+0=0, \quad 0+1=1+0=1, \quad 1+1=0.$$

It is easy to verify that all columns are linear combinations of four columns: col 2, col 3, col 4, col 5, say. Indeed, we have

col 1 = col 2 + col 3 + col 4,   col 6 = col 3 + col 4 + col 5,

col 16 = col 4 + col 5,   col 9 = col 4 + col 4, etc.

Hence, our code is characterized by linearity and by 4 independent columns of 8 co-ordinates each. In other terms, our code is a 4-dimensional subspace of the 8-dimensional vector space with binary co-ordinates.

From the present code we obtain a linear (7, 4) code by deleting from the matrix G^T any row, say the first row. The 16 columns of length 7 thus obtained are still linear combinations of 4 columns. Since any two columns differ in at least 3 co-ordinates, the (7, 4) code is 1-error-correcting. In addition, this code has the following remarkable property. The number of columns, not belonging to the code, which are at distance 1 from any given code word, equals 7. There are 16 code words, hence $16(1 + 7)$ columns of length 7 at distance at most 1 from any code word. But with these columns all 128 binary columns of length 7 are exhausted. Hence, in geometric language, the "spheres" of radius 1 about the code words of the 4-dimensional subspace constitute a packing of the 7-dimensional vector space, without any overlap and without holes. Such a code is called a *perfect* code.

A block design. Once again we consider the matrix G^T of page 274. We delete the first column and the ninth so as to obtain the 8×14 matrix M, which has the property that each column has exactly 4 ones:

$$M = \begin{bmatrix} 1 & 1 & 1 & 1 & 1 & 1 & 1 & 0 & 0 & 0 & 0 & 0 & 0 & 0 \\ 1 & 1 & 1 & 0 & 0 & 0 & 0 & 0 & 0 & 1 & 1 & 1 & 1 \\ 1 & 0 & 0 & 1 & 1 & 0 & 0 & 0 & 1 & 1 & 0 & 0 & 1 & 1 \\ 1 & 0 & 0 & 0 & 0 & 1 & 1 & 0 & 1 & 1 & 1 & 1 & 0 & 0 \\ 0 & 1 & 0 & 1 & 0 & 1 & 0 & 1 & 0 & 1 & 0 & 1 & 0 & 1 \\ 0 & 1 & 0 & 0 & 1 & 0 & 1 & 1 & 0 & 1 & 1 & 0 & 1 & 0 \\ 0 & 0 & 1 & 1 & 0 & 0 & 1 & 1 & 1 & 0 & 0 & 1 & 1 & 0 \\ 0 & 0 & 1 & 0 & 1 & 1 & 0 & 1 & 1 & 0 & 1 & 0 & 0 & 1 \end{bmatrix}$$

It is observed that each row of M has 7 ones, and that each pair of rows has 3 ones in common. The matrix M is the point-block incidence matrix of a *block design* with

$$v = 8, \ b = 14, \ k = 4, \ r = 7, \ \lambda = 3; \ JM = 4J, \ MJ = 7J, \ MM^T = 4I + 3J.$$

This means the following. We have a set of $v = 8$ points and a collection of $b = 14$ subsets called blocks. Each block contains $k = 4$ points. Each point is in $r = 7$ blocks. Any pair of points is in $\lambda = 3$ blocks. In addition, any triple of points is in one block.

We exemplify this block design as follows. Let the 8 points be represented by the 8 vertices of a cube. Let the 14 subsets of 4 points each be represented by the 6 faces of the cube, its 6 diagonal planes, and the 2 regular tetrahedra formed by its vertices. We obtain a better representation if we calculate modulo 2, that is, according to the binary addition rules. Then

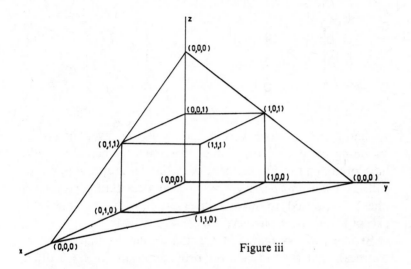

Figure iii

also the tetrahedra become planes, as may be seen from figure iii. The blocks are represented by the 14 planes with equations

$$x=0, \quad y=0, \quad z=0, \quad x+y=0, \quad x+z=0, \quad y+z=0, \quad x+y+z=0,$$

$$x=1, \quad y=1, \quad z=1, \quad x+y=1, \quad x+z=1, \quad y+z=1, \quad x+y+z=1,$$

which are to be interpreted as congruences modulo 2.

The block design discussed in the present section may be considered as an extension of the projective plane of page 271. Indeed, in the representation using the cube modulo 2 there are 7 lines and 7 planes passing through any point, $(0, 0, 0)$ say. Their incidence relations are the same as those of the 7 points and the 7 lines of the projective plane of order 2. This may clarify the resemblance between figure ii and figure iii.

We have observed that in the 8×14 matrix M (page 276) any pair of rows have 3 ones in common. We now consider the columns in pairs. Clearly, any column has 2 ones in common with all but one other column, and has no ones in common with exactly one column. We represent the 14 columns by the 14 vertices of a graph, and we call any 2 vertices adjacent whenever the corresponding columns have no one in common. Then the so-called *ladder graph* (figure iv) is obtained. If instead

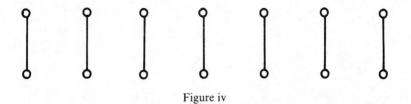

Figure iv

we had called any two vertices adjacent whenever the corresponding columns have 2 ones in common, then we would have obtained the complement of the ladder graph, sometimes called the *cocktail-party graph* (since at a cocktail party everyone is supposed to have conversation with everyone else except for his own partner).

Steiner triple systems.   Returning to our original problem, we recall that the lady's table provides room for $k = 3$ other persons, and that each pair of friends meets at $\lambda = 1$ dinner. Under these conditions, what arrangements could the lady make for more than 7 friends?

The projective plane of order 2 is a block design with $v = 7$ points, $b = 7$ blocks (the lines), any point being in $r = 3$ blocks, any block having $k = 3$ points, and any pair of points being in $\lambda = 1$ block. What other block designs exist with $k = 3$ and $\lambda = 1$? Thus we ask for a set containing v points, and a collection of triples of these points, such that any pair of points is in exactly one triple. Such a collection of triples, if it exists, is called a *Steiner triple system* of order v, for short $S(v)$, after Jacob Steiner (1796–1863),* who, as we shall see, was not the first to ask such a question.

Suppose a Steiner triple system of order v exists. Then any given point occurs once in a triple with each of the $v - 1$ other points, hence it occurs in $\frac{1}{2}(v - 1)$ triples. The total number of triples, being $\frac{1}{3}v$ times this number, is $\frac{1}{6}v(v - 1)$. Since both numbers are integers, we have as a necessary condition for the existence of $S(v)$ that v is congruent to 1 or 3 modulo 6, that is, $v = 7, 9, 13, 15, \ldots$. E.H. Moore proved† in 1893 that

---

*J. Steiner, *Journal für die reine und angewandte Mathematik*, 1853, vol. XLV, pp. 181–182.

†Marshall Hall, Jr., *Combinatorial Theory*, Waltham, Mass., 1967, p. 239.

this condition is also sufficient; hence the existence problem for Steiner triple systems is solved. (We have already considered the case $v = 7$.)

There exists* a unique Steiner triple system of order 9. It has 9 points, $a, b, c, d, e, f, g, h, i$, say, and 12 triples, as indicated in figure v. Thus the twelve·triples are: the rows abc, def, ghi;

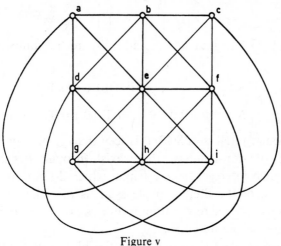

Figure v

the columns adg, beh, cfi; the positive diagonals aei, bfg, cdh; and the negative diagonals afh, bdi, ceg (as in the evaluation of a 3×3 determinant).

Any point is in 4 triples, any pair of points is in 1 triple. There are 4 families, consisting of 3 triples each, of "parallel lines," whose members have no point in common. The mutual relations of the 12 triples are governed by the graph on 12 vertices shown in figure vi, adjacency being defined by non-intersection.

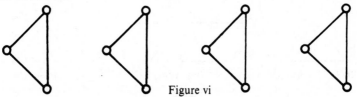

Figure vi

*Cf. G.A. Miller, H.F. Blichfeldt, and L.E. Dickson, *Theory and Applications of Finite Groups*, Dover, New York, 1961, p. 335.

The next case in which Steiner triple systems exist corresponds to the parameters

$$v = 13, \quad b = 26, \quad r = 6, \quad k = 3, \quad \lambda = 1.$$

One of the 2 existing systems is determined by its 13×26 point-block incidence matrix N of the form

$$N = [N_1 \; N_2],$$

where the 13×13 matrices N_1 and N_2 are the following circulants:

$$N_1 = \text{circ}(1\ 0\ 1\ 0\ 0\ 0\ 0\ 0\ 1\ 0\ 0\ 0\ 0),$$

$$N_2 = \text{circ}(0\ 0\ 0\ 0\ 0\ 0\ 1\ 1\ 0\ 0\ 1\ 0\ 0).$$

From the parameters of $S(13)$ it is seen that, given any point P and triple l, non-incident, there exist through P exactly 3 triples which intersect l and 3 triples which do not intersect l. In the case of the Steiner triple systems of orders 9 and 7, respectively, the corresponding numbers are 3, 1 and 3, 0, respectively. (See figure vii.)

This property of the points and the blocks of $S(13)$, $S(9)$, $S(7)$ agrees with the basic axiom for the points and the lines of the hyperbolic (non-Euclidean),* affine (or Euclidean), and projective (or elliptic) plane geometries, respectively, the first of which was discovered independently by Gauss (1777–1855), Lobachevsky (1793–1856), and Bolyai (1802–1860). Accordingly, we call $S(13)$ a finite *hyperbolic* plane, $S(9)$ a finite *affine* plane (cf. page 283), and $S(7)$ a finite *projective* plane.

Let us investigate some further properties of the points and the lines (= blocks) of these finite planes. Since they are Steiner

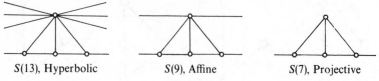

S(13), Hyperbolic   S(9), Affine   S(7), Projective

Figure vii

*Cf. Coxeter, *Introduction to Geometry*, 2nd ed., New York, 1969, chapters 15, 16.

triple systems, any 2 distinct points determine a unique line. Furthermore, these finite planes are *transitive* on points, that is, given any 2 distinct points P and Q, there exists a permutation of the set of points which maps P onto Q and lines onto lines, preserving the incidence of points and lines. For $S(7)$ and $S(9)$ this is seen immediately from figure ii and figure v, respectively. For $S(13)$, and also for $S(7)$, it follows from the circulant properties of the point-block incidence matrix N. (It should be remarked that, given any pair of lines l and m, there does not always exist an incidence-preserving permutation of the points which maps l onto m.) We mentioned earlier that our $S(13)$ is one of two Steiner systems of order 13; the other one is less interesting because it is not transitive on the points.

The next case in which Steiner triple systems exist corresponds to the parameters

$$v = 15, \quad b = 35, \quad r = 7, \quad k = 3, \quad \lambda = 1.$$

These systems were completely enumerated by Cole, White, and Cummings.* There are 80 non-isomorphic such systems $S(15)$. Many of them fail to be transitive on points. However, among the transitive ones there are 7 which have a remarkable additional property. They are Kirkman systems, a notion which will be discussed on page 288.

As for the higher Steiner triple systems $S(v)$, their number increases very rapidly. For instance, there are more than 2.10^{15} non-isomorphic $S(31)$.

Finite geometries. The finite projective plane introduced on page 271 satisfies the following axioms:

(1) for any 2 distinct points there exists a unique line incident with both of them,

(2) for any 2 distinct lines there exists a unique point incident with both of them,

(3) there exist 4 points no 3 of which are incident with the same line,

(4) the number of points is finite.

*F.N. Cole, A.S. White, and L.D. Cummings, *Memoirs of the National Academy of Sciences (U.S.A.)*, 1925, vol. xiv, Second Memoir, p. 89.

More generally, a *finite projective plane* is a system of points and lines, and an incidence relation which satisfies the above axioms. Let us consider 4 points P, Q, R, S, no 3 on a line, whose existence is guaranteed by (3). The line RS contains a finite number of points, $q+1$ say. Then any point non-incident with RS is on at least $q+1$ lines by (1), and on at most $q+1$ lines by (2), thus on exactly $q+1$ lines. In particular this holds for P and Q. This implies that every line determined by P, Q, R, S contains $q+1$ points. Therefore, every point of the plane is on $q+1$ lines, and every line of the plane contains $q+1$ points. The total number of points in the plane (the points on all lines through any point) equals $1+(q+1)q=q^2+q+1$. This is also the total number of lines in the plane.

The above observation, and the definition of a finite projective plane of order q, go back to von Staudt.[*] Finite projective planes were actually constructed by Fano[†] (for prime q) and by Veblen and Bussey[‡] (for prime power q). Projective planes have been constructed for prime power q only. It is known (compare page 192) that there does not exist a plane of order 6. The existence of a plane of order 10, the next composite number which is not a prime power, is unknown.

Now we give a construction for a finite projective plane of order $q=p^n$, p prime. Our point of departure is the Galois field $GF(p^n)$ with p^n elements, which was discussed on page 74. We consider[§] two types of ordered triads of elements of the Galois field, namely

$$(x_1, x_2, x_3) \text{ and } [X_1, X_2, X_3],$$

where $x_1, x_2, x_3, X_1, X_2, X_3$ are elements of $GF(p^n)$. We exclude $(0,0,0)$ and $[0,0,0]$, and regard two triads of the same kind as being equivalent (that is, geometrically indistinguishable) if

---

[*]K.G.C. von Staudt, *Beiträge zur Geometrie der Lage*, vol. I, Nürnberg, 1856.

[†]G. Fano, *Giornale di Matimatiche*, 1892, vol. XXX, pp. 114–124.

[‡]O. Veblen and W. Bussey, *Transactions of the American Mathematical Society*, 1906, vol. VII, pp. 241–259.

[§]H.S.M. Coxeter, *Projective Geometry*, University of Toronto Press, 1974, p. 112.

they are proportional. A point is the set of all triads

$$(\lambda x_1, \lambda x_2, \lambda x_3), \quad \lambda \neq 0, \quad \lambda \; \varepsilon \; \mathrm{GF}(p^n)$$

equivalent to a given triad (x_1, x_2, x_3). A line is the set of all triads

$$[\Lambda X_1, \Lambda X_2, \Lambda X_3], \quad \Lambda \neq 0, \quad \Lambda \; \varepsilon \; \mathrm{GF}(p^n)$$

equivalent to $[X_1, X_2, X_3]$. The point-line incidence is defined by

$$x_1 X_1 + x_2 X_2 + x_3 X_3 = 0.$$

It is easy to see that this system of points and lines, with the incidence relation, satisfies our axioms. Indeed, two lines $[A_1, A_2, A_3]$ and $[B_1, B_2, B_3]$ have a unique point of intersection (x_1, x_2, x_3) determined by $x_1 A_1 + x_2 A_2 + x_3 A_3 = 0$, $x_1 B_1 + x_2 B_2 + x_3 B_3 = 0$; two points (a_1, a_2, a_3) and (b_1, b_2, b_3) are connected by a unique line $[X_1, X_2, X_3]$ determined by $a_1 X_1 + a_2 X_2 + a_3 X_3 = 0, b_1 X_1 + b_2 X_2 + b_3 X_3 = 0$; the number of points is finite; the points $(1, 0, 0)$, $(0, 1, 0)$, $(0, 0, 1)$, $(1, 1, 1)$ satisfy (3). Thus we have obtained* a finite projective plane of order p^n, which is denoted by PG$(2, p^n)$. The case $p^n = 2$ brings us back to the projective plane of order 2, as explained on page 281 and in figure ii. In the case $p^n = 3$ the Galois field is $\{0, 1, -1\}$ with the ordinary multiplication and with $1 + 1 = -1$, $-1 - 1 = 1$, $1 - 1 = 0$. Writing $+$ for 1 and $-$ for -1, we have the 13 points and 13 lines indicated in figure viii.

A *finite affine plane* over GF(p^n), to be denoted by AG$(2, p^n)$, is the system of points and lines which is obtained from PG$(2, p^n)$ by deleting any one line and its points. For instance, AG$(2, 3)$ (figure v) is obtained from PG$(2, 3)$ by deleting the line $[0, 0, 1]$ and its points. The 9 points of AG$(2, 3)$ thus obtained may be indicated by their first two co-ordinates alone, since the third co-ordinate can be normalized to 1. For any $p^n = q$, the q^2 points and the $q^2 + q$ lines of AG$(2, p^n)$ may be

---

*By use of more sophisticated algebraic systems it is possible to construct other projective planes. Compare M. Hall, *Theory of Groups*, New York, 1959, p. 346.

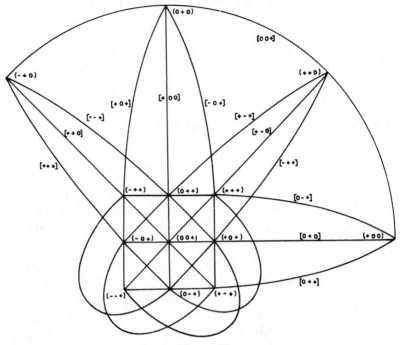

Figure viii

indicated by the ordered pairs (ξ, η) and by the equations $\alpha\xi + \beta\eta + \gamma = 0$, respectively, where ξ, η, α, β, γ run through $GF(p^n)$.

By use of Galois fields, higher-dimensional geometries may also be constructed. For three dimensions, two types of ordered tetrads, (x_1, x_2, x_3, x_4) and $[X_1, X_2, X_3, X_4]$, of elements of $GF(p^n)$ are considered. The proportional tetrads of each of these types yield the points and the planes, respectively, of the *finite projective space* $PG(3, p^n)$ of dimension 3. The point-plane incidence is given by the relation

$$x_1 X_1 + x_2 X_2 + x_3 X_3 + x_4 X_4 = 0.$$

The lines of $PG(3, p^n)$ are the intersections of pairs of planes or, equivalently, the joins of pairs of points, that is, all sets of points

$$(\lambda a_1 + \mu b_1,\ \lambda a_2 + \mu b_2,\ \lambda a_3 + \mu b_3,\ \lambda a_4 + \mu b_4) \neq (0,\ 0,\ 0,\ 0),$$

where (a_1, a_2, a_3, a_4) and (b_1, b_2, b_3, b_4) are distinct points while λ and μ run through $GF(p^n)$.

As an example we consider the *binary projective space* PG(3, 2). It has 15 points, namely all ordered quadruples of zeros and ones, except for $(0, 0, 0, 0)$. It also has 15 planes, and each plane forms a PG(2, 2) with 7 points and 7 lines. PG(3, 2) has 35 lines, each consisting of any two points and their sum modulo 2. In fact, the points and the lines of PG(3, 2) form one of the 80 Steiner triple systems of order 15 (compare page 281).

A better way to represent PG(3, 2), with its 15 points, is by the Galois field $GF(2^4)$ with its 15 non-zero elements, that is, the 15 polynomials in x modulo $x^4 + x + 1$ with coefficients 0 and 1 (compare page 74). Since $x^4 = x + 1$, all polynomials are powers of x. We refer to the following "logarithmic" table, where each polynomial is represented by its detached coefficients, so that $10011 = 0$:

$$
\begin{aligned}
x^0 &= & & & 1 &= 0\ 0\ 0\ 1 \\
x^1 &= & & x & &= 0\ 0\ 1\ 0 \\
x^2 &= & x^2 & & &= 0\ 1\ 0\ 0 \\
x^3 &= x^3 & & & &= 1\ 0\ 0\ 0 \\
x^4 &= & & x+1 & &= 0\ 0\ 1\ 1 \\
x^5 &= & x^2+x & & &= 0\ 1\ 1\ 0 \\
x^6 &= x^3+x^2 & & & &= 1\ 1\ 0\ 0 \\
x^7 &= x^3 & & +x+1 & &= 1\ 0\ 1\ 1 \\
x^8 &= & x^2 & +1 & &= 0\ 1\ 0\ 1 \\
x^9 &= x^3 & & +x & &= 1\ 0\ 1\ 0 \\
x^{10} &= & x^2+x+1 & & &= 0\ 1\ 1\ 1 \\
x^{11} &= x^3+x^2+x & & &= 1\ 1\ 1\ 0 \\
x^{12} &= x^3+x^2+x+1 & & &= 1\ 1\ 1\ 1 \\
x^{13} &= x^3+x^2 & +1 & &= 1\ 1\ 0\ 1 \\
x^{14} &= x^3 & & +1 & &= 1\ 0\ 0\ 1 \\
\end{aligned}
$$

Any 2 of the 15 points determine a line which consists of those 2 points and their sum. For instance, the line joining

x^4 and x^5 contains as a third point $x^4 + x^5 = 0011 + 0110 = 0101 = x^8$, or, more simply, $x^4 + x^5 = x^4(1 + x) = x^4 x^4 = x^8$. Hence, the operation of multiplying by x is a *collineation* of period 15, mapping points into points, lines into lines, and (consequently) planes into planes. The cyclic group generated by this collineation clearly is transitive on the 15 points x^0, $x^1, \ldots, x^{14}$; but it is not transitive on the 35 lines. It yields a partition of the set of these lines into three sets. One set contains 15 lines such as $\{x^0, x^1, x^4\}$ (arising from the equation $x^4 + x + 1 = 0$); another set contains 15 lines such as $\{x^0, x^2, x^8\}$; the last set contains 5 lines such as $\{x^0, x^5, x^{10}\}$. This partition* will be used in the next section for the solution of Kirkman's school-girl problem.

The *finite affine space* AG$(3, p^n)$ of dimension 3 is obtained from PG$(3, p^n)$ by deleting any one plane with its points and lines. Thus, its $(p^n)^3$ points are given by the ordered triples out of GF(p^n). But again we can do better. For example, the 27 points of the *ternary affine space* AG$(3, 3)$ may be described by the elements of GF(3^3), that is, by 0 and the 26 powers of a primitive root x, which satisfies $x^3 = x - 1$. Then modulo 3 we have $x^{13} = x^4(x - 1)^3 = x^4(x^3 - 1) = x^4(x + 1) = x(x^2 - 1) = -1$. Thus the 27 points are as indicated by their exponents in figure ix:

$$x^0 = 0\,0+, \quad x^1 = 0+0, \quad x^2 = +0\,0, \quad x^3 = 0+-,$$
$$x^4 = +-0, \quad \ldots, \quad x^{13} = 0\,0-, \quad \ldots$$

and $x^\infty = 000$ by definition. Again any 3 points are collinear if

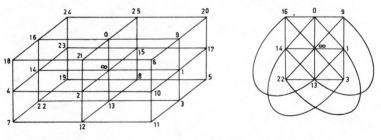

Figure ix

their sum is zero, for instance $\{x^\infty, x^0, x^{13}\}$ and $\{x^0, x^4, x^5\}$. Multiplication by x is now an *affinity* leaving the origin fixed. It maps any line $\{x^\infty, x^i, x^{i+13}\}$ and its 8 parallels onto the line $\{x^\infty, x^{i+1}, x^{i+14}\}$ and its 8 parallels. Therefore, the cyclic group generated by this affinity partitions the 117 lines of AG(3, 3) into 13 bundles of 9 parallel lines each, as follows:

$(\infty, 3, 16)$	$(0, 1, 22)$	$(4, 6, 12)$	$(7, 10, 21)$	$(11, 18, 2)$
$(\infty, 4, 17)$	$(1, 2, 23)$	$(5, 7, 13)$	$(8, 11, 22)$	$(12, 19, 3)$
$(\infty, 5, 18)$	$(2, 3, 24)$	$(6, 8, 14)$	$(9, 12, 23)$	$(13, 20, 4)$
etc.	etc.	etc.	etc.	etc.
	$(13, 14, 9)$	$(17, 19, 25)$	$(20, 23, 8)$	$(24, 5, 15)$
	$(14, 15, 10)$	$(18, 20, 0)$	$(21, 24, 9)$	$(25, 6, 16)$
	$(15, 16, 11)$	$(19, 21, 1)$	$(22, 25, 10)$	$(0, 7, 17)$
	etc.	etc.	etc.	etc.

Analogously, the points of the ternary affine geometry AG(n, 3) of dimension n are described by the 3^n elements of GF(3^n). The number b_n of lines in AG(n, 3) is calculated by arranging the points in the 3 hyperplanes $x_n = 1, x_n = 0, x_n = -1$, each having 3^{n-1} points. It follows that $b_n = 3^{2(n-1)} + 3b_{n-1}$, whence $b_n = \frac{1}{2} 3^{n-1}(3^n - 1)$. Every line has 3 points, every point is on $\frac{1}{2}(3^n - 1)$ lines, and every pair of points is on one line. Therefore, AG(n, 3) contains $\frac{1}{2}(3^n - 1)$ bundles of parallel lines.

Kirkman's school-girl problem. "A school-mistress is in the habit of taking her girls for a daily walk. The girls are fifteen in number, and are arranged in five rows of three each, so that each girl might have two companions. The problem is to dispose them so that for seven consecutive days no girl will walk with any of her school-fellows in any triplet more than once. In the general problem we require to arrange n girls, where n is an odd multiple of 3, in triplets to walk out for $\frac{1}{2}(n-1)$ days, so that no girl will walk with any of her school-fellows in any triplet more than once."

In this formulation the subject of the present section was introduced in the original chapter x of this book, which carries the same title. That chapter, written in 1892 and revised a number of times, consisted of a variety of solutions and methods of solutions to the problem, without solving the general case. In fact, it was only in 1969 that the complete solution was given by D.K. Ray-Chaudhuri and R.M. Wilson.

The special case of 15 girls was first enunciated by T.P. Kirkman* in 1847, thus anticipating Steiner. The general case asks for a Steiner triple system $S(v)$ with *parallelism*, that is, with a partition of the collection of triples in such a way that, for any point P and any triple l, there is a unique triple in the part containing l, which contains P and does not intersect l. In the example it is not sufficient to provide 35 triples out of the 15 girls such that any pair of girls occurs once in a triple. In addition, on each of the 7 days the 15 girls should be arranged in 5 triples. So, the 35 triples should be partitioned into 7 parallel classes of 5 triples each. The general school-girl problem asks for a *Kirkman system*, that is, a Steiner triple system with parallelism, having the parameters $v = 6t+3$, $b = (3t+1)(2t+1)$, $r = 3t+1$, $k = 3$, $\lambda = 1$.

Ray-Chaudhuri and Wilson† showed that this problem admits at least one solution for every non-negative integer t. They develop composition theorems by use of which a design of larger order is constructed from a number of designs of smaller order. To that end block designs with unequal block size, and other types of combinatorial designs, are used. In addition, they construct so many designs of small order that by application of the composition theorems all orders $v \equiv 3$ modulo 6 are covered. Earlier editions of the present book contain direct constructions of Kirkman systems for orders $9 \leqq v \leqq 99$.

Although the existence problem for Kirkman systems was solved in 1969, it remains to be determined how many different Kirkman systems the various orders $v \equiv 3$ modulo 6 admit. The only known results are that for $v = 9$ there is essentially 1, and for $v = 15$ there are‡ essentially 7 such systems.

*T. Kirkman, *Cambridge and Dublin Mathematical Journal*, 1847, vol. ii, pp. 191–204.

†D.K. Ray-Chaudhuri and R.M. Wilson, *Proceedings Symposia Pure Mathematics, Combinatorics*, 1971, vol. lxx, pp. 187–203.

‡P. Mulder, *Kirkman-Systemen* (Groningen Dissertation, Leiden, 1917); F.N. Cole, *Bulletin of the American Mathematical Society*, 1922, vol. xxviii, pp. 435–437.

We shall now give a solution for $v = 15$, and one for $v = 27$. We recall (see page 285) that the binary projective space PG(3, 2) has 15 points which are represented by the powers of a primitive element x of GF(2^4). Multiplication by x partitions the 35 lines into 3 classes, of sizes 5, 15, 15, viz.

$$\{x^0, x^5, x^{10}\}, \{x^1, x^6, x^{11}\}, \{x^2, x^7, x^{12}\}, \{x^3, x^8, x^{13}\},$$
$$\{x^4, x^9, x^{14}\},$$

$$\{x^0, x^1, x^4\}, \{x^1, x^2, x^5\}, \{x^2, x^3, x^6\}, \text{ etc.},$$

$$\{x^0, x^2, x^8\}, \{x^1, x^3, x^9\}, \{x^2, x^4, x^{10}\}, \text{ etc.},$$

Interpreting the 15 points as the girls, and the 35 lines as the rows, we have to find 7 sets of 5 lines, each of which contains all 15 points. Any such set, which is called a *spread* of lines, consists of 5 mutually skew lines, since any line contains 3 points. Now the first class constitutes one spread. The second class does not contain a spread, nor does the third class. However, the second class does contain 4 mutually skew lines and it so happens that the remaining 3 points, which are not covered by these lines, are on a line of the third class. After this observation a simple inspection leads to the following solution, where the girls are denoted by i instead of by x^i:

Sun.	Mon.	Tues.	Wed.	Thurs.	Fri.	Sat.
0, 5, 10	0, 1, 4	1, 2, 5	4, 5, 8	2, 4, 10	4, 6, 12	10, 12, 3
1, 6, 11	2, 3, 6	3, 4, 7	6, 7, 10	3, 5, 11	5, 7, 13	11, 13, 4
2, 7, 12	7, 8, 11	8, 9, 12	11, 12, 0	6, 8, 14	8, 10, 1	14, 1, 7
3, 8, 13	9, 10, 13	10, 11, 14	13, 14, 2	7, 9, 0	9, 11, 2	0, 2, 8
4, 9, 14	12, 14, 5	13, 0, 6	1, 3, 9	12, 13, 1	14, 0, 3	5, 6, 9

Proceeding with Kirkman's problem for 3^n girls we consider the affine geometry AG($n, 3$) of page 287. Interpreting the 3^n points as the girls and the $\frac{1}{2} 3^{n-1}(3^n - 1)$ lines as the rows, we have to find $\frac{1}{2}(3^n - 1)$ sets of 3^{n-1} lines, each of which contains all points. A solution is obtained by the partition of the set of all lines into $\frac{1}{2}(3^n - 1)$ bundles of mutually parallel lines. For $n = 3$ this solution is explicitly given on page 287.

Latin squares.   We recall from page 189 that a *Latin square* of order n is a square matrix of order n, every row and every column of which is a permutation of n letters. Two Latin squares of order n are *orthogonal* (Eulerian) if in their superposition all of the n^2 ordered pairs of letters occur exactly once. The first three of the following four Latin squares of order 4 are mutually orthogonal:

$$\begin{bmatrix} a & b & c & d \\ b & a & d & c \\ c & d & a & b \\ d & c & b & a \end{bmatrix}, \begin{bmatrix} a & b & c & d \\ c & d & a & b \\ d & c & b & a \\ b & a & d & c \end{bmatrix}, \begin{bmatrix} a & b & c & d \\ d & c & b & a \\ b & a & d & c \\ c & d & a & b \end{bmatrix}, \begin{bmatrix} a & b & c & d \\ b & c & d & a \\ c & d & a & b \\ d & a & b & c \end{bmatrix}.$$

However, the fourth Latin square is inextensible, having no orthogonal mate.

The notions of Latin square, and of mutually orthogonal Latin squares, admit various recreational, mathematical, and applied interpretations. We mention five examples, all concerning Latin squares of order 4.

How can one arrange 16 girls, on each of x days, in 4 subsets of 4 girls each, in such a way that no 2 girls will meet together in a subset on more than one day? For $x=3$, an answer is given by any Latin square of order 4 by interpreting the entries of the matrix as the girls, to be arranged on the first day according to the rows, on the second day according to the columns, and on the third day according to the letters. For $x=4$ a solution is provided by any pair of orthogonal Latin squares, and for $x=5$ by the 3 mutually orthogonal Latin squares of order 4.

Suppose we wish to compare the yields of 4 varieties of wheat in an agricultural experiment. In order to eliminate unavoidable variations in the structure and the fertility of the soil, the field of the experiment is subdivided into 16 plots, arranged in 4 rows and 4 columns. Then the 4 varieties are distributed over these plots according to a Latin square, so that each variety is planted once in each row and once in each column. Thus in comparing the varieties, systematic changes in soil fertility along rows and columns are eliminated. In the same experiment 4 different methods of manuring can be investigated provided they are applied according to a Latin square orthogonal to the previous one, so that each method of manuring is applied once to each variety of wheat.

If the resistance to wear of 4 brands of automobile tires is to be compared, it is obviously desirable to use one tire of each brand on the 4 wheels of one car. But the loads and hence the amount of wear may differ in these 4 positions and may also vary from week to week owing to different weather conditions. Hence, a suitable experiment will be to use the 4 tires for 4 weeks and to interchange the 4 positions from week to week according to a Latin square.

In any pair of orthogonal Latin squares of order 4 we indicate the rows, the columns, the Latin letters, and the Greek letters, by the same set of symbols $\{1, 2, 3, 4\}$. The pair of orthogonal Latin squares may be seen as a set of 16 ordered quadruples out of the symbol set $\{1, 2, 3, 4\}$ in such a way that for each pair of co-ordinates every pair of symbols occurs exactly once. Therefore, any two distinct quadruples differ in at least 3 co-ordinates. So we have a code, consisting of 16 code words with 4 co-ordinates each, using a set of 4 symbols, with minimal distance 3; hence the code is 1-error-correcting.

Our final example deals with experiments involving television screens. The cells of an $n \times n$ square grid are to be coloured with n different shades of grey such that they are contrasted in pairs as ordered neighbours, both horizontally and vertically. The following examples, for $n = 4$ and for $n = 6$, use Latin squares:

$$
\begin{bmatrix} a & b & c & d \\ c & a & d & b \\ b & d & a & c \\ d & c & b & a \end{bmatrix}
\qquad
\begin{bmatrix} a & b & c & d & e & f \\ b & d & f & a & c & e \\ c & f & b & e & a & d \\ d & a & e & b & f & c \\ e & c & a & f & d & b \\ f & e & d & c & b & a \end{bmatrix}
$$

These examples serve to establish a better understanding of the notion of a Latin square. Indeed, in several cases the special role of rows, columns, and letters is irrelevant, and the only property used is that we dispose of n^2 ordered triples out of a set of n symbols such that for each pair of co-ordinates every pair of symbols occurs once. If this property is taken as the definition, then for the number m of Latin squares of order n we have

n	2	3	4	5	6	7	8
m	1	1	2	2	12	147	$>250{,}000$

Under the original definition there are 110,592 Latin squares of order 4. If permutations of rows, of columns, of letters, and an interchange of rows and columns are permitted, then this number equals 4. Under the new definition this number is 2, the types being represented by an inextensible and an extensible Latin square.

The new definition also applies to orthogonal Latin squares. A set of $k-2$ mutually orthogonal Latin squares of order n then is a set of n^2 ordered k-tuples out of n symbols such that for each pair of co-ordinates every pair of symbols occurs once. Thus, any pair of k-tuples has either one or no co-ordinates in common.

As we remarked on page 192, Euler conjectured in 1782 that no pair of mutually orthogonal squares of order $n=4k+2$ exists. This conjecture was proved for $n=6$ by Tarry, and disproved for all $n=4k+2$, $k>1$, by Bose, Parker, and Shrikhande.* The reader should realize that the conjecture, which stood firm for 177 years, turned out to be true only in one case, and false in all other cases. Once known, it is easy to give the following example of 2 orthogonal Latin squares of order $n=10$, which immediately generalizes to $n=3m+1$:

```
0 6 5 4 7 8 9|1 2 3   0 9 8 7 1 3 5|2 4 6
9 1 0 6 5 7 8|2 3 4   6 1 9 8 7 2 4|3 5 0
8 9 2 1 0 6 7|3 4 5   5 0 2 9 8 7 3|4 6 1
7 8 9 3 2 1 0|4 5 6   4 6 1 3 9 8 7|5 0 2
1 7 8 9 4 3 2|5 6 0   7 5 0 2 4 9 8|6 1 3
3 2 7 8 9 5 4|6 0 1   8 7 6 1 3 5 9|0 2 4
5 4 3 7 8 9 6|0 1 2   9 8 7 0 2 4 6|1 3 5
----------------      ----------------
2 3 4 5 6 0 1|7 8 9   1 2 3 4 5 6 0|7 8 9
4 5 6 0 1 2 3|9 7 8   2 3 4 5 6 0 1|8 9 7
6 0 1 2 3 4 5|8 9 7   3 4 5 6 0 1 2|9 7 8
```

*R.C. Bose, S.S. Shrikhande, and E.T. Parker, *Canadian Journal of Mathematics*, 1960, vol. XII, pp. 189–203. The frontispiece shows a modified version.

Denoting the blocks by

$$\frac{A\ \mid B}{C\ \mid D}, \quad \frac{A^T\mid C^T}{B^\Gamma\mid E}$$

we explain the construction* for $n = 3m + 1$ (so read $m = 3$ in the example) as follows. The main diagonal of A consists of the symbols $0, 1, 2, \ldots, 2m$. Half of the $2m$ remaining parallel diagonals of A consist of the same symbols, but shifted, beginning with $2m, 2m - 1, \ldots, m + 1$, respectively. The other half are diagonals with the constant elements $2m + 1, 2m + 2, \ldots, 3m$, respectively. The columns of B [rows of C] contain the symbols in cyclic order, beginning with $1, 2, \ldots, m$ [with $2, 4, \ldots, 2m$], respectively. D and E stand for two mutually orthogonal Latin squares on the symbols $2m + 1, 2m + 2, \ldots, 3m$. Apart from E the second square of order $3m + 1$ is obtained from the first by transposition. In the case when $m = 3$, the partition of 10 into $7 + 3$ can be disguised so that the Eulerian square appears in the more symmetrical form shown in the frontispiece.

The story of the falsity of Euler's conjecture begins in 1922 with MacNeish, who constructed t mutually orthogonal Latin squares of order n, where t is one less than the smallest prime power in the prime power decomposition of n. Furthermore, MacNeish conjectured that, for every n, no more than t such squares can exist. Thus, for $n = 4k + 2 = 2(2k + 1)$, the case under consideration, there would be no Eulerian square. However, Parker (1959) disproved MacNeish's conjecture by constructing 3 mutually orthogonal Latin squares of order 21, by use of a projective plane $PG(2, 2^2)$. Then Bose and Shrikhande, by using a generalization of block designs, with unequal block size, constructed 5 mutually orthogonal Latin squares of order 50, thus disproving Euler's conjecture. By using a modified Kirkman system they also constructed an Eulerian square of order 22. Then Parker found an Eulerian square of

*P. Kesava Menon, *Sankhya*, 1961, vol. A XXIII, pp. 281–282.

order 10. In the subsequent weeks the three mathematicians settled all other cases with the result stated above.

The news that Euler's conjecture was disproved hit the headlines of the world press. The *New York Times* of April 26, 1959, contained an extensive article and a picture. Although only part of the Eulerian square was visible in the picture, dozens of schoolboys managed to fill in the missing part.

We conclude this section by a simple construction of Latin squares which are orthogonal to their transpose. Let GF(q) be any Galois field, $q \neq 2, 3$ (for instance, the residue classes modulo a prime number greater than 3). Let λ be a fixed element of GF(q), with $\lambda \neq 0$, $\lambda \neq 1$, $\lambda \neq \frac{1}{2}$. Consider the matrix

$$[\lambda a + (1-\lambda)b],$$

where a and b run through GF(q). By hypothesis, no row or column contains an element of the field twice. So we have a Latin square. This matrix and its transpose

$$[\lambda b + (1-\lambda)a]$$

form an Eulerian square. Indeed, $\lambda a + (1-\lambda)b = \lambda c + (1-\lambda)d$ and $\lambda b + (1-\lambda)a = \lambda d + (1-\lambda)c$ imply $a + b = c + d$ and, by substitution,

$$(1-2\lambda)b = (1-2\lambda)d,$$

whence $b = d$ and $a = c$. By standard methods this construction is extended to any order n whose prime factorization does not contain 2 or 3 to the first power. Here is an example, based on GF(2^3):

$$
\begin{bmatrix}
0 & 3 & 5 & 6 & 7 & 1 & 4 & 2 \\
2 & 1 & 4 & 7 & 6 & 3 & 5 & 0 \\
4 & 7 & 2 & 1 & 3 & 6 & 0 & 5 \\
5 & 6 & 0 & 3 & 1 & 7 & 2 & 4 \\
3 & 0 & 6 & 5 & 4 & 2 & 7 & 1 \\
7 & 4 & 1 & 2 & 0 & 5 & 3 & 6 \\
1 & 2 & 7 & 4 & 5 & 0 & 6 & 3 \\
6 & 5 & 3 & 0 & 2 & 4 & 1 & 7
\end{bmatrix}
$$

The cube and the simplex.   In the following sections we shall take as our point of departure the ordinary cube in 3-space (figure x). For the co-ordinates of its vertices we shall use sometimes 0 and 1 (with the origin O at a vertex and edge 1), and sometimes $+1$ and -1 (with the origin O at the centre and edge 2). The cube-graph, as indicated in the third part of figure

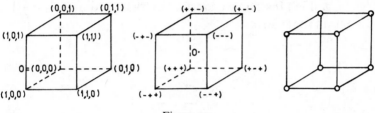

Figure x

x, is the set of the 8 vertices together with the 12 adjacencies (edges) between vertices which differ in exactly one co-ordinate.

The *n-cube* has 2^n vertices in *n*-space, namely the set V of points $(x_1, x_2, \ldots, x_n)$, where each $x_i = 0$ or 1. Its edge set E consists of the pairs of vertices that differ in exactly one co-ordinate.

The *n-simplex* in *n*-space is the *n*-polytope that has $n+1$ vertices whose mutual distances are all equal. It is the generalization of the equilateral triangle $(n=2)$ and the regular tetra-

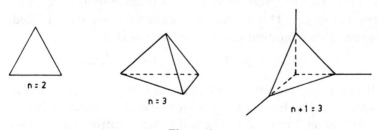

Figure xi

hedron $(n=3)$. The *n*-simplex also can be defined by the $(n+1)$-tuples $(1, 0, 0, \ldots, 0), (0, 1, 0, \ldots, 0), \ldots, (0, 0, 0, \ldots, 1)$, that is, by $n+1$ points in $(n+1)$-space which are in the *n*-dimensional subspace $x_1 + \ldots + x_{n+1} = 1$ (as in the last part of figure xi).

Hadamard matrices.   Given the 8 vertices of a cube in 3-space (as in the second part of figure x), can we select 4 of them so as to form a regular tetrahedron? An affirmative answer to this question is provided by the vectors

$$(+ \; + \; +), (+ \; - \; -), (- \; + \; -), (- \; - \; +),$$

which have length $\sqrt{3}$ and inner product -1 (for each pair). (See page 134.) Inserting a constant extra co-ordinate $x_0 = +1$, we obtain the matrix

$$H_4 = \begin{bmatrix} + & + & + & + \\ + & + & - & - \\ + & - & + & - \\ + & - & - & + \end{bmatrix},$$

whose pairs of rows have inner product 0. Therefore H_4 is an Hadamard matrix of order 4. On page 274 we constructed an Hadamard matrix of order 8, and on page 302 we shall construct one of order 12.

An Hadamard matrix H_r is a square matrix of order r, with elements $+1$ and -1, whose rows are orthogonal in pairs:

$$H_r H_r^T = rI.$$

The problem of constructing Hadamard matrices H_r of order r is equivalent* to the problem of selecting, in $(r-1)$-space, r out of the 2^{r-1} vertices of the $(r-1)$-cube, so as to form an $(r-1)$-simplex. This is obviously impossible for $r=3$. Indeed, a necessary condition for the existence of H_r is

$$r=2 \text{ or } r=4s, \quad s \text{ a positive integer.}$$

It has been conjectured that the necessary condition is also sufficient. This has been verified for all $r < 188$ and for infinitely many other values of r. The following iterative construction for an infinite series of Hadamard matrices of order 2^t $(t = 1, 2, 3, \ldots)$ goes back to Sylvester (see page 107):

*H.S.M. Coxeter, *Journal of Mathematics and Physics*, 1933, vol. XII, pp. 334–345.

$$H_2 = \begin{bmatrix} + & + \\ + & - \end{bmatrix}, \quad H_4 = \begin{bmatrix} H_2 & H_2 \\ H_2 & -H_2 \end{bmatrix}, \quad H_8 = \begin{bmatrix} H_4 & H_4 \\ H_4 & -H_4 \end{bmatrix}, \ldots .$$

The H_4 constructed above from the cube coincides with this H_4. The Hadamard matrix constructed on page 274 is, upon rearrangement, the same as this H_8.

From the H_{32}, obtained by the iterative construction, we form the 32×64 matrix

$$[H_{32} \quad -H_{32}].$$

Referring again to page 274, where the case $r = 8$ was considered, we observe that the columns of this matrix form a linear (32, 6) code which is 7-error-correcting. In March 1969 the Mariner 1969 spacecraft was launched toward the planet Mars. The pictures of Mars were received on earth five months later. The code used in the Mariner 1969 High Rate Telemetry System was based on the (32, 6) code described above.*

Picture transmission. Suppose we have to assign the integers $0, 1, 2, \ldots, 2^n - 1$ to the vertices of the n-cube graph, with vertex set V and edge set E, in a one-to-one way:

$$\phi : V \leftrightarrow \{0, 1, 2, \ldots, 2^n - 1\}.$$

For $n = 3$ there are $(2^n)! = 40320$ possible assignments, 3 of which are given in figure xii. We are interested in the question which of these ϕ minimizes the value of

$$\Phi = \sum_{(v,w) \, \varepsilon \, E} [\phi(v) - \phi(w)]^2.$$

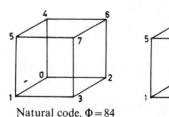
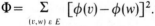
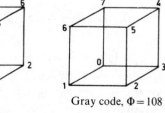

| Natural code, $\Phi = 84$ | $\Phi = 86$ | Gray code, $\Phi = 108$ |

Figure xii

*Edward C. Posner, "Combinatorial Structures in Planetary Reconnaissance," in H.B. Mann, *Error Correcting Codes*, New York, 1968; see fig. III on p. 23.

In the examples we have $\Phi = 84, 86, 108$, respectively, so the first assignment is the best one of the three. We shall prove that, for $n = 3$, this assignment essentially is best of all, without calculating Φ in all 40320 cases.

Let a vertex v be called odd or even according to the parity of the number of ones among its co-ordinates. Let v^c be the vertex opposite to v, that is, the member of the vertex set V which differs from v in every co-ordinate, e.g. $(0, 1, 0)^c = (1, 0, 1)$. Then, for $n = 3$, we have

$$\sum_{(v,w)\, \varepsilon\, E} [\phi(v) - \phi(w)]^2 = 3 \sum_{v \varepsilon V} \phi^2(v) - 2 \sum_{(v,w)\, \varepsilon\, E} \phi(v)\phi(w)$$

$$= 3 \sum_{v\, \varepsilon\, V} \phi^2(v) - 2 \sum_{v \text{ odd}} \phi(v) \sum_{w \text{ even}} \phi(w) + 2 \sum_{v, v^c\, \varepsilon\, V} \phi(v)\phi(v^c).$$

The first term is independent of ϕ. The second term is maximized if

$$\sum_{v \text{ odd}} \phi(v) = \sum_{w \text{ even}} \phi(w) = 14,$$

since $\qquad \sum_{v\, \varepsilon\, V} \phi(v) = 0 + 1 + \ldots + 7 = 28.$

For the same reason the last term is minimized if the summands are $0 \times 7, 1 \times 6, 2 \times 5, 3 \times 4$. The ϕ defined by

$$\phi(v) = \sum_{i=0}^{n-1} v_i 2^i$$

is easily seen to satisfy both of these conditions, and so minimizes Φ. Since the only solution of

$$\sum_{v \text{ odd}} \phi(v) = n_1 + n_2 + n_3 + 7 = 14, \quad 1 \leq n_1 < n_2 < n_3 \leq 6,$$

is $n_1 = 1$, $n_2 = 2$, $n_3 = 4$, our ϕ is unique in this respect. The solution thus obtained is the *natural one*: it assigns to each vertex the integer of which its co-ordinates form the binary expansion.

This mathematical problem has practical consequences. In engineering terms, the structure of a binary symmetric channel, with block length n, with low probability of error, in which only single errors are likely, is approximated by the n-cube graph. Suppose numerical data points $0, 1, 2, \ldots, 2^n - 1$ are to be transmitted, for instance the 2^n brightness levels of one of the thousand parts of a picture. Then the assignment ϕ becomes a *code* and

$$\sum_{(v,w)\, \varepsilon\, E} [\phi(v) - \phi(w)]^2$$

is proportional to the *noise power*, which it is obviously desirable to minimize. Both the natural and the Gray code are known to minimize the "average absolute error"

$$\sum_{(v,w)\, \varepsilon\, E} |\phi(v) - \phi(w)|,$$

but within that restriction the natural code minimizes the noise power, and the Gray code maximizes it. Consequently, the subjective effect of the errors is greater with the Gray code. A solution, for all n, of a more general problem has recently been given.*

Equiangular lines in 3-space. We take another look at the cube $(ADBC)$ $(B'C'A'D')$ with centre O (figure xiii). For the angles between the vectors OA, OB, OC, OD we have

$$\cos AOD = \cos DOB = \cos BOC = \cos COA = \tfrac{1}{3},$$

$$\cos AOB = \cos COD = -\tfrac{1}{3},$$

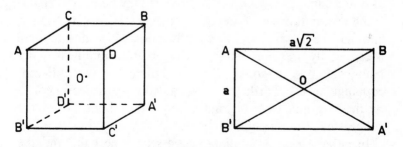

Figure xiii

*IEEE Transactions, Information Theory, 1969, vol. IT–15, pp. 72–78.

as we easily verify by considering a diagonal plane. It follows that the pairs of the 4 diagonals AA', BB', CC', DD' of the cube all have the same angle, if we define the angle of a pair of lines to be the smaller of the two supplementary angles formed by the lines. Calling a set of lines *equiangular* whenever each pair in the set has the same angle, we have found an equiangular set of 4 lines.

We can do better. Indeed, consider the 6 lines which connect the pairs of antipodal vertices of an icosahedron (figure xiv).

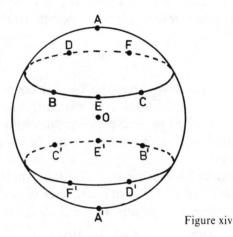

Figure xiv

By elementary symmetry arguments it is easily seen that the six diagonals AA', BB', CC', DD', EE', FF' of the icosahedron also form an equiangular set of lines. It may be proved that 6 is the maximal order of an equiangular set of lines in 3-space, and that the present set is essentially unique. In addition, there is essentially one equiangular set of order 5, obtained by deleting any one diagonal. However, there are two different equiangular sets of order 4, one obtained from the icosahedron by deleting any two diagonals, and one obtained from the cube, as has been explained above.*

In either case, the equiangular 4-set of the cube and the equiangular 6-set of the icosahedron, we take unit vectors

---

*The lines of these two equiangular sets are parallel to the edges of the two kinds of rhombic dodecahedron: Bilinski's (see page 143) and the classical one.

$p_1, p_2, p_3, \ldots$ along the lines, from O in the unprimed directions A, B, C, $\ldots$. Since all inner products of these unit vectors satisfy $|p_i . p_j| = \cos \phi$, where ϕ is the angle between the lines, the matrix P of the inner products is as follows:

$$P = [(p_i . p_j)] = \begin{bmatrix} 1 & & & & \pm\cos\phi \\ & 1 & & & \\ & & \cdot & & \\ & & & \cdot & \\ & & & & \cdot \\ \pm\cos\phi & & & & 1 \end{bmatrix}$$

By taking out the diagonal and dividing by $\cos \phi$, we arrive at the matrix

$$A = [P - I] \sec \phi,$$

which, in our examples, has the form:

$$A_4 = \begin{bmatrix} 0 & - & + & + \\ - & 0 & + & + \\ + & + & 0 & - \\ + & + & - & 0 \end{bmatrix}, \quad A_6 = \begin{bmatrix} 0 & + & + & + & + & + \\ + & 0 & - & + & + & - \\ + & - & 0 & - & + & + \\ + & + & - & 0 & - & + \\ + & + & + & - & 0 & - \\ + & - & + & + & - & 0 \end{bmatrix}.$$

These matrices indicate which of the pairs of vectors among OA, OB, OC, $\ldots$ have an acute angle (positive cosine), and which have an obtuse angle (negative cosine). Each of these matrices may be considered as the adjacency matrix of a graph, by taking any pair of vertices A, B, C, $\ldots$ adjacent whenever the corresponding matrix element is -1, and non-adjacent otherwise (see figure xv). These matrices, and the corresponding

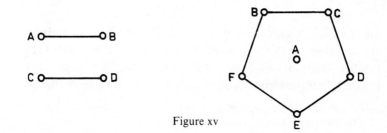

Figure xv

graphs, provide a basis for our further considerations. By a simple verification it is observed that the matrices A_4 and A_6 satisfy

$$(A_4 - 1)(A_4 + 3I) = 0, \quad A_4 J = J; \quad A_6{}^2 = 5I.$$

Hence A_6 is an orthogonal matrix: each pair of its rows has inner product zero. We note three consequences of this observation.

First, since the smallest *eigenvalue* (or *characteristic root*) of A_6 is $-\sqrt{5}$ and the corresponding P_6 is non-negative definite of rank 3, the angle ϕ between the diagonals of the icosahedron satisfies $-\sec \phi = -\sqrt{5}$, whence $\cos \phi = 1/\sqrt{5}$. Secondly, the matrix

$$H_{12} = \begin{bmatrix} A_6 + I_6 & A_6 - I_6 \\ A_6 - I_6 & -A_6 - I_6 \end{bmatrix}$$

is an Hadamard matrix of order 12.

For the third application,* the 12 columns $c_1, c_2, \ldots, c_{12}$ of the 6×12 matrix

$$[A_6 \quad I_6]$$

are taken as 6-dimensional vectors over GF(3). By inspection it is observed that no 5 of these columns are dependent. This means that, in every set of numbers $\alpha_1, \alpha_2, \ldots, \alpha_{12}$ out of $\{0, 1, -1\}$ such that

$$\alpha_1 c_1 + \alpha_2 c_2 + \ldots + \alpha_{12} c_{12} = 0,$$

at least 6 numbers are non-zero. Now such sets of numbers are easily provided; namely, by the 6 rows of the matrix

$$[I_6 \quad -A_6]$$

and by all 3^6 linear combinations of these rows. This 6-dimensional subspace of the 12-dimensional vector space over GF(3) is (in a terminology analogous to that used on page 275) a linear (12, 6) ternary code. Since all its vectors have distance

*R.C. Bose, *Bulletin de l'Institut International de Statistique*, 1961, vol. XXXVIII, pp. 257–271.

≥ 6, the code is 2-error-correcting and 3-error-detecting. By deleting any one co-ordinate we obtain an $(11,6)$ code of minimal distance 5. Therefore, the spheres of radius 2 about any 2 vectors of the 6-dimensional subspace of the 11-dimensional ternary vector space are disjoint. Since there are 3^{11} vectors in total, 3^6 vectors in the subspace, and $1+22+220$ vectors in each sphere, it follows that these spheres exhaust the vector space. Therefore the linear $(11, 6)$ ternary 2-error-correcting code is a perfect code. This code has been discovered by Golay,[*] together with a further linear $(23, 12)$ binary 3-error-correcting perfect code. They provide the only such codes[†] which correct more than one error.

Lines in higher-dimensional space. We make an excursion into higher dimensions. Which sets of equiangular lines exist in 7-space? An answer to this question is obtained by consideration of the 7-simplex whose 8 vertices are represented by the 8 vectors

$$(8, 0, 0, \ldots, 0), (0, 8, 0, \ldots, 0), \text{ etc.}$$

in 8-space. The hyperplane in 8-space with the equation $x_1 + x_2 + \ldots + x_8 = 8$, which contains these 8 vertices, is shifted parallel to itself into $x_1 + x_2 + \ldots + x_8 = 0$. Then the vertices become

$$(7, -1, -1, \ldots, -1), (-1, 7, -1, \ldots, -1), \text{ etc.,}$$

and the 28 mid-points $P_{h,i}$ $(h < i; h, i = 1, \ldots, 8)$ of the edges of the 7-simplex become

$$P_{1,2} = (3, 3, -1, -1, -1, -1, -1, -1),$$

$$P_{1,3} = (3, -1, 3, -1, -1, -1, -1, -1),$$

$$P_{3,4} = (-1, -1, 3, 3, -1, -1, -1, -1), \text{ etc.}$$

Now, the 28 lines which connect the origin O with the points

[*]M. Golay, *Proceedings of the Institute of Radio Engineers*, 1949, vol. XXXVII, p. 637.

[†]E.R. Berlekamp, *Algebraic Coding Theory*, New York, 1968; J.H. von Lint, *Coding Theory*, Lecture Notes in Mathematics 201, Springer, Berlin, 1971.

$P_{h,i}$ constitute an equiangular set with cos $\phi = \frac{1}{3}$. Indeed, the vectors $OP_{h,i}$ have length $\sqrt{24}$, and have in pairs inner product ± 8. The angle between the vectors $OP_{h,i}$ and $OP_{j,k}$ is acute if the index pairs have one number in common, and is obtuse if they have no number in common. The graph which belongs (in the sense of page 301) to this configuration is the complement of the triangular graph $T(8)$ (figure xvi). The *triangular*

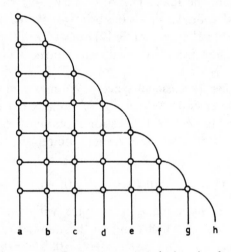

Figure xvi. $T(8)$ on the symbols $\{a, b, c, d, e, f, g, h\}$

graph $T(n)$ of order n, $n > 3$, consists of the $\frac{1}{2}n(n-1)$ unordered pairs out of n symbols, any two pairs being adjacent if and only if they have one symbol in common. For instance, in $T(8)$ the vertex ab is adjacent to the 12 vertices ij which contain a or b.

After having constructed the set $\{P_{h,i}, h < i; h, i = 1, 2, \ldots, 8\}$ of 28 equiangular lines* in 7-space, we go down to 6-space. The subset of $\{P_{h,i}\}$ which satisfies

$$x_1 + x_2 + \ldots + x_8 = 0, \quad x_1 = x_2,$$

*These lines connect the pairs of antipodal vertices of the polytope 3_{21}; see H.S.M. Coxeter, *Regular Polytopes*, 3rd edition, New York, 1973, p. 203. See also *Proceedings of the Cambridge Philosophical Society*, 1928, vol. XXIV, pp. 1–9.

contains the 16 points $P_{1,2}$ and $P_{j,k}$ ($j<k$; j, $k=3$, ..., 8). The lines connecting the origin with these points span 6-space and constitute a set of 16 equiangular lines with angle cos $\phi=\frac{1}{3}$. There is an easier way to obtain this set. To that end we again consider the set S of 3-vectors

$$(+\ +\ +)\,,\ (+\ -\ -)\,,\ (-\ +\ -)\,,\ (-\ -\ +)$$

used on page 296. The 16 vectors in 6-space, whose first 3 co-ordinates and last 3 co-ordinates both run through S, have inner product ± 2 and thus provide the required set. The graph, which belongs to this configuration of 16 equiangular lines, is the complement of the lattice graph $L(4)$ (figure xvii).

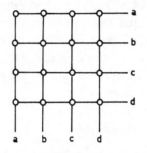

Figure xvii. $L(4)$ on the symbols $\{a, b, c, d\}$

The *lattice graph* $L(n)$ of order n, $n>1$, consists of the n^2 ordered pairs out of n symbols, any two pairs being adjacent if and only if they have one symbol in common. For instance, in $L(4)$ the vertex (a, a) is adjacent to the 6 vertices (a, b), (a, c), (a, d), (b, a), (c, a), (d, a).

Going down to 5-space we observe that the subset of the 28-set $\{P_{h,i}\}$, which satisfies $x_1=x_2=x_3$, consists of the 10 points $P_{j,k}$ ($j<k$; j, $k=4$, ..., 8) and leads to 10 equiangular lines in 5-space with cos $\phi=\frac{1}{3}$. The graph belonging to this configuration is the complement of the triangular graph $T(5)$, which is called the Petersen graph (figure xviii; see also page 225). These graphs are closely related to the *Desargues configuration* (figure xix). This configuration is defined by 10 points and 10 lines which are situated in a plane in the follow-

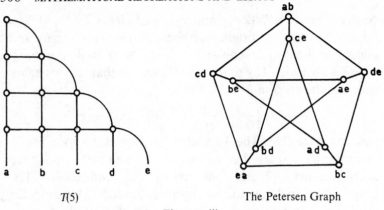

T(5)   The Petersen Graph

Figure xviii

ing way. The triangles *ac, ad, ae,* and *bc, bd, be,* are perspective from the centre *ab,* and their sides intersect in the collinear points *cd, de, ce.* The configuration may also be viewed as the intersection figure of 5 planes *a, b, c, d, e* in 3-space. (We have denoted each point of intersection by the pair of planes not passing through that point.)

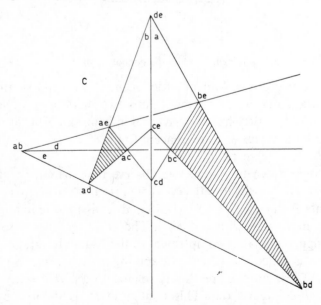

Figure xix.   The Desargues Configuration

The Desargues configuration becomes the triangular graph [the Petersen graph] if any two of its points are called adjacent whenever they are [are not] on a line of the configuration.

In order to derive equiangular sets in 4-space and in 3-space from the 28-set $\{P_{h,i}\}$, we put $x_1=x_2=x_3=x_4$, and $x_1=x_2=x_3=x_4=x_5$, respectively. The first case yields 6 equiangular lines in 4-space, whose graph is the complement of $T(4)$, that is, the ladder graph with 3 steps (figure xx). The second case

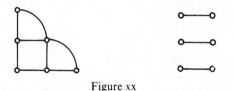

Figure xx

yields 4 equiangular lines in 3-space, whose graph is the complement of $L(2)$, that is, the ladder graph with 2 steps (figure xxi). This brings us back to where we started on page 300: the 4 diagonals of the 3-cube.

Figure xxi

The examples given above are optimal in the following sense. Let $v(n)$ be the maximum number of equiangular lines in n-space. Then it is known* that $v(2)=3$, $v(3)=v(4)=6$, $v(5)=10$, $v(6)=16$, $v(7)=28$, $v(15)=36$, $v(22)=176$, $v(23)=276$. The 28-set with $\cos \phi=\frac{1}{3}$ in 7-space is inextensible.

On the other hand, there do exist sets of lines which admit two angles one of which is arccos $\frac{1}{3}$. So, in 24-space there exists a two-angle set of 2048 lines with $\cos \phi=\frac{1}{3}$ and $\cos \psi=0$. This set is closely related to the perfect Golay code (23, 12) mentioned on page 303. In addition, this set is related to a close packing of spheres in 24-space. A still closer sphere

*J.H. van Lint and J.J. Seidel, *Koninklijke Nederlandsche Akademie van Wetenschappen te Amsterdam, Proceedings*, 1966, vol. A LXIX, pp. 335–348.

packing* of 24-space corresponds to a three-angle set of 98280 $=\binom{28}{5}$ lines with $\cos\phi$ =¼, ½, or 0. It is the automorphism group of this set which plays an important role in the theory of finite simple groups.†

C-matrices

$$C_6 = \begin{bmatrix} 0 & + & + & + & + & + \\ + & 0 & + & - & - & + \\ + & + & 0 & + & - & - \\ + & - & + & 0 & + & - \\ + & - & - & + & 0 & + \\ + & + & - & - & + & 0 \end{bmatrix} \qquad C_4 = \begin{bmatrix} 0 & + & + & + \\ - & 0 & - & + \\ - & + & 0 & - \\ - & - & + & 0 \end{bmatrix}$$

These matrices, the first of which resembles the icosahedron graph of page 300, are examples of C-matrices: they are orthogonal since the inner product of any pair of rows vanishes. Any symmetric or skew matrix C of order v with diagonal elements 0 and other elements $+1$ and -1, satisfying

$$CC^T = (v-1)I,$$

is called a C-matrix. Here are two applications.

The v directors of a company wish to have their conferences by telephone, in such a way that any director can speak to any colleague and that all others can follow their discussions. The construction of such conference-networks (linear, lossless, reciprocal v-ports, independent of frequency, with uniform distribution and zero reflection) is equivalent to the construction of symmetric C-matrices.‡

Which weighing design is "best" in order to weigh v objects in v weighings (under certain conditions and with a definition of "best")? The strategy for the weighing is described by the matrix C defined by its elements c_{ij}:

$c_{ij} = 1$, if the object j at the weighing i is on the left scale,
$c_{ij} = -1$, if the object j at the weighing i is on the right scale,
$c_{ij} = 0$, if the object j at the weighing i is not involved.

*J. Leech, *Canadian Journal of Mathematics*, 1967, vol. LXX, pp. 251–267.

†J.H. Conway, *Bulletin of the London Mathematical Society*, 1969, vol. I, pp. 79–88.

‡V. Belevitch, *Annales de la Société scientifique de Bruxelles*, 1968, vol. LXXXII, pp. 13–32.

For $v \equiv 0$ (mod 4) the best weighing design is provided by an Hadamard matrix, and for $v \equiv 2$ (mod 4) by a symmetric C-matrix.*

A skew C-matrix can exist only if its order v is divisible by 4. A symmetric C-matrix can exist only if $v-2$ is divisible by 4 and if $v-1$ is the sum of two squares of integers. But an integer n is the sum of two squares if and only if the square-free part of n has no prime factor $\equiv 3$ (mod 4). Thus, there does not exist a symmetric C-matrix of order 22.

We have encountered already C-matrices of orders 4, 6, 10 (the adjacency matrix of Petersen's graph). By way of example we now give a construction for a C-matrix of order 8. The method, due to Paley,† may be extended to all odd prime powers $v-1$.

The residue classes modulo 7 are the multiples of 7 plus 0, 1, 2, 3, 4, 5, 6, respectively. They are denoted by $a_0, a_1, a_2, a_3, a_4, a_5, a_6$. Clearly a_1, a_2, a_4 are squares, and a_3, a_5, a_6 are non-squares. To any class $a_i - a_j$ $(i, j = 0, 1, \ldots, 6)$, there is attached the number

$$0, \text{ if } a_i - a_j = a_0,$$
$$+1, \text{ if } a_i - a_j \text{ is a non-zero square,}$$
$$-1, \text{ if } a_i - a_j \text{ is a non-square.}$$

The matrix of these "Legendre symbols" is a circulant matrix of order 7, with first row $(0, -, -, +, -, +, +)$. Bordering this matrix we obtain the skew C-matrix

$$C_8 = \begin{bmatrix} 0 & + & + & + & + & + & + & + \\ - & 0 & - & - & + & - & + & + \\ - & + & 0 & - & - & + & - & + \\ - & + & + & 0 & - & - & + & - \\ - & - & + & + & 0 & - & - & + \\ - & + & - & + & + & 0 & - & - \\ - & - & + & - & + & + & 0 & - \\ - & - & - & + & - & + & + & 0 \end{bmatrix}$$

*D. Raghavarao, *Constructions and Combinatorial Problems in Design of Experiments*, New York, 1971.
†See the paper cited on page 108.

Projective planes. The 7 vertices of a 6-simplex are represented by the following 7 points in 7-space:

$$(1, 0, 0, 0, 0, 0, 0), \quad (0, 1, 0, 0, 0, 0, 0), \quad \ldots, \quad (0, 0, 0, 0, 0, 0, 1).$$

The 35 centres of the triangles of the 6-simplex have the coordinates

$$\tfrac{1}{3} (1, 1, 1, 0, 0, 0, 0), \quad \tfrac{1}{3} (1, 1, 0, 1, 0, 0, 0), \text{ etc.}$$

We consider the following question. Can 7 of such centres be selected so as to form again a 6-simplex? Then these centres should have all their distances equal. This implies that each pair of the required centres should have one co-ordinate 1 in common. This brings us back to our point of departure on page 271. Indeed, a solution is provided by the 7 rows of the incidence matrix of the projective plane PG(2, 2).

Let us review formally and briefly some of the combinatorial designs which we have encountered in this chapter.

Let V be a finite set consisting of v elements. Let k and λ be integers such that $0 < k < v - 1$ and $0 < \lambda$. A (balanced incomplete) *block design* (v, k, λ) is a collection of k-subsets of V such that each 2-subset of V is contained in λ of these k-subsets. Examples are the Steiner triple systems of order v, with $k = 3$ and $\lambda = 1$.

A block design is *symmetric* if the number of k-subsets in the collection equals the number of elements in V. An example: any normalized Hadamard matrix of order $4t \geq 8$ is equivalent to a symmetric block design with

$$v = 4t - 1, \quad k = 2t - 1, \quad \lambda = t - 1.$$

A *finite projective plane* of order n is a symmetric block design with

$$v = n^2 + n + 1, \quad k = n + 1, \quad \lambda = 1.$$

The elements of V are the points, the k-subsets are the lines of the plane. A complete set of $n - 1$ mutually orthogonal Latin squares of order n is equivalent to a projective plane of order n.

It is by no means true that, given any v, k, λ, the corresponding block design can be constructed. Certain non-existence theorems are known, certain series of block designs have been constructed, but much remains unknown. Finite projective planes have been constructed only for prime power orders; sometimes several non-isomorphic copies exist for the same order. There is one non-existence theorem* known, due to Bruck and Ryser: if n satisfies $n \equiv 1$ or 2 (mod 4), $n \neq a^2 + b^2$, a and b integers, then there exists no projective plane of order n. This excludes $n = 6$. The first undecided case is $n = 10$. For further details the reader is referred to the literature.†

*R.H. Bruck and H.J. Ryser, *Canadian Journal of Mathematics*, 1949, vol. I, pp. 88–93.

†P. Dembowski, *Finite Geometries*, Berlin, 1968; H.J. Ryser, *Combinatorial Mathematics*, Mathematical Association of America, 1963.

CHAPTER XI

MISCELLANEOUS PROBLEMS

I propose to discuss in this chapter the mathematical theory of a few common mathematical amusements and games. I might have dealt with them in the first four chapters, but, since most of them involve mixed geometry and algebra, it is rather more convenient to deal with them apart from the problems and puzzles which have been described already; the arrangement is, however, based on convenience rather than on any logical distinction.

The majority of the questions here enumerated have no connection one with another, and I jot them down almost at random.

I shall discuss in succession the *Fifteen Puzzle*, the *Tower of Hanoï*, *Chinese Rings*, and some miscellaneous *Problems connected with a Pack of Cards*.

THE FIFTEEN PUZZLE*

Some years ago the so-called *Fifteen Puzzle* was on sale in all toy-shops. It consists of a shallow wooden box – one side being marked as the top – in the form of a square, and contains fifteen square blocks or counters numbered 1, 2, 3, . . . up to 15. The box will hold just sixteen such counters, and, as it contains only fifteen, they can be moved about in the box relatively to one another. Initially they are put in the box in any order, but leaving the sixteenth cell or small square empty; the puzzle is to move them so that finally they occupy the position shown in the first of the annexed figures.

*There are two articles on the subject in the *American Journal of Mathematics*, 1879, vol. II, by Professors Woolsey Johnson and Storey; but the whole theory is deducible immediately from the proposition I give in the text.

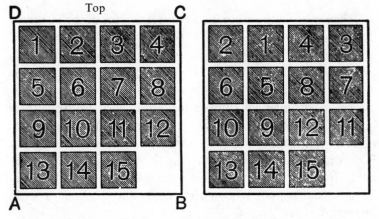

We may represent the various stages in the game by supposing that the blank space, occupying the sixteenth cell, is moved over the board, ending finally where it started.

The route pursued by the blank space may consist partly of tracks followed and again retraced, which have no effect on the arrangement, and partly of closed paths travelled round, which necessarily are cyclic permutations of an odd number of counters. No other motion is possible.

Now, a cyclic permutation of n letters is equivalent to $n-1$ simple interchanges or *transpositions*; accordingly the cyclic permutation of an odd number of letters is the product of an even number of transpositions. Hence, if we move the counters so as to bring the blank space back into the sixteenth cell, the new order must differ from the initial order by an even number of transpositions. If therefore the order we want to get can be obtained from this initial order only by an odd number of transpositions, the problem is incapable of solution; if it can be obtained by an even number, the problem can be solved.*

The order in the second of the diagrams given above is deducible from that in the first diagram by six transpositions – namely, by interchanging the counters 1 and 2, 3 and 4, 5 and 6, 7 and 8, 9 and 10, 11 and 12. Hence the one can be deduced from the other by moving the counters about in the box.

*See H.V. Mallison's article in the *Mathematical Gazette*, 1940, vol. XXIV, p. 119 (Note 1454).

If, however, in the second diagram the order of the last three counters had been 13, 15, 14, then it would have required seven transpositions of counters to bring them into the order given in the first diagram. Hence in this case the problem would be insoluble.

The easiest way of finding the number of transpositions necessary in order to obtain one given arrangement from another is to make the transformation by a series of cycles. For example, suppose that we take the counters in the box in any definite order, such as taking the successive rows from left to right, and suppose the original order and the final order to be respectively

$$1, 13, 2, 3, 5, 7, 12, 8, 15, \quad 6, \quad 9, \quad 4, 11, 10, 14,$$

and $11, \quad 2, 3, 4, 5, 6, \quad 7, 1, \quad 9, 10, 13, 12, \quad 8, 14, 15.$

We can deduce the second order from the first by 12 transpositions. The simplest way of seeing this is to arrange the process in three separate cycles as follows:

$$1, 11, 8; \mid 13, 2, 3, \quad 4, 12, 7, \quad 6, 10, 14, 15, \quad 9; \mid 5.$$
$$11, \quad 8, 1; \mid 2, 3, 4, 12, \quad 7, 6, 10, 14, 15, \quad 9, 13; \mid 5.$$

If in the first row of figures 11 is substituted for 1, then 8 for 11, then 1 for 8, we have made a cyclic permutation of 3 numbers, which is equivalent to two transpositions (namely, interchanging 1 and 11, and then 1 and 8). Thus the whole process is equivalent to one cyclic permutation of 3 numbers, another of 11 numbers, and another of 1 number. Hence it is equivalent to $(2+10+0)$ transpositions. This is an even number, and thus one of these orders can be deduced from the other by moving the counters about in the box.

It is obvious that, if the initial order is the same as the required order except that the last three counters are in the order 15, 14, 13, it would require one transposition to put them in the order 13, 14, 15; hence the problem is insoluble.

If, however, the box is turned through a right angle, so as to make AD the top, this rotation will be equivalent to 13 transpositions. For, if we keep the sixteenth square always blank, then such a rotation would change any order such as

1, 2, 3, 4,   5,   6, 7, 8,   9, 10, 11, 12, 13, 14, 15,

o   13, 9, 5, 1, 14, 10, 6, 2, 15, 11,   7,   3, 12,   8,   4,

which is equivalent to 13 transpositions. Hence it will change the arrangement from one where a solution is impossible to one where it is possible, and vice versa.

Again, even if the initial order is one which makes a solution impossible, yet if the first cell and not the last is left blank, it will be possible to arrange the fifteen counters in their natural order. For, if we represent the blank cell by b, this will be equivalent to changing the order

1, 2, 3, 4, 5, 6, 7, 8, 9, 10, 11, 12, 13, 14, 15, b,

to b, 1, 2, 3, 4, 5, 6, 7, 8,   9, 10, 11, 12, 13, 14, 15;

this is a cyclic permutation of 16 things, and therefore is equivalent to 15 transpositions. Hence it will change the arrangement from one where a solution is impossible to one where it is possible, and vice versa.

So, too, if it were permissible to turn the 6 and the 9 upside down, thus changing them to 9 and 6 respectively, this would be equivalent to one transposition, and therefore would change an arrangement where a solution is impossible to one where it is possible.

It is evident that the above principles are applicable equally to a rectangular box containing mn cells or spaces and $mn-1$ counters which are numbered. Of course, m may be equal to n. When m and n are both even (and in certain other cases, such as $m=3$, $n=5$), the operation of turning the box through a right angle is equivalent to an odd number of transpositions, and thus will change an impossible position to a possible one, and vice versa. Similarly, if m and n are not both odd, and it is impossible to solve the problem when the last cell is left blank, then it will be possible to solve it by leaving the first cell blank.

The problem may be made more difficult by limiting the

possible movements by fixing bars inside the box which will prevent the movements of a counter transverse to their directions. Piet Hien has invented a similar cubical puzzle called Bloxbox.*

<h3 align="center">THE TOWER OF HANOÏ</h3>

I may mention next the ingenious puzzle known as the *Tower of Hanoi*. It was brought out in 1883 by M. Claus (Lucas).

It consists of three pegs fastened to a stand, and of eight circular discs of wood or cardboard, each of which has a hole in the middle through which a peg can be passed. These discs are of different radii, and initially they are placed all on one peg, so that the biggest is at the bottom, and the radii of the successive discs decrease as we ascend: thus the smallest disc is at the top. This arrangement is called the *Tower*. The problem is to shift the discs from one peg to another in such a way that a disc shall never rest on one smaller than itself, and finally to transfer the tower (i.e. all the discs in their proper order) from the peg on which they initially rested to one of the other pegs.

The method of effecting this is as follows. (i) If initially there are n discs on the peg A, the first operation is to transfer gradually the top $n-1$ discs from the peg A to the peg B, leaving the peg C vacant: suppose that this requires x separate transfers. (ii) Next, move the bottom disc to the peg C. (iii) Then, reversing the first process, transfer gradually the $n-1$ discs from B to C, which will necessitate x transfers. Hence, if it requires x transfers of single discs to move a tower of $n-1$ discs, then it will require $2x+1$ separate transfers of single discs to move a tower of n sides. Now, with 2 discs it requires 3 transfers, i.e. 2^2-1 transfers; hence with 3 discs the number of transfers required will be $2(2^2-1)+1$, that is, 2^3-1. Proceeding in this way, we see that with a tower of n discs it will require 2^n-1 transfers of single discs to effect the complete transfer. Thus the eight discs of the puzzle will require 255 single transfers. It will be noticed that every alternate move

*Scientific American, 1973, vol. CCXXVII, no. 2, p. 109.

consists of a transfer of the smallest disc from one peg to another, the pegs being taken in cyclic order; further, if the discs be numbered consecutively 1, 2, 3, ... beginning with the smallest, all those with odd numbers rotate in one direction, and all those with even numbers in the other direction.

Obviously, the discs may be replaced by cards numbered 1, 2, 3, ..., n; and if n is not greater than 10, playing-cards may be conveniently used.

De Parville gave an account of the origin of the toy which is a sufficiently pretty idea to deserve repetition.* In the great temple at Benares, says he, beneath the dome which marks the centre of the world, rests a brass plate in which are fixed three diamond needles, each a cubit high and as thick as the body of a bee. On one of these needles, at the creation, God placed sixty-four discs of pure gold, the largest disc resting on the brass plate, and the others getting smaller and smaller up to the top one. This is the Tower of Bramah. Day and night unceasingly the priests transfer the discs from one diamond needle to another according to the fixed and immutable laws of Bramah, which require that the priest on duty must not move more than one disc at a time and that he must place this disc on a needle so that there is no smaller disc below it. When the sixty-four discs shall have been thus transferred from the needle on which at the creation God placed them to one of the other needles, tower, temple, and Brahmins alike will crumble into dust, and with a thunderclap the world will vanish.

The number of separate transfers of single discs which the Brahmins must make to effect the transfer of the tower is $2^{64} - 1$, that is, 18,446744,073709,551615: a number which will require so long to carry out that, when the final thunderclap occurs, the universe will be a thousand times as old as it is now.

*La Nature, Paris, 1884, part I, pp. 285-286.

CHINESE RINGS*

A somewhat more elaborate toy, known as *Chinese Rings*, which is on sale in most English toy-shops, is represented in the accompanying figure. It consists of a number of rings hung upon a bar in such a manner that the ring at one end (say *A*)

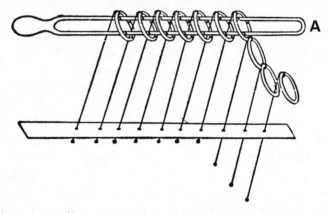

can be taken off or put on the bar at pleasure; but any other ring can be taken off or put on only when the one next to it towards *A* is on, and all the rest towards *A* are off the bar. The order of the rings cannot be changed.

Only one ring can be taken off or put on at a time. (In the toy, as usually sold, the first two rings form an exception to the rule. Both these can be taken off or put on together. To simplify the discussion I shall assume at first that only one ring is taken off or put on at a time.) I proceed to show that, if there are *n* rings, then, in order to disconnect them from the bar, it will be necessary to take a ring off or to put a ring on either $\frac{1}{3}(2^{n+1}-1)$ times or $\frac{1}{3}(2^{n+1}-2)$ times, according as *n* is odd or even.

Let the taking a ring off the bar or putting a ring on the bar be called a *step*. It is usual to number the rings from the free end *A*. Let us suppose that we commence with the first *m*

*This was described by Cardan in 1550 in his *De Subtilitate*, bk. xv, paragraph 2, ed. Sponius, vol. iii, p. 587; by Wallis in his *Algebra*, Latin edition, 1693, *Opera*, vol. ii, chap. cxi, pp. 472–478; and allusion is made to it also in Ozanam's *Récréations*, 1723 edition, vol. iv, p. 439.

rings off the bar and all the rest on the bar; and suppose that then it requires $x-1$ steps to take off the next ring, that is, it requires $x-1$ additional steps to arrange the rings so that the first $m+1$ of them are off the bar and all the rest are on it. Before taking these steps we can take off the $(m+2)$th ring, and thus it will require x steps from our initial position to remove the $(m+1)$th and $(m+2)$th rings.

Suppose that these x steps have been made, and that thus the first $m+2$ rings are off the bar and the rest on it, and let us find how many additional steps are now necessary to take off the $(m+3)$th and $(m+4)$th rings. To take these off, we begin by taking off the $(m+4)$th ring: this requires 1 step. Before we can take off the $(m+3)$th ring we must arrange the rings so that the $(m+2)$th ring is on and the first $m+1$ rings are off: to effect this, (i) we must get the $(m+1)$th ring on and the first m rings off, which requires $x-1$ steps, (ii) then we must put on the $(m+2)$th ring, which requires 1 step, (iii) and lastly we must take the $(m+1)$th ring off, which requires $x-1$ steps: these movements require in all $\{2(x-1)+1\}$ steps. Next we can take the $(m+3)$th ring off, which requires 1 step; this leaves us with the first $m+1$ rings off, the $(m+2)$th on, the $(m+3)$th and $(m+4)$th off, and all the rest on. Finally, to take off the $(m+2)$th ring, (i) we get the $(m+1)$th ring on and the first m rings off, which requires $x-1$ steps, (ii) we take off the $(m+2)$th ring, which requires 1 step, (iii) we take the $(m+1)$th ring off, which requires $x-1$ steps: these movements require $\{2(x-1)+1\}$ steps.

Therefore, if when the first m rings are off it requires x steps to take off the $(m+1)$th and $(m+2)$th rings, then the number of additional steps required to take off the $(m+3)$th and $(m+4)$th rings is $1+\{2(x-1)+1\}+1+\{2(x-1)+1\}$, that is, $4x$.

To find the whole number of steps necessary to take off an odd number of rings we proceed as follows.

To take off the first ring requires 1 step;
∴ to take off the first 3 rings requires 4 additional steps;
∴ 　　 ,, 　　 ,, 　　 5 　,, 　　 ,, 　4^2　 ,, 　　 ,,

In this way we see that the number of steps required to take off the first $2n+1$ rings is $1+4+4^2+\ldots+4^n$, which is equal to $\frac{1}{3}(2^{2n+2}-1)$.

To find the number of steps necessary to take off an even number of rings we proceed in a similar manner.

To take off the first 2 rings requires 2 steps;

$\therefore$ to take off the first 4 rings requires 2×4 additional steps;

$\therefore$ „ „ „ 6 „ „ 2×4^2 „ „

In this way we see that the number of steps required to take off the first $2n$ rings is $2+(2\times4)+(2\times4^2)+\ldots+(2\times4^{n-1})$, which is equal to $\frac{1}{3}(2^{2n+1}-2)$.

If we take off or put on the first two rings in one step instead of two separate steps, these results become respectively 2^{2n} and $2^{2n-1}-1$.

I give the above analysis because it is the direct solution of a problem attacked unsuccessfully by Cardan in 1550 and by Wallis in 1693, and which at one time attracted some attention.

I proceed next to give another solution, more elegant, though rather artificial. This, which is due to Monsieur Gros,* depends on a convention by which any position of the rings is denoted by a certain number expressed in the binary scale of notation in such a way that a step is indicated by the addition or subtraction of unity.

Let the rings be indicated by circles: if a ring is on the bar, it is represented by a circle drawn above the bar; if the ring is off the bar, it is represented by a circle below the bar. Thus figure i below represents a set of seven rings, of which the first two are off the bar, the next three are on it, the sixth is off it, and the seventh is on it.

Denote the rings which are on the bar by the digits 1 or 0 alternately, reckoning from left to right, and denote a ring which is off the bar by the digit assigned to that ring on the bar which is nearest to it on the left of it, or by a 0 if there is no ring to the left of it.

---

*Théorie du Baguenodier, by L. Gros, Lyons, 1872. I take the account of this from Lucas, vol. i, part 7.

Thus the three positions indicated below are denoted respectively by the numbers written below them. The position represented in figure ii is obtained from that in figure i by putting the first ring on to the bar, while the position represented in figure iii is obtained from that in figure i by taking the fourth ring off the bar.

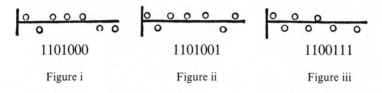

1101000	1101001	1100111
Figure i	Figure ii	Figure iii

It follows that every position of the rings is denoted by a number expressed in the binary scale: moreover, since in going from left to right every ring on the bar gives a variation (that is, 1 to 0, or 0 to 1) and every ring off the bar gives a continuation, the effect of a step by which a ring is taken off or put on the bar is either to subtract unity from this number or to add unity to it. For example, the number denoting the position of the rings in figure ii is obtained from the number denoting that in figure i by adding unity to it. Similarly the number denoting the position of the rings in figure iii is obtained from the number denoting that in figure i by subtracting unity from it.

The position when all the seven rings are off the bar is denoted by the number 0000000: when all of them are on the bar, by the number 1010101. Hence to change from one position to the other requires a number of steps equal to the difference between these two numbers in the binary scale. The first of these numbers is 0: the second is equal to $2^6 + 2^4 + 2^2 + 1$, that is, to 85. Therefore 85 steps are required. In a similar way we may show that to put on a set of $2n+1$ rings requires $(1 + 2^2 + \ldots + 2^{2n})$ steps, that is, $\frac{1}{3}(2^{2n+2} - 1)$ steps, and to put on a set of $2n$ rings requires $(2 + 2^3 + \ldots + 2^{2n-1})$ steps, that is, $\frac{1}{3}(2^{2n+1} - 2)$ steps.

I append a table indicating the steps necessary to take off the first four rings from a set of five rings. The diagrams in the middle column show the successive position of the rings

Initial position		10101
After 1st step		10110 ⎫
” 2nd ”		10111 ⎬
” 3rd ”		11000
” 4th ”		11001 ⎫
” 5th ”		11010 ⎬
” 6th ”		11011
” 7th ”		11100
” 8th ”		11101
” 9th ”		11110 ⎫
” 10th ”		11111 ⎬

after each step. The number following each diagram indicates that position, each number being obtained from the one above it by the addition of unity. The steps which are bracketed together can be made in one movement, and, if thus effected, the whole process is completed in 7 movements instead of 10 steps: this is in accordance with the formula given above.

Gros asserted that it is possible to take from 64 to 80 steps a minute, which in my experience is a rather high estimate. If we accept the lower of these numbers, it would be possible to take off 10 rings in less than 8 minutes; to take off 25 rings would require more than 582 days, each of ten hours' work; and to take off 60 rings would necessitate no less than 768614,336404,564650 steps, and would require nearly 55000,000000 years' work – assuming, of course, that no mistakes were made.

PROBLEMS CONNECTED WITH A PACK OF CARDS

An ordinary pack of playing-cards can be used to illustrate many questions depending on simple properties of numbers, or involving the relative position of the cards. In problems of this kind, the principle of solution generally consists in re-arranging the pack in a particular manner so as to bring the card into some definite position. Any such rearrangement is a species of shuffling.

I shall treat in succession of problems connected with *Shuffling a Pack*, *Arrangements by Rows and Columns*, the *Determination of a Pair out of* $\frac{1}{2}n(n+1)$ *Pairs*, *Gergonne's Pile Problem*, the *Window Reader*, and the game known as the *Mouse Trap*.

SHUFFLING A PACK

Any system of *shuffling a pack* of cards, if carried out consistently, leads to an arrangement which can be calculated; but tricks that depend on it generally require considerable technical skill.

Suppose, for instance, that a pack of n cards is shuffled, as is not unusual, by placing the second card on the first, the third below these, the fourth above them, and so on. The theory of this system of shuffling is due to Monge.* The following are some of the results, and are not difficult to prove directly.

One shuffle of a pack of $2p$ cards will move the card which was in the x_0th place to the x_1th place, where $x_1 = \frac{1}{2}(2p + x_0 + 1)$ if x_0 is odd, and $x_1 = \frac{1}{2}(2p - x_0 + 2)$ if x_0 is even. For instance, if a complete pack of 52 cards is shuffled as described above, the eighteenth card will remain the eighteenth card. If an écarté pack of 32 cards is so shuffled, the seventh and the twentieth cards will change places.

Again, in any pack of n cards, after a certain number of shufflings, not greater than n, the cards will return to their primitive order. This will always be the case as soon as the

*Monge's investigations are printed in the *Mémoires de l'Académie des Sciences*, Paris, 1773, pp. 390–412. Among those who have studied the subject afresh I may in particular mention V. Bouniakowski, *Bulletin physico-mathématique de St. Pétersbourg*, 1857, vol. XV, pp. 202–205, summarized in the *Nouvelles Annales de Mathématiques*, 1858, *Bulletin*, pp. 66–67; T. de St. Laurent, *Mémoires de l'Académie de Gard*, 1865; L. Tanner, *Educational Times Reprints*, 1880, vol. XXXIII, pp. 73–75; M.J. Bourget, *Liouville's Journal*, 1882, pp. 413–434; H.F. Baker, *Transactions of the British Association* for 1910, pp. 526–528; and P.H. Cowell, *The Field*, 2 April 1921, p. 444.

original top card occupies that position again. To determine the number of shuffles required for a pack of $2p$ cards, it is sufficient to put $x_m = x_0$ and find the smallest value of m which satisfies the resulting equation for all values of x_0 from 1 to $2p$.

The result can, however, be obtained more easily if the cards are numbered from the bottom of the original pack. Doing this, we can show that if after s shuffles a card is in the rth place from the bottom, its original number from the bottom was the difference between $2^s \times r$ and the nearest multiple of $4p + 1$. Hence, if m shuffles are required to restore the original order, m is the least number for which $2^m + 1$ or $2^m - 1$ is divisible by $4p + 1$. The number for a pack of $2p + 1$ cards is the same as that for a pack of $2p$ cards. With an écarté pack of 32 cards, six shuffles are sufficient; with a pack of 2^n cards, $n + 1$ shuffles are sufficient; with a full pack of 52 cards, twelve shuffles are sufficient; with a pack of 13 cards, ten shuffles are sufficient; while with a pack of 50 cards, fifty shuffles are required; and so on.

In general, for a pack of n cards, whatever the law of shuffling may be, if the same shuffle is repeated sufficiently often, the cards will ultimately fall into their original positions. In fact, W.H.H. Hudson[*] has shown that this will always happen before the number of shuffles exceeds the greatest L.C.M. of all sets of numbers whose sum is n.

If $P_1, \ldots, P_n$ represent the n positions that the cards may occupy, then every shuffle S can be represented by the product of a number of disjoint "subshuffles" or *cycles*:

$$S = (P_a P_b \ldots P_i)(P_j \ldots P_m) \ldots (P_u \ldots P_z),$$

where the card in position P_a is moved to position P_b and the card in the last position of a given cycle (such as P_m) is moved to the first position (such as P_j) of that cycle. Since the subshuffles are independent, any change in their order will not

[*]*Educational Times Reprints*, London, 1865, vol. II, p. 105. See also E. Landau, *Archiv der Mathematik und Physik*, 1903, series 3, vol. v, pp. 92–103; W. Feller, *An Introduction to Probability Theory*, New York, 1950, pp. 335–341.

alter the whole shuffle; because of this, S performed twice may be written

$$S^2 = (P_a P_b \ldots P_i)^2 (P_j \ldots P_m)^2 \ldots (P_u \ldots P_z)^2.$$

One can check, however, that every card whose position belongs to a cycle of period (or "length") r will return to its original place after the rth shuffle. If, for a particular shuffle, m is the L.C.M. of the periods of the cycles, then this shuffle, performed m times, will restore all the cards to their original positions. Thus in a pack of n cards we have Hudson's result. For instance, when $n = 52$, we know that the period of any shuffle will never exceed 180180.

ARRANGEMENTS BY ROWS AND COLUMNS

A not uncommon trick, which rests on a species of shuffling, depends on the obvious fact that if n^2 cards are arranged in the form of a square of n rows, each containing n cards, then any card will be defined if the row and the column in which it lies are mentioned.

This information is generally elicited by first asking in which row the selected card lies, and noting the extreme left-hand card of that row. The cards in each column are then taken up, face upwards, one at a time, beginning with the lowest card of each column and taking the columns in their order from right to left – each card taken up being placed on the top of those previously taken up. The cards are then dealt out again in rows, from left to right, beginning with the top left-hand corner, and a question is put as to which row contains the card. The selected card will be that card in the row mentioned which is in the same vertical column as the card which was originally noted.

The trick is improved by allowing the pack to be cut as often as is liked before the cards are re-dealt, and then giving one cut at the end so as to make the top card in the pack one of those originally in the top row. For instance, take the case of 16 cards. The first and second arrangements may be represented by figures i and ii. Suppose we are told that in figure i the card

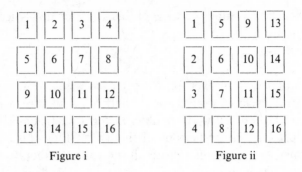

1	2	3	4
5	6	7	8
9	10	11	12
13	14	15	16

Figure i

1	5	9	13
2	6	10	14
3	7	11	15
4	8	12	16

Figure ii

is in the third row, it must be either 9, 10, 11, 12: hence, if we know in which row of figure ii it lies, it is determined. If we allow the pack to be cut between the deals, we must secure somehow that the top card is either 1, 2, 3, or 4, since that will leave the cards in each row of figure ii unaltered though the positions of the rows will be changed.

DETERMINATION OF A SELECTED PAIR OF CARDS OUT OF $\frac{1}{2}n(n+1)$ GIVEN PAIRS*

Another common trick is to throw twenty cards on to a table in ten couples, and ask someone to select one couple. The cards are then taken up, and dealt out in a certain manner into four rows each containing five cards. If the rows which contain the given cards are indicated, the cards selected are known at once.

This depends on the fact that the number of monomials of degree two which can be formed out of four symbols is 10. Hence the products formed out of four symbols can be used to define ten things.

1	2	3	5	7
4	9	10	11	13
6	12	15	16	17
8	14	18	19	20

*Bachet, problem XVII, avertissement, pp. 146 *et seq.*

Suppose that ten couples of cards are placed on a table and someone selects one couple. Take up the cards in their couples. Then the first two cards form the first couple, the next two the second couple, and so on. Deal them out in four rows each containing five cards according to the scheme shown above.

The first couple (1 and 2) are in the first row. Of the next couple (3 and 4), put one in the first row and one in the second. Of the next couple (5 and 6), put one in the first row and one in the third, and so on, as indicated in the diagram. After filling up the first row proceed similarly with the second row, and so on.

Enquire in which rows the two selected cards appear. If only one line, the mth, is mentioned as containing the cards, then the required pair of cards are the mth and $(m+1)$th cards in that line. These occupy the *clue* squares of that line. Next, if two lines are mentioned, then proceed as follows. Let the two lines be the pth and the qth and suppose $q > p$. Then that one of the required cards which is in the qth line will be the $(q-p)$th card which is below the first of the clue squares in the pth line. The other of the required cards is in the pth line, and is the $(q-p)$th card to the right of the second of the clue squares.

Bachet's rule, in the form in which I have given it, is applicable to a pack of $n(n+1)$ cards divided into couples, and dealt in n rows each containing $n+1$ cards; for there are $\frac{1}{2}n(n+1)$ such couples, also there are $\frac{1}{2}n(n+1)$ homogeneous products of two dimensions which can be formed out of n things. Bachet gave the diagrams for the cases of 20, 30, and 42 cards: these the reader will have no difficulty in constructing for himself, and I have enunciated the rule for 20 cards in a form which covers all the cases.

I have seen the same trick performed by means of a sentence and not by numbers. If we take the case of ten couples, then, after collecting the pairs, the cards must be dealt in four rows each containing five cards, in the order indicated by the sentence *Mutus dedit nomen Cocis*. This sentence must be imagined as written on the table, each word forming one line. The first card is dealt on the M. The next card (which is the pair of the first) is placed on the second m in the sentence, that is, third in

the third row. The third card is placed on the *u*. The fourth
card (which is the pair of the third) is placed on the second
u, that is, fourth in the first row. Each of the next two cards
is placed on a *t*, and so on. Enquire in which rows the two
selected cards appear. If two rows are mentioned, the two cards
are on the letters common to the words that make these rows.
If only one row is mentioned, the cards are on the two letters
common to that row.

The reason is obvious: let us denote each of the first pair by
a *u*, and similarly each of any of the other pairs by an *e, i, o,
c, d, m, n, s,* or *t* respectively. Now the sentence *Mutus dedit
nomen Cocis* contains four words each of five letters; ten letters
are used, and each letter is repeated only twice. Hence, if two
of the words are mentioned, they will have one letter in common, or, if one word is mentioned, it will have two like letters.

To perform the same trick with any other number of cards
we should require a different sentence.

The number of homogeneous products of three dimensions
which can be formed out of four things is 20, and of these the
number consisting of products in which three things are alike
and those in which three things are different is 8. This leads
to a trick with 8 trios of things, which is similar to that last
given – the cards being arranged in the order indicated by the
sentence *Lanata levete livini novoto.*

I believe that these arrangements by sentences are well-
known, but I am not aware who invented them.

GERGONNE'S PILE PROBLEM

Before discussing Gergonne's theorem I will describe the
familiar three-pile problem, the theory of which is included in
his results.

The three-pile problem.* This trick is usually performed as
follows. Take 27 cards and deal them into three piles, face
upwards. By "dealing" is to be understood that the top card
is placed as the bottom card of the first pile, the second card in

---

*The trick is mentioned by Bachet, problem XVIII, p. 143, but his analysis
of it is insufficient.

the pack as the bottom card of the second pile, the third card as the bottom card of the third pile, the fourth card on the top of the first one, and so on: moreover, I assume that throughout the problem the cards are held in the hand face upwards. The result can be modified to cover any other way of dealing.

Request a spectator to note a card, and remember in which pile it is. After finishing the deal, ask in which pile the card is. Take up the three piles, placing that pile between the other two. Deal again as before, and repeat the question as to which pile contains the given card. Take up the three piles again, placing the pile which now contains the selected card between the other two. Deal again as before, but in dealing note the middle card of each pile. Ask again, for the third time, in which pile the card lies, and you will know that the card was the one which you noted as being the middle card of that pile. The trick can be finished then in any way that you like. The usual method – but a very clumsy one – is to take up the three piles once more, placing the named pile between the other two as before, when the selected card will be the middle one in the pack – that is, if 27 cards are used it will be the 14th card.

The trick is often performed with 15 cards or with 21 cards, in either of which cases the same rule holds.

Gergonne's generalization.   The general theory for a pack of m^m cards was given by M. Gergonne.* Suppose the pack is arranged in m piles, each containing m^{m-1} cards, and that, after the first deal, the pile indicated as containing the selected card is taken up ath; after the second deal, is taken up bth; and so on; and finally after the mth deal, the pile containing the card is taken up kth. Then when the cards are collected after the mth deal, the selected card will be nth from the top where

if m is even, $n = km^{m-1} - jm^{m-2} + \ldots + bm - a + 1$,
if m is odd, $n = km^{m-1} - jm^{m-2} + \ldots - bm + a$.

For example, if a pack of 256 cards (i.e. $m=4$) was given, and anyone selected a card out of it, the card could be deter-

---

*Gergonne's *Annales de Mathématiques*, Nîmes, 1813–1814, vol. IV, pp. 276–283.

mined by making four successive deals into four piles of 64 cards each, and after each deal asking in which pile the selected card lay. The reason is that after the first deal you know it is one of 64 cards. In the next deal these 64 cards are distributed equally over the four piles, and therefore, if you know in which pile it is, you will know that it is one of 16 cards. After the third deal you know it is one of 4 cards. After the fourth deal you know which card it is.

Moreover, if the pack of 256 cards is used, it is immaterial in what order the pile containing the selected card is taken up after a deal. For, if after the first deal it is taken up ath, after the second bth, after the third cth, and after the fourth dth, the card will be the $(64d - 16c + 4b - a + 1)$th from the top of the pack, and thus will be known. We need not take up the cards after the fourth deal, for the same argument will show that it is the $(64 - 16c + 4b - a + 1)$th in the pile then indicated as containing it. Thus if $a = 3$, $b = 4$, $c = 1$, $d = 2$, it will be the 62nd card in the pile indicated after the fourth deal as containing it, and will be the 126th card in the pack as then collected.

In exactly the same way a pack of 27 cards may be used, and three successive deals, each into three piles of 9 cards, will suffice to determine the card. If after the deals the pile indicated as containing the given card is taken up ath, bth, and cth, respectively, then the card will be the $(9c - 3b + a)$th in the pack or will be the $(9 - 3b + a)$th card in the pile indicated after the third deal as containing it.

The method of proof will be illustrated sufficiently by considering the usual case of a pack of 27 cards, for which $m = 3$, which are dealt into three piles each of 9 cards.

Suppose that, after the first deal, the pile containing the selected card is taken up ath: then (i) at the top of the pack there are $a - 1$ piles each containing 9 cards; (ii) next there are 9 cards, of which one is the selected card; and (iii) lastly there are the remaining cards of the pack. The cards are dealt out now for the second time: in each pile the bottom $3(a - 1)$ cards will be taken from (i), the next 3 cards from (ii), and the remaining $9 - 3a$ cards from (iii).

Suppose that the pile now indicated as containing the selected card is taken up bth: then (i) at the top of the pack are $9(b-1)$ cards; (ii) next are $9-3a$ cards; (iii) next are 3 cards, of which one is the selected card; and (iv) lastly are the remaining cards of the pack. The cards are dealt out now for the third time: in each pile the bottom $3(b-1)$ cards will be taken from (i), the next $3-a$ cards will be taken from (ii), the next card will be one of the 3 cards in (iii), and the remaining $8-3b+a$ cards are from (iv).

Hence, after this deal, as soon as the pile is indicated, it is known that the card is the $(9-3b+a)$th from the top of that pile. If the process is continued by taking up this pile as cth, then the selected card will come out in the place $9(c-1)+(8-3b+a)+1$ from the top, that is, will come out as the $(9c-3b+a)$th card.

Since, after the third deal, the position of the card in the pile then indicated is known, it is easy to notice the card, in which case the trick can be finished in some way more effective than dealing again.

If we put the pile indicated always in the middle of the pack, we have $a=2$, $b=2$, $c=2$, hence $n=9c-3b+a=14$, which is the form in which the trick is usually presented, as was explained above on page 328.

I have shown that if a, b, c are known, then n is determined. We may modify the rule so as to make the selected card come out in any assigned position – say the nth. In this case we have to find values of a, b, c which will satisfy the equation $n=9c-3b+a$, where a, b, c can have only the values 1, 2, or 3.

Hence, if we divide n by 3 and the remainder is 1 or 2, this remainder will be a; but, if the remainder is 0, we must decrease the quotient by unity so that the remainder is 3, and this remainder will be a. In other words, a is the smallest positive number (exclusive of zero) which must be subtracted from n to make the difference a multiple of 3.

Next let p be this multiple – i.e. p is the next lowest integer to $n/3$: then $3p=9c-3b$, therefore $p=3c-b$. Hence b is the smallest positive number (exclusive of zero) which must be added to p to make the sum a multiple of 3, and c is that multiple.

A couple of illustrations will make this clear. Suppose we wish the card to come out 22nd from the top, therefore $22 = 9c - 3b + a$. The smallest number which must be subtracted from 22 to leave a multiple of 3 is 1, therefore $a = 1$. Hence $22 = 9c - 3b + 1$, therefore $7 = 3c - b$. The smallest number which must be added to 7 to make a multiple of 3 is 2, therefore $b = 2$. Hence $7 = 3c - 2$, therefore $c = 3$. Thus $a = 1$, $b = 2$, $c = 3$.

Again, suppose the card is to come out 21st. Hence $21 = 9c - 3b + a$. Therefore a is the smallest number which subtracted from 21 makes a multiple of 3, therefore $a = 3$. Hence $6 = 3c - b$. Therefore b is the smallest number which added to 6 makes a multiple of 3, therefore $b = 3$. Hence $9 = 3c$, therefore $c = 3$. Thus $a = 3$, $b = 3$, $c = 3$.

If any difficulty is experienced in this work, we can proceed thus. Let $a = x + 1$, $b = 3 - y$, $c = z + 1$; then x, y, z may have only the value 0, 1, or 2. In this case Gergonne's equation takes the form $9z + 3y + x = n - 1$. Hence, if $n - 1$ is expressed in the ternary scale of notation, x, y, z will be determined, and therefore a, b, c will be known.

The rule in the case of a pack of m^m cards is just the same. We want to make the card come out in a given place. Hence, in Gergonne's formula, we are given n and we have to find a, b, ..., k. We can effect this by dividing n continually by m, with the convention that the remainders are to be alternately positive and negative, and that their numerical values are to be not greater than m or less than unity.

An analogous theorem with a pack of lm cards can be constructed. C.T. Hudson and L.E. Dickson[*] have discussed the general case where such a pack is dealt n times, each time into l piles of m cards; and they have shown how the piles must be taken up in order that after the nth deal the selected card may be rth from the top.

Educational Times Reprints, 1868, vol. IX, pp. 89–91; and *Bulletin of the American Mathematical Society*, New York, April 1895, vol. I, pp. 184–186.

The principle will be sufficiently illustrated by one example treated in a manner analogous to the cases already discussed. For instance, suppose that an écarté pack of 32 cards is dealt into four piles each of 8 cards, and that the pile which contains some selected card is picked up ath. Suppose that on dealing again into four piles, one pile is indicated as containing the selected card; the selected card cannot be one of the bottom $2(a-1)$ cards, or of the top $8-2a$ cards, but must be one of the intermediate 2 cards, and the trick can be finished in any way, as for instance by the common conjuring ambiguity of asking someone to choose one of them, leaving it doubtful whether the one he takes is to be rejected or retained.

M.R. Goormaghtigh* has determined the condition which l and m must satisfy in order that a selected card may be located by dealing three times into l piles of m, the indicated pile being taken up second every time. The condition is $[(m+h)/l] = [(m+k)/l]$ $(=p$, say), where $h = [m/l]$ and $k = [(2m-1)/l]$. The selected card is then the $(p+1)$th in the pile indicated after the third deal. For an ordinary pack of 52 cards, we may take $l=4$, $m=13$; then $h=3$, $k=6$, and $p=4$, so that after three deals the selected card will be the fifth in the indicated pile.

THE WINDOW READER

Some years ago a set of 8 numbered and perforated cards was brought out which enabled an operator to state a number chosen by a spectator. Of the 8 cards each of the first 7 was pierced with window-like openings, each of the last 7 contained some of the numbers less than 100 headed by the word *Yes*, each of the last 3 had also certain digits on its back, and the first of the cards was headed with the word *Top*. Each card if turned upside down bore on what was then its top the word *No*.

The cards were employed to determine any number less than 100 chosen by someone – say, A. They were used by B thus. B first laid on the table the card numbered 1 with the

*Sphinx, 1936, pp. 113–115.

side marked *Top* uppermost. *B* then took the second card, and asked *A* if the chosen number was on it; if *A* said yes, *B* placed the card 2 on the top of the card 1 with the *Yes* uppermost; and if *A* said no, *B* turned the card round and placed it with the *No* uppermost. *B* then asked if the chosen number was on the third card, and placed it on the top of 2 with the appropriate end uppermost; and so on with the rest of the cards 4 to 8. Finally, on turning the whole pile over, the chosen number was seen through the windows.

The puzzle must have been widely circulated. It was sold in Italy and Germany as well as in London. The method used is fairly obvious, and I will leave to any reader, sufficiently interested, the task of constructing cards suitable for the purpose.

Evidently, however, any number not exceeding 128 can be determined by only 7 cards, each bearing 64 selected numbers. For the first card serves to divide the numbers into two sets of 64 numbers, numbers on the second card can be chosen so as to divide each of these into two sets of 32 cards, the third divides each of these into two sets of 16 cards, and so on. The numbers must be written on the cards and the windows cut so that after arranging the cards in their proper order, and turning the pack over, the chosen number appears on the back of the seventh card as seen through the windows cut in the first 6 cards. To arrange the numbers in this way presents no difficulty, but the geometrical problem of cutting the windows is less easy. I give one solution.

If we work with 7 cards, one way of preparing them is as follows. We write under the word *Yes* on the first card, the 64 numbers 1 to 32, 65 to 96; on the second card, the numbers 1 to 16, 33 to 48, 65 to 80, 97 to 112; on the third card, the numbers in four arithmetical progressions starting from 1, 2, 5, 6, each of 16 terms, with 8 as difference; on the fourth card, the numbers in eight arithmetical progressions starting from 1, 2, 3, 4, 5, 6, 7, 8, each of 8 terms, with 16 as difference; on the fifth card, the odd numbers from 1 to 127; on the sixth card, the numbers in four arithmetical progressions starting from 1, 2, 3, 4, each of 16 terms, with 8 as difference; and on the seventh card the consecutive numbers from 1 to 64; on this card, however, the *No* must be written on the left-hand edge of the card, and not on its lowest edge. On the back of the last card we must now write the numbers from 1 to 128 in their natural order, 1 to 32 occupying the first quarter, 33 to 64 the fourth quarter, 65 to 96 the second quarter rotated through a right angle, and 97 to 128 the third quarter also rotated through a right angle. The spacing of the figures requires care, but is not difficult if the principle of construction is grasped and squared paper used.

The windows should be cut as follows. We will suppose that we use squared paper dividing each card into four equal quadrants with margins. In the first card, we form the window by cutting out the whole of the second quadrant. In the second card, we get two windows by cutting out the top half of the second quadrant and the top half of the third quadrant. In the third card, we get two windows by cutting out the right half of the second quadrant and the right half of the third quadrant. In the fourth card, we divide the second and third quadrants into four equal horizontal strips, and from each of these quadrants cut out the first and third strips. In the fifth card, we divide the second and third quadrants into four equal vertical strips, and from each of these quadrants cut out the second and fourth of these strips. In the sixth card, we divide the second and third quadrants into eight horizontal strips, and from each of these quadrants cut out the first, third, fifth, and seventh strips.

It will be noticed that no windows are cut in the first or fourth quarter of any card; hence they are free for insertion of the 64 numbers written on the face of each card. The construction here given is due to my friend R.A.L. Cole.

Possibly the puzzle is better presented by omitting all numbers exceeding 100, for the introduction of 128 at once suggests the method of construction. With that restriction I think the use of only 7 cards is better and more elegant than the form in which I have seen it on sale.

THE MOUSE TRAP. TREIZE

I will conclude this chapter with the bare mention of another game of cards, known as the *Mouse Trap*, the discussion of which involves some rather difficult considerations.

It is played as follows. A set of cards, marked with the numbers 1, 2, 3, ..., n, is dealt in any order, face upwards, in the form of a circle. The player begins at any card and counts round the circle always in the same direction. If the kth card has the number k on it – which event is called a *hit* – the player takes up the card and begins counting afresh. According to Cayley, the player wins if he thus takes up all the cards, and the cards win if at any time the player counts up to n without being able to take up a card.

For example, if a pack of only 4 cards is used, and these cards come in the order 3214, then the player would obtain the second card 2 as a hit, next he would obtain 1 as a hit, but if he went on for ever, he would not obtain another hit. On the other hand, if the cards in the pack were initially in the order 1423, the player would obtain successively all four cards in the order 1, 2, 3, 4,

The problem may be stated as the determination of what hits and how many hits can be made with a given number of cards; and what permutations will give a certain number of hits in a certain order.

Cayley* showed that there are 9 arrangements of a pack of 4 cards in which no hit will be made, 6 arrangements in which only one hit will be made, 3 arrangements in which only two hits will be made, and 6 arrangements in which four hits will be made.

Prof. Steen† has investigated the general theory for a pack of n cards. He has shown how to determine the number of arrangements in which x is the first hit [Arts. 3–5]; the number of arrangements in which 1 is the first hit and x is the second hit [Art. 6]; and the number of arrangements in which 2 is

---

Quarterly Journal of Mathematics, 1878, vol. xv, pp. 8–10.
†*Ibid.*, vol. xv, pp. 230–241.

the first hit and x the second hit [Arts. 7–8]; but beyond this point the theory has not been carried. It is obvious that, if there are $n-1$ hits, the nth hit will necessarily follow.

The French game of *treize* is very similar. It is played with a full pack of fifty-two cards (knave, queen, and king counting as 11, 12, and 13 respectively). The dealer calls out 1, 2, 3, ..., 13, as he deals the 1st, 2nd, 3rd, ..., 13th cards respectively. At the beginning of a deal the dealer offers to lay or take certain odds that he will make a hit in the thirteen cards next dealt. (Cf. page 47.)

CHAPTER XII

THREE CLASSICAL GEOMETRICAL PROBLEMS

Among the more interesting geometrical problems of antiquity are three questions which attracted the special attention of the early Greek mathematicians. Our knowledge of geometry is derived from Greek sources, and thus these questions have attained a classical position in the history of the subject. The three questions to which I refer are: (i) the duplication of a cube – that is, the determination of the side of a cube whose volume is double that of a given cube; (ii) the trisection of an angle; and (iii) the squaring of a circle – that is, the determination of a square whose area is equal to that of a given circle; each problem to be solved by a geometrical construction involving the use of straight lines and circles only – that is, by Euclidean geometry.

This limitation to the use of straight lines and circles implies that the only instruments available in Euclidean geometry are compasses and rulers. But the compasses must be capable of opening as wide as is desired, and the ruler must be of unlimited length. Further, the ruler must not be graduated, for if there were two fixed marks on it we could obtain constructions equivalent to those obtained by the use of the conic sections.

With the Euclidean restriction all three problems are insoluble.* To duplicate a cube the length of whose side is a, we have to find a line of length x, such that $x^3 = 2a^3$. Again, to trisect a given angle, we may proceed to find the sine of the angle, say a, then, if x is the sine of an angle equal to one-

*See F.C. Klein, *Vorträge über ausgewählte Fragen der Elementargeometrie*, Leipzig, 1895; and F.G. Texeira, *Sur les Problèmes célèbres de la Géomètrie Élémentaire non resolubles avec la Règle et le Compas*, Coimbra, 1915. It is said that the earliest rigorous proof that the problems were insoluble by Euclidean geometry was given by P.L. Wantzel in 1837.

third of the given angle, we have $4x^3 = 3x - a$. Thus the first and second problems, when considered analytically, require the solution of a cubic equation; and since a construction by means of circles (whose equations are of the form $x^2 + y^2 + ax + by + c = 0$) and straight lines (whose equations are of the form $\alpha x + \beta y + \gamma = 0$) cannot be equivalent to the solution of a cubic equation, it is inferred that the problems are insoluble if in our constructions we are restricted to the use of circles and straight lines. If the use of the conic sections is permitted, both of these questions can be solved in many ways. The third problem is different in character, but under the same restrictions it also is insoluble.

I propose to give some of the constructions which have been proposed for solving the first two of these problems. To save space I shall not draw the necessary diagrams, and in most cases I shall not add the proofs: the latter present but little difficulty. I shall conclude with some historical notes on approximate solutions of the quadrature of the circle.

THE DUPLICATION OF THE CUBE*

The problem of the duplication of the cube was known in ancient times as the Delian problem, in consequence of a legend that the Delians had consulted Plato on the subject. In one form of the story, which is related by Philoponus,† it is asserted that the Athenians in 430 B.C. when suffering from the plague of eruptive typhoid fever, consulted the oracle at Delos as to how they could stop it. Apollo replied that they must double the size of his altar, which was in the form of a cube. To the unlearned suppliants nothing seemed more easy, and a new altar was constructed either having each of its edges double that of the old one (from which it followed that the volume was increased eight-fold) or by placing a similar cube altar next to the old one. Whereupon, according to the

---

*See *Historia Problematis de Cubi Duplicatione* by N.T. Reimer, Göttingen, 1798; and *Historia Problematis Cubi Duplicandi* by C.H. Biering, Copenhagen, 1844; also *Das Delische Problem*, by A. Sturm, Linz, 1895–7. Some notes on the subject are given in my *History of Mathematics*.

†*Philoponus ad Aristotelis Analytica Posteriora*, bk. I, chap. vii.

legend, the indignant god made the pestilence worse than before, and informed a fresh deputation that it was useless to trifle with him, as his new altar must be a cube and have a volume exactly double that of his old one. Suspecting a mystery, the Athenians applied to Plato, who referred them to the geometricians. The insertion of Plato's name is an obvious anachronism. Eratosthenes* relates a somewhat similar story, but with Minos as the propounder of the problem.

In an Arab work, the Greek legend was distorted into the following extraordinarily impossible piece of history, which I cite as a curiosity of its kind. "Now in the days of Plato," says the writer, "a plague broke out among the children of Israel. Then came a voice from heaven to one of their prophets, saying, 'Let the size of the cubic altar be doubled, and the plague will cease'; so the people made another altar like unto the former, and laid the same by its side. Nevertheless the pestilence continued to increase. And again the voice spake unto the prophet, saying, 'They have made a second altar like unto the former, and laid it by its side, but that does not produce the duplication of the cube.' Then applied they to Plato, the Grecian sage, who spake to them, saying, 'Ye have been neglectful of the science of geometry, and therefore hath God chastised you, since geometry is the most sublime of all the sciences.' Now, the duplication of a cube depends on a rare problem in geometry, namely..." And then follows the solution of Apollonius, which is given later.

If a is the length of the side of the given cube and x that of the required cube, we have $x^3 = 2a^3$, that is, $x:a = \sqrt[3]{2}:1$. It is probable that the Greeks were aware that the latter ratio is irrational, in other words, that no two integers can be found whose ratio is the same as that of $\sqrt[3]{2}:1$, but it did not therefore follow that they could not find the ratio by geometry: in fact, the side and diagonal of a square are instances of lines whose numerical measures are incommensurable.

I proceed now to give some of the geometrical constructions

*Archimedis Opera cum Eutocii Commentariis, ed. Torelli, Oxford, 1792, p. 144; ed. Heiberg, Leipzig, 1880–1, vol. III, pp. 104–107.

which have been proposed for the duplication of the cube.*
With one exception, I confine myself to those which can be
effected by the aid of the conic sections.

Hippocrates† (*circ.* 420 B.C.) was perhaps the earliest mathematician who made any progress towards solving the problem.
He did not give a geometrical construction, but he reduced
the question to that of finding two means between one line
segment (*a*), and another twice as long (2*a*). If these means are *x*
and *y*, we have $a:x=x:y=y:2a$, from which it follows that
$x^3 = 2a^3$. It is in this form that the problem is always presented
now. Formerly any process of solution by finding these means
was called a *mesolabum.*

One of the first solutions of the problem was that given by
Archytas‡ in or about the year 400 B.C. His construction is
equivalent to the following. On the diameter *OA* of the base
of a right circular cylinder describe a semicircle whose plane is
perpendicular to the base of the cylinder. Let the plane containing this semicircle rotate round the generator through *O*;
then the surface traced out by the semicircle will cut the
cylinder in a tortuous curve. This curve will itself be cut by
a right cone, whose axis is *OA* and semi-vertical angle is (say)
60°, in a point *P*, such that the projection of *OP* on the base
of the cylinder will be to the radius of the cylinder in the ratio
of the side of the required cube to that of the given cube. Of
course, the proof given by Archytas is geometrical; and it is
interesting to note that in it he shows himself familiar with
the results of the propositions Euc. III, 18; III, 35; and XI, 19.
To show analytically that the construction is correct, take *OA*
as the axis of *x*, and the generator of the cylinder drawn through
O as axis of *z*; then with the usual notation, in polar co-
ordinates, if *a* is the radius of the cylinder, we have for the

*On the application to this problem of the traditional Greek methods of
analysis by Hero and Philo (leading to the solution by the use of Apollonius's
circle), by Nicomedes (leading to the solution by the use of the conchoid),
and by Pappus (leading to the solution by the use of the cissoid), see *Geometrical
Analysis* by J. Leslie, Edinburgh, second edition, 1811, pp. 247–250, 453.

†Proclus, ed. Friedlein, pp. 212–213.

‡*Archimedis Opera*, ed. Torelli, p. 143; ed. Heiberg, vol. III, pp. 98–103.

equation of the surface described by the semicircle $r = 2a \sin \theta$; for that of the cylinder $r \sin \theta = 2a \cos \phi$; and for that of the cone $\sin \theta \cos \phi = \frac{1}{2}$. These three surfaces cut in a point such that $\sin^3 \theta = \frac{1}{2}$, and therefore $(r \sin \theta)^3 = 2a^3$. Hence the volume of the cube whose side is $r \sin \theta$ is twice that of the cube whose side is a.

The construction attributed to Plato* (*circ.* 360 B.C.) depends on the theorem that, if CAB and DAB are two right-angled triangles, having one side, AB, common, their other sides, AD and BC, parallel, and their hypotenuses, AC and BD, at right angles, then if these hypotenuses cut in P, we have $PC:PB = PB:PA = PA:PD$. Hence, if such a figure can be constructed having $PD = 2PC$, the problem will be solved. It is easy to make an instrument by which the figure can be drawn.

The next writer whose name is connected with the problem is Menaechmus,† who in or about 340 B.C. gave two solutions of it.

In the first of these he pointed out that two parabolas having a common vertex, axes at right angles, and such that the latus rectum of the one is double that of the other, will intersect in another point whose abscissa (or ordinate) will give a solution. If we use analysis, this is obvious; for, if the equations of the parabolas are $y^2 = 2ax$ and $x^2 = ay$, they intersect in a point whose abscissa is given by $x^3 = 2a^3$. It is probable that this method was suggested by the form in which Hippocrates had cast the problem, namely, to find x and y so that $a:x = x:y = y:2a$, whence we have $x^2 = ay$ and $y^2 = 2ax$.

The second solution given by Menaechmus was as follows. Describe a parabola of latus rectum l. Next describe a rectangular hyperbola, the length of whose real axis is $4l$, and having for its asymptotes the tangent at the vertex of the parabola and the axis of the parabola. Then the ordinate and the abscissa of the point of intersection of these curves are the mean proportionals between l and $2l$. This is at once obvious by analysis. The curves are $x^2 = ly$ and $xy = 2l^2$. These cut in

*Archimedis Opera, ed. Torelli, p. 135; ed. Heiberg, vol. III, pp. 66–71.
†Ibid., ed. Torelli, pp. 141–143; ed. Heiberg, vol. III, pp. 92–99.

a point determined by $x^3 = 2l^3$ and $y^3 = 4l^3$. Hence

$$l : x = x : y = y : 2l.$$

The solution of Apollonius,* which was given about 220 B.C., was as follows. The problem is to find two mean proportionals between two given lines. Construct a rectangle $OADB$, of which the adjacent sides OA and OB are respectively equal to the two given lines. Bisect AB in C. With C as centre describe a circle cutting OA produced in a and cutting OB produced in b, so that aDb shall be a straight line. If this circle can be so described, it will follow that $OA : Bb = Bb : Aa = Aa : OB$, that is, Bb and Aa are the two mean proportionals between OA and OB. It is impossible to construct the circle by Euclidean geometry, but Apollonius gave a mechanical way of describing it.

The only other construction of antiquity to which I will refer is that given by Diocles and Sporus.† It is as follows. Take two sides of a rectangle OA, OB, equal to the two lines between which the means are sought. Suppose OA to be the greater. With centre O and radius OA describe a circle. Let OB produced cut the circumference in C and let AO produced cut it in D. Find a point E on BC so that if DE cuts AB produced in F and cuts the circumference in G, then $FE = EG$. If E can be found, then OE is the first of the means between OA and OB. Diocles invented the cissoid in order to determine E, but it can be found equally conveniently by the aid of conics.

In more modern times several other solutions have been suggested. I may allude in passing to three given by Huygens,‡ but I will enunciate only those proposed respectively by Vieta, Descartes, Gregory of St. Vincent, and Newton.

Vieta's construction is as follows.§ Describe a circle, centre O, whose radius is equal to half the length of the larger of the two given lines. In it draw a chord AB equal to the smaller

*Archimedis Opera, ed. Torelli, p. 137; ed. Heiberg, vol. III, pp. 76–79. The solution is given in my History of Mathematics, London, 1901, p. 84.
†Ibid., ed. Torelli, pp. 138, 139, 141; ed. Heiberg, vol. III, pp. 78–84, 90–93.
‡Opera Varia, Leyden, 1724, pp. 393–396.
§Opera Mathematica, ed. Schooten, Leyden, 1646, prop. V, pp. 242–243.

of the two given lines. Produce AB to E so that $BE = AB$. Through A draw a line AF parallel to OE. Through O draw a line $DOCFG$, cutting the circumference in D and C, cutting AF in F, and cutting BA produced in G, so that $GF = OA$. If this line can be drawn, then $AB : GC = GC : GA = GA : CD$.

Descartes pointed out* that the curves

$$x^2 = ay \quad \text{and} \quad x^2 + y^2 = ay + bx$$

cut in a point (x, y) such that $a : x = x : y = y : b$. Of course, this is equivalent to the first solution given by Menaechmus, but Descartes preferred to use a circle rather than a second conic.

Gregory's construction was given in the form of the following theorem.† The hyperbola drawn through the point of intersection of two sides of a rectangle so as to have the two other sides for its asymptotes meets the circle circumscribing the rectangle in a point whose distances from the asymptotes are the mean proportionals between two adjacent sides of the rectangle. This is the geometrical expression of the proposition that the curves $xy = ab$ and $x^2 + y^2 = ay + bx$ cut in a point (x, y) such that $a : x = x : y = y : b$.

One of Newton's constructions is as follows.‡ Let OA be the greater of two given line segments. Bisect OA in B. With centre O and radius OB describe a circle. Take a point C on the circumference so that BC is equal to the other of the two given line segments. From O draw ODE cutting AC produced in D, and BC produced in E, so that the intercept $DE = OB$. Then $BC : OD = OD : CE = CE : OA$. Hence OD and CE are two mean proportionals between any two line segments BC and OA.

THE TRISECTION OF AN ANGLE §

The trisection of an angle is the second of these classical prob-

---

*Geometria, bk. III, ed. Schooten, Amsterdam, 1659, p. 91.

†Gregory of St. Vincent, Opus Geometricum Quadraturae Circuli, Antwerp, 1647, bk. VI, prop. 138, p. 602.

‡Arithmetica Universalis, Ralphson's (second) edition, 1728, p. 242; see also pp. 243, 245.

§On the bibliography of the subject see the supplements to L Intermédiaire des Mathématiciens, Paris, May and June 1904.

lems, but tradition has not enshrined its origin in romance. The following two constructions are among the oldest and best known of those which have been suggested; they are quoted by Pappus,* but I do not know to whom they were due originally.

The first of them is as follows. Let AOB be the given angle. From any point P in OB draw PM perpendicular to OA. Through P draw PR parallel to OA. On MP take a point Q so that if OQ is produced to cut PR in R, then $QR = 2 . OP$. If this construction can be made, then $AOR = \frac{1}{3}AOB$. The solution depends on determining the position of R. This was effected by a construction which may be expressed analytically thus. Let the given angle be $\tan^{-1}(b/a)$. Construct the hyperbola $xy = ab$, and the circle $(x-a)^2 + (y-b)^2 = 4(a^2 + b^2)$. Of the points where they cut, let x be the abscissa which is greatest, then $PR = x - a$, and $\tan^{-1}(b/x) = \frac{1}{3}\tan^{-1}(b/a)$.

The second construction is as follows. Let AOB be the given angle. Take $OB = OA$, and with centre O and radius OA describe a circle. Produce AO indefinitely and take a point C on it external to the circle, so that if CB cuts the circumference in D, then CD shall be equal to OA. Draw OE parallel to CDB. Then, if this construction can be made, $AOE = \frac{1}{3}AOB$. The ancients determined the position of the point C by the aid of the conchoid: it could be also found by the use of the conic sections.

I proceed to give a few other solutions, confining myself to those effected by the aid of conics.

Among other constructions given by Pappus† I may quote the following. Describe a hyperbola whose eccentricity is two. Let its centre be C and its vertices A and A'. Produce CA' to S so that $A'S = CA'$. On AS describe a segment of a circle to contain the given angle. Let the orthogonal bisector of AS cut this segment in O. With centre O and radius OA or OS

*Pappus, *Mathematicae Collectiones*, bk. IV, props. 32, 33 (ed. Commandino, Bonn, 1670, pp. 97–99). On the application to this problem of the traditional Greek methods of analysis see *Geometrical Analysis*, by J. Leslie, Edinburgh, second edition, 1811, pp. 245–247.

†Pappus, *Mathematicae Collectiones*, bk. IV, prop. 34, pp. 99–104.

describe a circle. Let this circle cut the branch of the hyperbola through A' in P. Then $SOP = \frac{1}{3}SOA$.

In modern times one of the earliest of the solutions by a direct use of conics was suggested by Descartes, who effected it by the intersection of a circle and a parabola. His construction* is equivalent to finding the points of intersection, other than the origin, of the parabola $y^2 = \frac{1}{4}x$ and the circle

$$x^2 + y^2 - \tfrac{13}{4}x + 4ay = 0.$$

The ordinates of these points are given by the equation $4y^3 = 3y - a$. The smaller positive root is the sine of one-third of the angle whose sine is a. The demonstration is ingenious.

One of the solutions proposed by Newton is practically equivalent to the third one which is quoted above from Pappus. It is as follows.† Let A be the vertex of one branch of a hyperbola whose eccentricity is two, and let S be the focus of the other branch. On AS describe the segment of a circle containing an angle equal to the supplement of the given angle. Let this circle cut the S branch of the hyperbola in P. Then PAS will be equal to one-third of the given angle.

The following elegant solution is due to Clairaut.‡ Let AOB be the given angle. Take $OA = OB$, and with centre O and radius OA describe a circle. Join AB, and trisect it in H, K, so that $AH = HK = KB$. Bisect the angle AOB by OC cutting AB in L. Then $AH = 2 \cdot HL$. With focus A, vertex H, and directrix OC, describe a hyperbola. Let the branch of this hyperbola which passes through H cut the circle in P. Draw PM perpendicular to OC and produce it to cut the circle in Q. Then by the focus and directrix property we have $AP : PM = AH : HL = 2 : 1$; therefore $AP = 2 \cdot PM = PQ$. Hence, by symmetry, $AP = PQ = QB$. Therefore $AOP = POQ = QOB$.

I may conclude by giving the solution which Chasles§ re-

*Geometria, bk. III, ed. Schooten, Amsterdam, 1659, p. 91.

†Arithmetica Universalis, problem XLII, Ralphson's (second) edition, London, 1728, p. 148; see also pp. 243–245.

‡I believe that this was first given by Clairaut, but I have mislaid my reference. The construction occurs as an example in the Geometry of Conics, by C. Taylor, Cambridge, 1881, no. 308, p. 126.

§Traité des sections coniques, Paris, 1865, art. 37, p. 36.

gards as the most fundamental. It is equivalent to the following proposition. If OA and OB are the bounding radii of a circular arc AB, then a rectangular hyperbola having OA for a diameter and passing through the point of intersection of OB with the tangent to the circle at A will pass through one of the two points of trisection of the arc.

Several instruments have been constructed by which mechanical solutions of the problem can be obtained.

THE QUADRATURE OF THE CIRCLE*

The object of the third of the classical problems was the determination of a side of a square whose area should be equal to that of a given circle.

The investigation of this question was of extreme interest to people throughout the ages, not only to mathematicians but often even more so to the general public. As a matter of fact, so great did the number of (erroneous) proofs become that in 1775 the Paris Academy found it necessary to pass a resolution which stated that no more solutions on the quadrature of the circle were to be examined. Nevertheless, the pursuit of the problem was fruitful in discoveries of allied theorems, but in more recent times it has been abandoned by those who are able to realize what is required. The history of this subject has been treated by competent writers in such detail that I shall content myself with a very brief allusion to it.

Archimedes showed † (what possibly was known before) that the problem is equivalent to finding the area of a right-angled triangle whose sides are equal respectively to the perimeter of the circle and the radius of the circle. Half the ratio of these lines is a number, usually denoted by π.

*See Montucla's *Histoire des Recherches sur la Quadrature du Cercle*, edited by P.L. Lacroíx, Paris, 1831; also various articles by A. De Morgan, and especially his *Budget of Paradoxes*, London, 1872. A popular sketch of the subject has been compiled by H. Schubert, *Die Quadratur des Zirkels*, Hamburg, 1889; and since the publication of the earlier editions of these *Recreations* Prof. F. Rudio of Zurich has given an analysis of the arguments of Archimedes, Huygens, Lambert, and Legendre on the subject, with an introduction on the history of the problem, Leipzig, 1892.

†*Archimedis Opera*, Κύκλου μέτρησις, prop. ι, ed. Torelli, pp. 203–205; ed. Heiberg, vol. ι, pp. 258–261, vol. ιιι, pp. 269–277.

The irrationality of this number was proved by Lambert* in 1761; he showed that if x ($\neq 0$) is rational, neither e^x nor $\tan x$ can be rational. Yet $\tan \frac{1}{4}\pi = 1$; thus $\frac{1}{4}\pi$ and hence π must be irrational.

The transcendental nature of π was demonstrated by Lindemann† in 1882. The proof leads to the conclusion that, if x is a root of a rational integral algebraic equation, then e^x cannot be rational: hence, if πi was the root of such an equation, $e^{\pi i}$ could not be rational; but $e^{\pi i}$ is equal to -1, and therefore is rational; hence πi cannot be the root of such an algebraical equation, and therefore neither can π.

Earlier, James Gregory ‡ tried to demonstrate the impossibility of quadrature, but the proof was not conclusive.

I need hardly add that, if π represented merely the ratio of the circumference of a circle to its diameter, the determination of its numerical value would have but slight interest. It is, however, a mere accident that π is defined usually in that way, and it really represents a certain number which would enter into analysis from whatever side the subject was approached, a typical example being the equation $e^{\pi i} + 1 = 0$. In reality, the fact that the ratio of the length of the circumference of a circle to its diameter is the number denoted by π does not afford the best analytical definition of π, and is only one of its properties. I recollect a distinguished professor explaining how different would be the ordinary life of a race of beings for whom the fundamental processes of arithmetic, algebra, and geometry were different from those which seem to us so evident; but, he added, it is impossible to conceive of a universe in which e and π should not exist.

The use of the symbol π to designate the number 3·14159265 3589793238462643383279502884197169399375105820974944 59230781640628620899862803482534211 70679... goes back

---

Mémoires de l'Académie de Berlin for 1761, Berlin, 1768, pp. 265–322.

†Ueber die Zahl π, *Mathematische Annalen*, Leipzig, 1882, vol. xx, pp. 213–225.

‡*Vera Circuli et Hyperbolae Quadratura*, Padua, 1668; this is reprinted in Huygens's *Opera Varia*, Leyden, 1724, pp. 405–462.

to 1647, when Oughtred used δ/π for the ratio of diameter to circumference; but in 1697 D. Gregory used π/ρ for the ratio of circumference to radius. The single symbol π was introduced about the beginning of the eighteenth century. William Jones* in 1706 represented it by π; a few years later† Johann Bernoulli denoted it by c. Euler in 1734 used p, and in 1736 used c; Christian Goldbach in 1742 used π; and after the publication of Euler's *Analysis* the symbol π was generally employed.

The numerical value of π can be determined by either of two methods with as close an approximation to the truth as is desired.

The first of these methods is geometrical. It consists in calculating the perimeters of polygons inscribed in and circumscribed about a circle, and assuming that the circumference of the circle is intermediate between these perimeters.‡ The approximation would be closer if the areas and not the perimeters were employed. The second and modern method rests on the determination of converging infinite series for π.

We may say that the π-calculators who used the first method regarded π as equivalent to a geometrical ratio, but those who adopted the modern method treated it as the symbol for a certain number which enters into numerous branches of mathematical analysis.

It may be interesting if I add here a list of some of the approximations to the value of π given by various writers.§ This will indicate incidentally those who have studied the subject to the best advantage.

Synopsis Palmariorum Matheseos, London, 1706, pp. 243, 263 *et seq.*

†See notes by G. Eneström in the *Bibliotheca Mathematica*, Stockholm, 1889, vol. III, p. 28; *ibid.*, 1890, vol. IV, p. 22.

‡The history of this method has been written by K.E.I. Selander, *Historik öfver Ludolphska Talet*, Upsala, 1868.

§ For the methods used in classical times and the results obtained, see the notices of their authors in M. Cantor's *Geschichte der Mathematik*, Leipzig, vol. I, 1880. For medieval and modern approximations, see the article by A. De Morgan on the Quadrature of the Circle in vol. XIX of the *Penny Cyclopaedia*, London, 1841; with the additions given by B. de Haan in the *Verhandelingen* of Amsterdam, 1858, vol. IV, p. 22: the conclusions were tabulated, corrected, and extended by Dr. J.W.L. Glaisher in the *Messenger of Mathematics*, Cambridge, 1873, vol. II, pp. 119–128; and *ibid.*, 1874, vol. III, pp. 27–46.

The Egyptians,* around 1700 B.C., took 256/81 as the value of π; this is equal to 3·1605. . . ; but the rougher approximation of 3 was used by the Babylonians† and by the Jews. ‡ It is not unlikely that these numbers were obtained empiricially.

We come next to a long roll of Greek mathematicians who attacked the problem. Whether the researches of the members of the Ionian School, the Pythagoreans, Anaxagoras, Hippias, Antipho, and Bryso led to numerical approximations for the value of π is doubtful, and their investigations need not detain us. The quadrature of certain lunes by Hippocrates of Chios is ingenious and correct, but a value of π cannot be thence deduced; and it seems likely that the later members of the Athenian School concentrated their efforts on other questions.

It is probable that Euclid, § the illustrious founder of the Alexandrian School, was aware that π was greater than 3 and less than 4, but he did not state the result explicitly.

The mathematical treatment of the subject began with Archimedes, who proved that π is less than $3\frac{1}{7}$ and greater than $3\frac{10}{71}$, that is, it lies between 3·1428... and 3·1408.... He established ‖ this by inscribing in a circle and circumscribing about it regular polygons of 96 sides, then determining by geometry the perimeters of these polygons, and finally assuming that the circumference of the circle was intermediate between these perimeters: this leads to a result from which he deduced the limits given above. This method is equivalent to using the proposition $\sin\theta < \theta < \tan\theta$, where $\theta = \pi/96$: the values of $\sin\theta$ and $\tan\theta$ were deduced by Archimedes from those of $\sin\frac{1}{3}\pi$ and $\tan\frac{1}{3}\pi$ by repeated bisections of the angle. With a polygon of n sides this process gives a value of π correct to at least the integral part of $(2\log n - 1\cdot 19)$ places of decimals. The result given by Archimedes is correct to two places of

*Ein mathematisches Handbuch der alten Aegypter (i.e. the Rhind papyrus), by A. Eisenlohr, Leipzig, 1877, arts. 100–109, 117, 124.

†Oppert, Journal Asiatique, August 1872, and October 1874.

‡1 Kings, ch. 7, ver. 23; 2 Chronicles, ch. 4, ver. 2.

§These results can be deduced from Euc. IV, 15, and IV, 8; see also book XII, prop. 16.

‖Archimedis Opera, Κύκλού μέτρησις, prop. III, ed. Torelli, Oxford, 1792, pp. 205–216; ed. Heiberg, Leipzig, 1880, vol. I, pp. 263–271.

decimals. His analysis leads to the conclusion that the perimeters of these polygons for a circle whose diameter is 4970
feet would lie between 15610 feet and 15620 feet; actually it
is about 15613 feet 9 inches.

Apollonius discussed these results, but his criticisms have
been lost.

Hero of Alexandria gave* the value 3, but he quoted † the
result 22/7: possibly the former number was intended only
for rough approximations.

The only other Greek approximation that I need mention
is that given by Ptolemy,‡ who asserted that $\pi = 3° \; 8' \; 30''$. This
is equivalent to taking $\pi = 3 + \frac{8}{60} + \frac{30}{3600} = 3\frac{17}{120} = 3\cdot1416$.

The Roman surveyors seem to have used 3, or sometimes 4,
for rough calculations. For closer approximations they often
employed $3\frac{1}{8}$ instead of $3\frac{1}{7}$, since the fractions then introduced
are more convenient in duodecimal arithmetic. On the other
hand, Gerbert§ (*circ.* A.D. 1000) recommended the use of 22/7.

Before coming to the medieval and modern European
mathematicians, it may be convenient to note the results arrived at in India and the East.

Baudhayana‖ took 49/16 as the value of π.

Arya-Bhata,¶ *circ.* 530, gave 62832/20000, which is equal
to $3\cdot1416$. He showed that, if a is the side of a regular polygon
of n sides inscribed in a circle of unit diameter, and if b is the
side of a regular inscribed polygon of $2n$ sides, then $b^2 = \frac{1}{2} - \frac{1}{2}(1 - a^2)^{1/2}$. From the side of an inscribed hexagon, he found
successively the sides of polygons of 12, 24, 48, 96, 192, and
384 sides. The perimeter of the last is given as equal to $\sqrt{9\cdot8694}$,
from which his result was obtained by approximation.

Brahmagupta,** *circ.* 650, gave $\sqrt{10}$, which is equal to

* *Mensurae*, ed. Hultsch, Berlin, 1864, p. 188.
† *Geometria*, ed. Hultsch, Berlin, 1864, pp. 115, 136.
‡ *Almagest*, bk. VI, chap. 7; ed. Halma, vol. I, p. 421.
§ *Œuvres de Gerbert*, ed. Olleris, Clermont, 1867, p. 453.
‖ The *Sulvasutras* by G. Thibaut, Asiatic Society of Bengal, 1875, arts. 26–28.
¶ *Leçons de calcul d'Aryabhata*, by L. Rodet in the *Journal Asiatique*, 1879,
series 7, vol. XIII, pp. 10, 21.
**Algebra...from Brahmegupta and Bhascara*, trans. by H.T. Colebrooke,
London, 1817, chap. XII, art. 40, p. 308.

3·1622. . . . He is said to have obtained this value by inscribing in a circle of unit diameter regular polygons of 12, 24, 48, and 96 sides, and calculating successively their perimeters, which he found to be $\sqrt{9\cdot65}$, $\sqrt{9\cdot81}$, $\sqrt{9\cdot86}$, $\sqrt{9\cdot87}$ respectively; and to have assumed that as the number of sides is increased indefinitely the perimeter would approximate to $\sqrt{10}$.

Bhaskara, *circ.* 1150, gave two approximations. One* (possibly copied from Arya-Bhata, but calculated afresh by Archimedes' method from the perimeters of regular polygons of 384 sides) is 3927/1250, which is equal to 3·1416.

The Chinese astronomer Tsu Ch'ung-chih (born 430 A.D.)† proved that π lies between 3·1415926 and 3·1415927, and deduded what he called the "accurate" value, 355/113.

Among the Arabs, J.M. al Kashî (*circ.* 1436) obtained the value

$$6\cdot2831853071795865$$

for 2π. This is correct in all its 16 decimal places. It was converted from a 9-sexagesimal place value that he had calculated earlier. It gave him the world record till 1596. Moreover, it is almost certainly the first instance of radix conversion for a fraction.‡

Returning to European mathematicians, we have the following successive approximations to the value of π: many of those prior to the eighteenth century having been calculated originally with the view of demonstrating the incorrectness of some alleged quadrature.

Leonardo of Pisa,§ in the thirteenth century, gave for π the value $1440/458\frac{1}{3}$, which is equal to 3·1418. . . . In the fifteenth century, Purbach‖ gave or quoted the value 62832/20000, which is equal to 3·1416; Cusa believed that the accurate value

Ibid., p. 87.

†E.W. Hobson, *Squaring the Circle*, Cambridge, 1913, p. 24.

‡D.E. Knuth, *The Art of Computer Programming*, vol. II, Reading, Massachusetts, 1969, pp. 165–167.

§Boncompagni's *Scritti di Leonardo*, vol. II (*Practica Geometriae*), Rome, 1862, p. 90. Leonardo was nicknamed Fibonacci (see page 56).

‖Appendix to the *De Triangulis* of Regiomontanus, Basle, 1541, p. 131.

was $\frac{3}{4}(\sqrt{3} + \sqrt{6})$, which is equal to $3 \cdot 1423 \ldots$; and, in 1464 Regiomontanus* is said to have given a value equal to $3 \cdot 14243$.

Vieta,† in 1579, showed that π was greater than $31415926535/10^{10}$, and less than $31415926537/10^{10}$. This was deduced from the perimeters of the inscribed and circumscribed polygons of 6×2^{16} sides, obtained by repeated use of the formula $2 \sin^2 \frac{1}{2}\theta = 1 - \cos\theta$. He also gave‡ a result equivalent to the formula

$$\frac{2}{\pi} = \frac{\sqrt{2}}{2} \frac{\sqrt{(2 + \sqrt{2})}}{2} \frac{\sqrt{\{2 + \sqrt{(2 + \sqrt{2})}\}}}{2} \ldots.$$

The father of Adrian Metius,§ in 1585, gave $355/113$, which is equal to $3 \cdot 14159292 \ldots$, and is correct to six places of decimals. This was a curious and lucky guess, for all that he proved was that π was intermediate between $377/120$ and $333/106$, whereon he jumped to the conclusion that he would obtain the true fractional value by taking the mean of the numerators and the mean of the denominators of these fractions.

In 1593 Adrian Romanus‖ calculated the perimeter of the inscribed regular polygon of $1,073,741824$ (i.e. 2^{30}) sides, from which he determined the value of π correct to 15 places of decimals.

Ludolph van Ceulen devoted a considerable part of his life to the subject. In 1596¶ he gave the result to 20 places of decimals: this was calculated by finding the perimeters of the inscribed and circumscribed regular polygons of 60×2^{33} sides, obtained by the repeated use of a theorem of his discovery equivalent to

---

*In his correspondence with Cardinal Nicholas de Cusa, *De Quadratura Circuli*, Nuremberg, 1533, wherein he proved that the cardinal's result was wrong. I cannot quote the exact reference, but the figures are given by competent writers and I have no doubt are correct.

†*Canon Mathematicus seu ad Triangula*, Paris, 1579, pp. 56, 66: probably this work was printed for private circulation only; it is very rare.

‡*Vietae Opera*, ed. Schooten, Leyden, 1646, p. 400.

§*Arithmeticae libri duo et Geometriae*, by A. Metius, Leyden, 1626, pp. 88–89. [Probably issued originally in 1611.]

‖*Ideae Mathematicae*, Antwerp, 1593: a rare work, which I have never been able to consult.

¶*Van den Circkel*, Delft, 1596, fol. 14, p. 1; or *De Circulo*, Leyden, 1619, p. 3.

the formula $1 - \cos A = 2 \sin^2 \frac{1}{2}A$. I possess a finely executed engraving of him of this date, with the result printed round a circle which is below his portrait. He died in 1610, and by his directions the result to 35 places of decimals (which was as far as he had calculated it) was engraved on his tombstone* in St. Peter's Church, Leyden. His posthumous arithmetic† contains the result to 32 places; this was obtained by calculating the perimeter of a polygon, the number of whose sides is 2^{62}, i.e. 4,611686,018427,387904. Van Ceulen also compiled a table of the perimeters of various regular polygons.

Willebrord Snell,‡ in 1621, obtained from a polygon of 2^{30} sides an approximation to 34 places of decimals. This is less than the numbers given by van Ceulen, but Snell's method was so superior that he obtained his 34 places by the use of a polygon from which van Ceulen had obtained only 14 (or perhaps 16) places. Similarly, Snell obtained from a hexagon an approximation as correct as that for which Archimedes had required a polygon of 96 sides, while from a polygon of 96 sides he determined the value of π correct to seven decimal places instead of the two places obtained by Archimedes. The reason is that Archimedes, having calculated the lengths of the sides of inscribed and circumscribed regular polygons of n sides, assumed that the length of $1/n$th of the perimeter of the circle was intermediate between them; whereas Snell constructed from the sides of these polygons two other lines which gave closer limits for the corresponding arc. His method depends on the theorem $3 \sin \theta /(2 + \cos \theta) < \theta < (2 \sin \frac{1}{3}\theta + \tan \frac{1}{3}\theta)$, by the aid of which a polygon of n sides gives a value of π correct to at least the integral part of $(4 \log n - \cdot 2305)$ places of decimals, which is more than twice the number given by the

---

*The inscription is quoted by Prof. de Haan in the *Messenger of Mathematics*, 1874, (N.S.) vol. III, p. 25.

†*De Arithmetische en Geometrische Fondamenten*, Leyden, 1615, p. 163; or p. 144 of the Latin translation by W. Snell, published at Leyden in 1615 under the title *Fundamenta Arithmetica et Geometrica*. This was reissued, together with a Latin translation of the *Van den Circkel*, in 1619, under the title *De Circulo*, in which see pp. 3, 29–32, 92.

‡*Cyclometricus*, Leyden, 1621, p. 55.

older rule. Snell's proof of his theorem is incorrect, though the result is true.

Snell also added a table* of the perimeters of all regular inscribed and circumscribed polygons, the number of whose sides is 10×2^n where n is not greater than 19 and not less than 3. Most of these were quoted from van Ceulen, but some were recalculated. This list has proved useful in refuting circle-squarers. A similar list was given by James Gregory.†

In 1630 Grienberger,‡ by the aid of Snell's theorem, carried the approximation to 39 places of decimals. He was the last mathematician who adopted the classical method of finding the perimeters of inscribed and circumscribed polygons. Closer approximations serve no useful purpose. Proofs of the theorems used by Snell and other calculators in applying this method were given by Huygens in a work § which may be taken as closing the history of this method.

In 1656 Wallis‖ proved that

$$\frac{\pi}{2} = \frac{2}{1}\ \frac{2}{3}\ \frac{4}{3}\ \frac{4}{5}\ \frac{6}{5}\ \frac{6}{7}\ \frac{8}{7} \cdots,$$

and quoted a proposition given a few years earlier by Viscount Brouncker to the effect that

$$4/\pi = 1 + 1^2/2 + 3^2/2 + 5^2/2 + 7^2/\ldots,$$

but neither of these theorems was used to any large extent for calculation.

Subsequent calculators have relied on converging infinite series, a method that was hardly practicable prior to the inven-

---

*It is quoted by Montucla, ed. 1831, p. 70.

†*Vera Circuli et Hyperbolae Quadratura*, prop. 29, quoted by Huygens, *Opera Varia*, Leyden, 1724, p. 447.

‡*Elementa Trigonometrica*, Rome, 1630, end of preface.

§*De Circula Magnitudine Inventa*, 1654; *Opera Varia*, pp. 351–387. The proofs are given in G. Pirie's *Geometrical Methods of Approximating to the Value of* π, London, 1877, pp. 21–23.

‖*Arithmetica Infinitorum*, Oxford, 1656, prop. 191. An analysis of the investigation by Wallis was given by Cayley, *Quarterly Journal of Mathematics*, 1889, vol. xxiii, pp. 165–169.

tion of the calculus, though Descartes* had indicated a geometrical process which was equivalent to the use of such a
series. The employment of infinite series was proposed by
James Gregory,† who established the theorem that

$$\theta = \tan\,\theta - \tfrac{1}{3}\tan^3\theta + \tfrac{1}{5}\tan^5\theta - \ldots,$$

the result being true only if θ lies between $-\tfrac{1}{4}\pi$ and $\tfrac{1}{4}\pi$.

The first mathematician to make use of Gregory's series
for obtaining an approximation to the value of π was Abraham
Sharp, ‡ who, in 1699, on the suggestion of Halley, determined
to it 72 places of decimals (71 correct). He obtained this value
by putting $\theta = \tfrac{1}{6}\pi$ in Gregory's series.

Machin, § earlier than 1706, gave the result to 100 places
(all correct). He calculated it by the formula

$$\tfrac{1}{4}\pi = 4\,\tan^{-1}\tfrac{1}{5} - \tan^{-1}\tfrac{1}{239}.$$

De Lagny, ‖ in 1719, gave the result to 127 places of decimals
(112 correct), calculating it by putting $\theta = \tfrac{1}{6}\pi$ in Gregory's
series.

Hutton,¶ in 1776, and Euler,** in 1779, suggested the use
of the formula $\tfrac{1}{4}\pi = \tan^{-1}\tfrac{1}{2} + \tan^{-1}\tfrac{1}{3}$ or $\tfrac{1}{4}\pi = 5\,\tan^{-1}\tfrac{1}{7} +$
$2\tan^{-1}\tfrac{3}{79}$, but neither carried the approximation as far as had
been done previously.

*See Euler's paper in the *Novi Commentarii Academiae Scientiarum*, Petro-
grad, 1763, vol. VIII, pp. 157–168.

†See the letter to Collins, dated Feb. 15, 1671, printed in the *Commercium
Epistolicum*, London, 1712, p. 25, and in the Macclesfield Collection, *Correspondence of Scientific Men of the Seventeenth Century*, Oxford, 1841, vol. II,
p. 216.

‡See *Life of A. Sharp*, by W. Cudworth, London, 1889, p. 170. Sharp's work
is given in one of the preliminary discourses (pp. 53 *et seq.*) prefixed to H. Sherwin's *Mathematical Tables*. The tables were issued at London in 1705:
probably the discourses were issued at the same time, though the earliest copies
I have seen were printed in 1717.

§W. Jones's *Synopsis Palmariorum*, London, 1706, p. 243; and Maseres,
Scriptores Logarithmici, London, 1796, vol. III, pp. vii–ix, 155–164.

‖*Histoire de l'Académie* for 1719, Paris, 1721, p. 144.

¶*Philosophical Transactions*, 1776, vol. LXVI, pp. 476–492.

**Nova Acta Academiae Scientiarum Petropolitanae* for 1793, Leningrad,
1798, vol. XI, pp. 133–149: the memoir was read in 1779.

Vega, in 1789,* gave the value of π to 143 places of decimals (126 correct); and, in 1794,† to 140 places (136 correct).

Towards the end of the eighteenth century F.X. von Zach saw in the Radcliffe Library, Oxford, a manuscript by an unknown author which gives the value of π to 154 places of decimals (152 correct).

In 1837, the result of a calculation of π to 154 places of decimals (152 correct) was published.‡

In 1841 Rutherford§ calculated it to 208 places of decimals (152 correct), using the formula

$$\tfrac{1}{4}\pi = 4 \tan^{-1} \tfrac{1}{5} - \tan^{-1} \tfrac{1}{70} + \tan^{-1} \tfrac{1}{99}.$$

In 1844 Dase ‖ calculated it to 205 places of decimals (200 correct), using the formula

$$\tfrac{1}{4}\pi = \tan^{-1} \tfrac{1}{2} + \tan^{-1} \tfrac{1}{5} + \tan^{-1} \tfrac{1}{8}.$$

In 1847 Clausen ¶ carried the approximation to 250 places of decimals (248 correct), calculating it independently by the formulae

$$\tfrac{1}{4}\pi = 2 \tan^{-1} \tfrac{1}{3} + \tan^{-1} \tfrac{1}{7}$$

and $\qquad\qquad \tfrac{1}{4}\pi = 4 \tan^{-1} \tfrac{1}{5} - \tan^{-1} \tfrac{1}{239}.$

In 1853 Rutherford** carried his former approximation to 440 places of decimals (all correct), and William Shanks prolonged the approximation to 530 places (527 correct).†† During the next twenty years Shanks attempted to carry the approxi-

---

*Nova Acta Academiae Scientiarum Petropolitanae 'for 1790, Leningrad, 1795, vol. IX, p. 41

†Thesaurus Logarithmorum (logarithmisch-trigonometrischer Tafeln), Leipzig, 1794, p. 633.

‡J.F. Callet's Tables, etc., Précis Élémentaire, Paris, tirage 1837. Tirage 1894, p. 96.

§Philosophical Transactions, 1841, p. 283.

‖ Crelle's Journal, 1844, vol. XXVII, p. 198.

¶ Schumacher, Astronomische Nachrichten, vol. XXV, col. 207.

**Proceedings of the Royal Society, Jan. 20, 1853, vol. VI, pp. 273–275.

††Contributions to Mathematics, W. Shanks, London, 1853, pp. 86–87.

mation further, but his mistake in the 528th place invalidated all that followed.*

In 1853 Richter, presumably in ignorance of what had been done in England, found the value of π to 333 places† of decimals (330 correct); in 1854 he carried the approximation to 400 places,‡ and in 1855 to 500 places. §

Of the series and formulae by which these approximations have been calculated, those used by Machin and Dase are perhaps the easiest to employ. Other series which converge rapidly are the following:

$$\frac{\pi}{6} = \frac{1}{2} + \frac{1}{2} \cdot \frac{1}{3 \cdot 2^3} + \frac{1 \cdot 3}{2 \cdot 4} \cdot \frac{1}{5 \cdot 2^5} + \cdots,$$

$$\frac{\pi}{4} = 22 \tan^{-1} \frac{1}{28} + 2 \tan^{-1} \frac{1}{443} - 5 \tan^{-1} \frac{1}{1393} - 10 \tan^{-1} \frac{1}{11018};$$

the latter of these is due to Mr. Escott. ‖

There is a tragic irony in the thought that this calculation, which occupied poor Mr. Shanks for a substantial portion of his life, can now be repeated (without his colossal mistake) in a few seconds, as a means of "warming up" an electronic computer.

As to those writers who believe that they have squared the circle, their number is legion and, in most cases, their ignorance profound, but their attempts are not worth discussing here. "Only prove to me that it is impossible," said one of them, "and I will set about it immediately"; and doubtless the statement that the problem is insoluble has attracted much attention to it.

Among the geometrical ways of approximating to the truth the following is one of the simplest. Inscribe in the given circle

---

*D.F. Ferguson, *Nature*, March 1946, vol. CLVII, p. 342.

†*Grünert's Archiv*, vol. XXI, p. 119.

‡*Ibid.*, vol. XXIII, p. 476: the approximation given in vol. XXII, p. 473, is correct only to 330 places.

§*Ibid.*, vol. XXV, p. 472; and *Elbinger Anzeigen*, No. 85.

‖ For a full discussion of such formulae, see John Todd, *American Mathematical Monthly*, 1949, vol. LVI, pp. 517–528.

a square, and to three times the diameter of the circle add a fifth of a' side of the square; the result will differ from the circumference of the circle by less than one-seventeen-thousandth part of it.

An approximate value of π has been obtained experimentally by the theory of probability. On a plane a number of equidistant parallel straight lines, distance apart a, are ruled; and a stick of length l, which is less than a, is dropped on the plane. The probability that it will fall so as to lie across one of the lines is $2l/\pi a$. If the experiment is repeated many hundreds of times, the ratio of the number of favourable cases to the whole number of experiments will be very nearly equal to this fraction: hence the value of π can be found. In 1855 Mr. A. Smith† of Aberdeen made 3204 trials, and deduced $\pi = 3\cdot1553$. A pupil of Prof. De Morgan,* from 600 trials, deduced $\pi = 3\cdot137$. In 1864 Captain Fox† made 1120 trials with some additional precautions, and obtained as the mean value $\pi = 3\cdot1419$.

Other similar methods of approximating to the value of π have been indicated. For instance, it is known that if two numbers are written down at random, the probability that they will be prime to each other is $6/\pi^2$. Thus, in one case‡ where each of 50 students wrote down 5 pairs of numbers at random, 154 of the pairs were found to consist of numbers prime to each other. This gives $6/\pi^2 = 154/250$, from which we get $\pi = 3\cdot12$.

*A. De Morgan, *Budget of Paradoxes*, London, 1872, pp. 171, 172.

†*Messenger of Mathematics*, Cambridge, 1873, vol. II, pp. 113, 114. For an amusing account of Captain Fox's "additional precautions" and the impossibility of obtaining really good approximations by such a method, see the section "Too good to be true?" in T.H. O'Beirne, *Puzzles and Paradoxes*, London, 1965, pp. 195–197.

‡Note on π by R. Chartres, *Philosophical Magazine*, London, series 6, vol. XXXIX, March 1904, p. 315.

CHAPTER XIII

CALCULATING PRODIGIES

At rare intervals there have appeared lads who possess extraordinary powers of mental calculation.* In a few seconds they gave the answers to questions connected with the multiplication of numbers and the extraction of roots of numbers, which an expert mathematician could obtain only in a longer time and with the aid of pen and paper. Nor were their powers always limited to such simple problems. More difficult questions, dealing for instance with factors, compound interest, annuities, the civil and ecclesiastical calendars, and the solution of equations, were solved by some of them with facility as soon as the meaning of what was wanted had been grasped. In most cases these lads were illiterate, and usually their rules of working were of their own invention.

The performances were so remarkable that some observers held that these prodigies possessed powers differing in kind from those of their contemporaries. For such a view there is no foundation. Any lad with an excellent memory and a natural turn for arithmetic can, if he continuously gives his undivided attention to the consideration of numbers and indulges in constant practice, attain great proficiency in mental arithmetic, and of course the performances of those that are specially gifted are exceptionally astonishing.

*Most of the facts about calculating prodigies have been collected by E.W. Scripture, *American Journal of Psychology*, 1891, vol. IV, pp. 1–59; by F.D. Mitchell, *ibid.*, 1907, vol. XVIII, pp. 61–143; and G.E. Müller, *Zur Analyse der Gedächtnistätigkeit und des Vorstellungsverlaufes*, Leipzig, 1911. I have used these papers freely, and in some cases where authorities are quoted of which I have no first-hand information have relied exclusively on them. These articles should be consulted for bibliographical notes on the numerous original authorities.

In this chapter I propose to describe briefly the doings of the more famous calculating prodigies. It will be seen that their performances were of much the same general character, though carried to different extents; hence in the later cases it will be enough to indicate briefly peculiarities of the particular calculators.

I confine myself to self-taught calculators, and thus exclude the consideration of a few public performers who by practice, arithmetical devices, and the tricks of the showman have simulated like powers. I also concern myself only with those who showed the power in youth. As far as I know, the only self-taught mathematician of advanced years whom I thus exclude is John Wallis, 1616–1703, the Savilian Professor at Oxford, who in middle-life developed, for his own amusement, his powers in mental arithmetic. As an illustration of his achievements, I note that on 22 December 1669, he, when in bed, occupied himself in finding (mentally) the integral part of the square root of 3×10^{40}; and several hours afterwards wrote down the result from memory. This fact having attracted notice, two months later he was challenged to extract the square root of a number of fifty-three digits; this he performed mentally, and a month later he dictated the answer, which he had not meantime committed to writing. Such efforts of calculation and memory are typical of calculating prodigies.

One of the earliest of these prodigies of whom we have records was *Jedediah Buxton*, who was born in or about 1707 at Elmton, Derbyshire. Although a son of the village schoolmaster, his education was neglected, and he never learned to write or cipher. With the exception of his power of dealing with large numbers, his mental faculties were of a low order: he had no ambition, and remained throughout his life a farm labourer, nor did his exceptional skill with figures bring him any material advantage other than that of occasionally receiving small sums of money from those who induced him to exhibit his peculiar gift. He died in 1772.

He had no recollection as to when or how he was first attracted by mental calculation, and of his performances in early life we have no reliable details. Mere numbers, however, seem always to have had a strange fascination for him. If the size of an object was stated, he began at once to compute how many inches or hair-breadths it contained; if a period of time was mentioned, he calculated the number of minutes in it; if he heard a sermon, he thought only of the number of words or syllables in it. No doubt his powers in these matters increased by incessant practice, but his ideas were childish, and do not seem to have gone beyond pride in being able to state accurately the results of such calculations. He was slow-witted, and took far longer to answer arithmetical questions than most of these prodigies. The only practical accomplishment to which his powers led him was the ability to estimate by inspection the acreage of a field of irregular shape.

His fame gradually spread through Derbyshire. Among many questions put to him by local visitors were the following, which fairly indicate his powers when a young man: How many acres are there in a rectangular field 351 yards long and 261 wide? answered in 11 minutes. How many cubic yards of earth must be removed in order to make a pond 426 feet long, 263 feet wide, and $2\frac{1}{2}$ feet deep? answered in 15 minutes. If sound travels 1,142 feet in one second, how long will it take to travel 5 miles? answered in 15 minutes. Such questions involve no difficulties of principle.

Here are a few of the harder problems solved by Buxton when his powers were fully developed. He calculated to what sum a farthing would amount if doubled 140 times: the answer is a number of pounds sterling which requires thirty-nine digits to represent it, with 2s. 8d. over. He was then asked to multiply this number of thirty-nine digits by itself: to this he gave the answer two and a half months later, and he said he had carried on the calculation at intervals during that period. In 1751 he calculated how many cubic inches there are in a right-angled block of stone 23,145,789 yards long, 5,642,732 yards wide, and 54,965 yards thick; how many grains of corn would be required to fill a cube whose volume is 202,680,000,360 cubic miles; and how many hairs one inch long would be required to fill the same space – the dimensions of a grain and a hair being given. These problems involve high numbers, but are not intrinsically difficult, though they could not be solved mentally unless the calculator had a phenomenally good memory. In each case he gave the correct answer, though only after considerable effort. In 1753 he was asked to give the dimensions of a cubical cornbin which holds exactly one quarter of malt. He recognized that to answer this required a process equivalent to the extraction of a cube root, which was a novel idea to him, but in an hour he said that the edge of the cube would be between $25\frac{1}{2}$ and 26 inches, which is correct: it has been suggested that he got this answer by trying various numbers.

Accounts of his performances were published, and his reputation reach London, which he visited in 1754. During his stay there he was examined by various members of the Royal Society, who were satisfied as to the genuineness of his performances. Some of his acquaintances took him to Drury Lane Theatre to see Garrick, being curious to see how a play would impress his imagination. He was entirely unaffected by the scene, but on coming out informed his hosts of the exact number of words uttered by the various actors, and of the number of steps taken by others in their dances.

It was only in rare cases that he was able to explain his methods of work, but enough is known of them to enable us to say that they were clumsy. He described the process by which he arrived at the product of 456 and 378; shortly it was as follows: If we denote the former of these numbers by a, he proceeded first to find $5a = $ (say) b; then to find $20b = $ (say) c; and then to find $3c = $ (say) d. He next formed $15b = $ (say) e, which he added to d. Lastly he formed $3a$, which, added to the sum last obtained, gave the result. This is equivalent to saying that he used the multiplier 378 in the form $(5 \times 20 \times 3) + (5 \times 15) + 3$. Mitchell suggests that this may mean that Buxton counted by multiples of 60 and of 15, and thus reduced the multiplication to addition. It may be so, for it is difficult to suppose that he did not realize that successive multiplications by 5 and 20 are equivalent to a multiplication by 100, of which the result can be at once obtained. Of billions, trillions, etc., he had never heard, and in order to represent the high numbers required in some of the questions proposed to him, he invented a notation of his own, calling 10^{18} a tribe and 10^{36} a cramp.

As in the case of all these calculators, his memory was exceptionally good, and in time he got to know a large number of facts (such as the products of certain constantly recurring numbers, the number of minutes in a year, and the number of hair-breadths in a mile), which greatly facilitated his calculations. A curious and perhaps unique feature in his case was that he could stop in the middle of a piece of mental calculation, take up other subjects, and after an interval, sometimes of weeks, could resume the consideration of the problem. He could answer simple questions when two or more were proposed simultaneously.

Another eighteenth-century prodigy was *Thomas Fuller*, a negro, born in 1710 in Africa. He was captured there in 1724, and exported as a slave to Virginia, U.S.A., where he lived till his death in 1790. Like Buxton, Fuller never learned to read or write, and his abilities were confined to mental arithmetic. He could multiply together two numbers, if each contained not more than nine digits, could state the number of seconds in a given period of time, the number of grains of corn in a given mass, and so on – in short, answer the stock problems commonly proposed to these prodigies, as long as they involved only multiplications and the solutions of problems by rule of three. Although more rapid than Buxton, he was a slow worker as compared with some of those whose doings are described below.

I mention next the case of two mathematicians of note who showed similar aptitude in early years. The first of these was *André Marie Ampère*, 1775–1836, who, when a child some four years old, was accustomed to perform long mental calculations, which he effected by means of rules learned from playing with arrangements of pebbles. But though always expert at mental arithmetic, and endowed with a phenomenal memory for figures, he did not specially cultivate this arithmetical power. On the other hand, *Carl Friedrich Gauss*, 1777–1855, spent much of his life doing prodigious calculations. Although these calculations relied, in .part, on his vast knowledge of the theory of numbers,* he did demonstrate his aptitude in early years. At the age of three, he astonished his father by correcting him in his calculations of certain payments for overtime; perhaps, however, this is only evidence of the early age at which his consummate abilities began to develop.

Another remarkable case is that of *Richard Whately*, 1787–1863, afterwards Archbishop of Dublin. When he was about five or six years old he showed considerable skill in mental arithmetic: it disappeared in about three years. "I soon," said he, "got to do the most difficult sums, always in my head, for I knew nothing of figures beyond numeration, nor had I any names for the different processes I employed. But I believe my sums were chiefly in multiplication, division, and the rule of three. . . . I did these sums much quicker than anyone could upon paper, and I never remember committing the smallest error. I was engaged either in calculating or in castle-building . . . morning, noon, and night. . . . When I went to school, at which time the passion was worn off, I was a perfect dunce at ciphering, and so have continued ever since." The archbishop's arithmetical powers were, however, greater in later life than he here allows.

The performances of *Zerah Colburn* in London, in 1812, were more remarkable. Colburn, born in 1804 at Cabot, Vermont, U.S.A., was the son of a small farmer. While still less than six years old he showed extraordinary powers of mental calculation, which were displayed in a tour in America. Two years later he was brought to England, where he was repeatedly examined by competent observers. He could instantly give the product of two numbers each of four digits, but hesitated if both numbers exceeded 10,000. Among questions asked him

*For a discussion of Gauss's techniques, see P. Maenncher, *Nachrichten der Königlichen Gessellschaft der Wissenschaft zu Göttingen*, 1918 (Beiheft 7), pp. 1–47.

at this time were to raise 8 to the 16th power; in a few seconds he gave the answer 281,474,976,710,656, which is correct. He was next asked to raise the numbers 2, 3, ..., 9 to the 10th power; and he gave the answers so rapidly that the gentleman who was taking them down was obliged to ask him to repeat them more slowly; but he worked less quickly when asked to raise numbers of two digits like 37 or 59 to high powers. He gave instantaneously the square roots and cube roots (when they were integers) of high numbers, e.g. the square root of 106,929 and the cube root of 268,336,125; such integral roots can, however, be obtained easily by various methods. More remarkable are his answers to questions on the factors of numbers. Asked for the factors of 247,483, he replied 941 and 263; asked for the factors of 171,395, he gave 5, 7, 59, and 83; asked for the factors of 36,083, he said there were none. He, however, found it difficult to answer questions about the factors of numbers higher than 1,000,000. His power of factorizing high numbers was exceptional, and depended largely on the method of two-digit terminals described below. Like all these public performers, he had to face buffoons who tried to make fun of him, but he was generally equal to them. Asked on one such occasion how many black beans were required to make three white ones, he is said to have at once replied, "Three, if you skin them" – this, however, has much the appearance of a prearranged show.

It was clear to observers that the child operated by certain rules, and during his calculations his lips moved as if he was expressing the process in words. Of his honesty there seems to have been no doubt. In a few cases he was able to explain the method of operation. Asked for the square of 4,395, he hesitated, but on the question being repeated, he gave the correct answer – namely, 19,316,025. Questioned as to the cause of his hesitation, he said he did not like to multiply four figures by four figures, but said he, "I found out another way; I multiplied 293 by 293 and then multiplied this product twice by the number 15." On another occasion when asked for the product of 21,734 by 543, he immediately replied 11,801,562;

and on being questioned explained that he had arrived at this by multiplying 65,202 by 181. These remarks suggest that whenever convenient he factorized the numbers with which he was dealing.

In 1814 he was taken to Paris, but amid the political turmoil of the time his exhibitions fell flat. His English and American friends, however, raised money for his education, and he was sent in succession to the Lycée Napoleon in Paris and Westminster School in London. With education his calculating powers fell off, and he lost the frankness which when a boy had charmed observers. His subsequent career was diversified and not altogether successful. He commenced with the stage, then tried schoolmastering, then became an itinerant preacher in America, and finally a "professor" of languages. He wrote his own biography, which contains an account of the methods he used. He died in 1840.

Contemporary with Colburn we find another instance of a self-taught boy, *George Parker Bidder*, who possessed quite exceptional powers of this kind. He is perhaps the most interesting of these prodigies, because he subsequently received a liberal education, retained his calculating powers, and in later life analysed and explained the methods he had invented and used.

Bidder was born in 1806 at Moreton Hampstead, Devonshire, where his father was a stone-mason. At the age of six he was taught to count up to 100, but though sent to the village school, learned little there, and at the beginning of his career was ignorant of the meaning of arithmetical terms and of numerical symbols. Equipped solely with this knowledge of counting, he taught himself the results of addition, subtraction, and multiplication of numbers (less than 100) by arranging and rearranging marbles, buttons, and shot in patterns. In later life he attached great importance to such concrete representations, and believed that his arithmetical powers were strengthened by the fact that at that time he knew nothing about the symbols for numbers. When seven years old he heard a dispute between two of his neighbours about the price of something which was being sold by the pound, and to their astonishment remarked that they were both wrong, mentioning the correct price. After this exhibition the villagers delighted in trying to pose him with arithmetical problems.

His reputation increased and, before he was nine years old, his father found it profitable to take him about the country to exhibit his powers. A couple of distinguished Cambridge graduates (Thomas Jephson, then tutor of St. John's, and John Herschel) saw him in 1817, and were so impressed by his general intelligence that they raised a fund for his education, and induced his father to give up the role of showman; but after a few months Bidder senior repented of his abandonment of money so easily earned, insisted on his son's return, and began again to make an exhibition of the boy's powers. In 1818, in the course of a tour, young Bidder was pitted against Colburn, and on the whole proved the abler calculator. Finally the father and son came to Edinburgh, where some members of that university intervened and persuaded his father to leave the lad in their care to be educated. Bidder remained with them, and in due course graduated at Edinburgh, shortly afterwards entering the profession of civil engineering, in which he rose to high distinction. He died in 1878.

With practice Bidder's powers steadily developed. His earlier performances seem to have been of the same type as those of Buxton and Colburn which have been already described. In addition to answering questions on products of numbers and the number of specified units in given quantities, he was, after 1819, ready in finding square roots, cube roots, etc., of high numbers, it being assumed that the root is an integer, and later explained his method, which is easy of application: this method is the same as that used by Colburn. By this time he was able also to give immediate solutions of easy problems on compound interest and annuities which seemed to his contemporaries the most astonishing of all his feats. In factorizing numbers he was less successful than Colburn, and was generally unable to deal at sight with numbers higher than 10,000. As in the case of Colburn, attempts to be witty at his expense were often made, but he could hold his own. Asked at one of his performances in London in 1818 how many bulls' tails were wanted to reach to the moon, he immediately answered one, if it is long enough.

Here are some typical questions put to and answered by him in his exhibitions during the years 1815–1819; they are taken from authenticated lists which comprise some hundreds of such problems: few, if any, are inherently difficult. His rapidity of work was remarkable, but the time limits given were taken by unskilled observers, and can be regarded as only approximately correct. Of course, all the calculations were mental without the aid of books, pencil, or paper. In 1815, being then nine years old, he was asked: If the moon be distant from the earth 123,256 miles, and sound travels at the rate of 4 miles a minute, how long would it be before the inhabitants of the moon could hear of the battle of Waterloo? answer, 21 days 9 hours 34 minutes, given in less than one minute. In 1816, being then ten years old, just learning to write, but unable to form figures, he answered questions such as the following: What is the interest on £11,111 for 11,111 days at 5 per cent a year? answer, £16,911 11s., given in one minute. How many hogsheads of cider can be made from a million of apples, if 30 apples make one quart? answer, 132 hogsheads 17 gallons 1 quart and 10 apples over, given in35 seconds. If a coach-wheel is 5 feet 10 inches in circumference, how many times will it revolve in running 800,000,000 miles? answer, 724,114,285,704 times and 20 inches remaining, given in 50 seconds. What is the square root of 119,550,669,121? answer 345,761, given in 30 seconds. In 1817, being then eleven years old, he was asked: How long would it take to fill a reservoir whose volume is one cubic mile if there flowed into it from a river 120 gallons of water a minute? answered in 2 minutes. Assuming that light travels from the sun to the earth in 8 minutes, and that the sun is 98,000,000 miles off, if light takes 6 years 4 months travelling from the nearest fixed star to the earth, what is the distance of that star, reckoning 365 days 6 hours to each year and 28 days to each month? – asked by Sir William Herschel: answer,

40,633,740,000,000 miles. In 1818, at one of his performances, he was asked: If the pendulum of a clock vibrates the distance of $9\frac{3}{4}$ inches in a second of time, how many inches will it vibrate in 7 years 14 days 2 hours 1 minute 56 seconds, each year containing 365 days 5 hours 48 minutes 55 seconds? answer, $2,165,625,744\frac{3}{4}$ inches, given in less than a minute. If I have 42 watches for sale and I sell the first for a farthing, and double the price for every succeeding watch I sell, what will be the price of the last watch? answer, £2,290,649,224 10s. 8d. If the diameter of a penny piece is $1\frac{3}{8}$ inches, and if the world is girdled with a ring of pence put side by side, what is their value sterling, supposing the distance to be 360 degrees, and a degree to contain 69·5 miles? answer, £4,803,340, given in one minute. Find two numbers whose difference is 12 and whose product, multiplied by their sum, is equal to 14,560? answer, 14 and 26. In 1819, when fourteen years old, he was asked: Find a number whose cube less 19 multiplied by its cube shall be equal to the cube of 6: answer, 3, given instantly. What will it cost to make a road for 21 miles 5 furlongs 37 poles 4 yards, at the rate of of £123 14s. 6d. a mile? answer, £2,688 13s. $9\frac{3}{4}d.$, given in 2 minutes. If you are now 14 years old and you live 50 years longer and spend half-a-crown a day, how many farthings will you spend in your life? answer, 2,805,120, given in 15 seconds. Mr. Moor contracted to illuminate the city of London with 22,965,321 lamps; the expense of trimming and lighting was 7 farthings a lamp, the oil consumed was $\frac{2}{3}$ths of a pint for every three lamps, and the oil cost 3s. $7\frac{1}{2}d.$ a gallon; he gained $16\frac{1}{2}$ per cent on his outlay: how many gallons of oil were consumed, what was the cost to him, and what was the amount of the contract? answer, he used 212,641 gallons of oil, the cost was £205,996 16s. $1\frac{3}{4}d.$, and the amount of the contract was £239,986 13s. 2d.

It should be noted that Bidder did not visualize a number like 984 in symbols, but thought of it in a concrete way as so many units which could be arranged in 24 groups of 41 each. It should also be observed that he, like Inaudi, whom I mention later, relied largely on the auditory sense to enable him to recollect numbers. "For my own part," he wrote, in later life, "though much accustomed to see sums and quantities expressed by the usual symbols, yet if I endeavour to get any number of figures that are represented on paper fixed in my memory, it takes me a much longer time and a very great deal more exertion than when they are expressed or enumerated verbally." For instance, suppose a question put to find the product of two numbers each of nine digits, if they were "read to me, I should not require this to be done more than once; but if they were represented in the usual way, and put into my hands, it would probably take me four times to peruse them before it would be in my power to repeat them, and after all they would not be impressed so vividly on my imagination."

Bidder retained his power of rapid mental calculation to the end of his life, and as a constant parliamentary witness in matters connected with engineering it proved a valuable accomplishment. Just before his death an illustration of his powers was given to a friend who, talking of then recent discoveries, remarked that if 36,918 waves of red light which occupy only one inch are required to give the impression of red, and if light travels at 190,000 miles a second, how immense must be the number of waves which must strike the eye in one second to give the impression of red. "You need not work it out," said Bidder; "the number will be 444,433,651,200,000."

Other members of the Bidder family have also shown exceptional powers of a similar kind as well as extraordinary memories. Of Bidder's elder brothers, one became an actuary, and on his books being burned in a fire, he rewrote them in six months from memory, but, it is said, died of consequent brain fever; another was a Plymouth Brother, and knew the whole Bible by heart, being able to give chapter and verse for any text quoted. Bidder's eldest son, a lawyer of eminence, was able to multiply together two numbers each of fifteen digits. Neither in accuracy nor rapidity was he equal to his father, but, then, he never steadily and continuously devoted himself to developing his abilities in this direction. He remarked that in his mental arithmetic, he worked with pictures of the figures, and said, "If I perform a sum mentally it always proceeds in a visible form in my mind; indeed I can conceive no other way possible of doing mental arithmetic": this, it will be noticed, is opposed to his father's method. Two of his children, one son and one daughter, representing a third generation, inherited analogous powers.

I mention next the names of *Henri Mondeux* and *Vito Mangiamele*. Both were born in 1826 in humble circumstances, were shepherds, and became when children noticeable for feats in calculation which deservedly procured for them local fame. In 1839 and 1840 respectively they were brought to Paris, where their powers were displayed in public, and tested by Arago, Cauchy, and others. Mondeux's performances were the more striking. One question put to him was to solve the equation $x^3 + 84 = 37x$: to this he at once gave the answer 3 and 4, but did not detect the third root, namely, -7. Another question asked was to find solutions of the indeterminate equation $x^2 - y^2 = 133$: to this he replied immediately 66 and 67; asked for a simpler solution he said after an instant 6 and 13. I do not, however, propose to discuss their feats in detail, for there was at least a suspicion that these lads were not frank, and that those who were exploiting them had taught them rules which enabled them to simulate powers they did not really possess. Finally both returned to farm work, and ceased to interest the scientific world. If Mondeux was self-taught, we must credit him with a discovery of some algebraic theorems which would entitle him to rank as a mathematical genius, but in that case it is inconceivable that he never did anything more, and that his powers appeared to be limited to the particular problems solved by him.

Johann Martin Zacharias Dase, whom I next mention, is a far more interesting example of these calculating prodigies. Dase was born in 1824 at Hamburg. He had a fair education, and was afforded every opportunity to develop his powers, but save in matters connected with reckoning and numbers he made little progress, and struck all observers as dull. Of geometry and any language but German he remained ignorant to the end of his days. He was trustworthy, and filled various small official posts in Germany. He gave exhibitions of his calculating powers in Germany, Austria, and England. He died in 1861.

When exhibiting in Vienna in 1840, he made the acquaintance of Strasznicky, who urged him to apply his powers to scientific purposes. This Dase gladly agreed to do, and so became acquainted with Gauss, Schumacher, Petersen, and Encke. To his contributions to science I allude later. In mental arithmetic the only problems to which I find allusions are straightforward examples like the following: Multiply 79,532,853 by 93,758,479: asked by Schumacher, answered in 54 seconds. In answer to a similar request to find the product of two numbers each of twenty digits, he took 6 minutes; to find the product of two numbers each of forty digits, he took 40 minutes; to find the product of two numbers each of a hundred digits, he took 8 hours 45 minutes. Gauss thought that perhaps on paper the last of these problems could be solved in half this time by a skilled computator. Dase once extracted the square root of a number of a hundred digits in 52 minutes. These feats far surpass all other records of the kind, the only calculations comparable to them being Buxton's squaring of a number of thirty-nine digits, and Wallis's extraction of the square root of a number of fifty-three digits. Dase's mental work, however, was not always accurate, and once (in 1845) he gave incorrect answers to every question put to him, but on that occasion he had a headache, and there is nothing astonishing in his failure.

Like all these calculating prodigies, he had a wonderful memory, and an hour or two after a performance could repeat all the numbers mentioned in it. He had also the peculiar gift of being able after a single glance to state the number (up to about 30) of sheep in a flock, of books in a case, and so on; and of visualizing and recollecting a large number of objects. For instance, after a second's look at some dominoes he gave the sum (117) of their points; asked how many letters were in a certain line of print chosen at random in a quarto page, he instantly gave the correct number (63); shown twelve digits, he had in half a second memorized them and their positions so as to be able to name instantly the particular digit occupying any assigned place. It is to be regretted that we do not know more of these performances. Those who are acquainted with the delightful autobiography of Robert-Houdin will recollect how he cultivated a similar power, and how valuable he found it in the exercise of his art.

Dase's calculations, when also allowed the use of paper and pencil, were almost incredibly rapid, and invariably accurate. When he was sixteen years old Strasznicky taught him the use of the familiar formula $\pi/4 = \tan^{-1}(\frac{1}{2}) + \tan^{-1}(\frac{1}{5}) + \tan^{-1}(\frac{1}{8})$, and asked him thence to calculate π. In two months he carried the approximation to 205 places of decimals, of which 200 are correct.* Dase's next achievement was to calculate the natural logarithms of the first 1,005,000 numbers to 7 places of decimals; he did this in his off-time from 1844 to 1847, when occupied by the Prussian survey. During the next two years he compiled in his spare time a hyperbolic table which was published by the Austrian Government in 1857. Later he offered to makes tables of the factors of all numbers from 7,000,000 to 10,000,000, and, on the recommendation of Gauss, the Hamburg Academy of Sciences agreed to assist him so that he might have leisure for the purpose, but he lived only long enough to finish about half the work.

Truman Henry Safford, born in 1836 at Royalton, Vermont, U.S.A., was another calculating prodigy. He was of a somewhat different type, for he received a good education, graduated in due course at Harvard, and ultimately took up astronomy, in which subject he held a professional post. I gather that though always a rapid calculator, he gradually lost the exceptional powers shown in his youth. He died in 1901.

Safford never exhibited his calculating powers in public, and I know of them only through the accounts quoted by Scripture and Mitchell, but they seem to have been typical of these calculators. In 1842, he amused and astonished his family by

*The result was published in *Crelle's Journal*, 1844, vol. XXVII, p. 198; on closer approximations and easier formulae, see chapter XII.

mental calculations. In 1846, when ten years old, he was examined, and here are some of the questions then put to him: Extract the cube root of a certain number of seven digits; answered instantly. What number is that which being divided by the product of its digits, the quotient is three, and if 18 be added the digits will inverted? answer 24, given in about a minute. What is the surface of a regular pyramid whose slant height is 17 feet, and the base a pentagon of which each side is 33·5 feet? answer 3354·5558 square feet, given in two minutes. Asked to square a number of eighteen digits, he gave the answer in a minute or less, but the question was made the more easy as the number consisted of the digits 365 repeated six times. Like Colburn, he factorized high numbers with ease. In such examples his processes were empirical; he selected (he could not tell how) likely factors and tested the matter in a few seconds by actual division.

More recently there have been four calculators of some note: *Ugo Zamebone*, an Italian, born in 1867; *Pericles Diamandi*, a Greek, born in 1868; *Carl Rückle*, a German; and *Jacques Inaudi*, born in 1867. The three first mentioned were of the normal type, and I do not propose to describe their performances, but Inaudi's performances merit a fuller treatment.

Jacques Inaudi* was born in 1867 at Onorato in Italy. He was employed in early years as a shepherd, and spent the long idle hours in which he had no active duties in pondering on numbers, but used for them no concrete representations such as pebbles. His calculating powers first attracted notice about 1873. Shortly afterwards his elder brother sought his fortune as an organ-grinder in Provence, and young Inaudi, accompanying him, came into a wider world, and earned a few coppers for himself by street exhibitions of his powers. His ability was exploited by showmen, and thus in 1880 he visited Paris, where he gave exhibitions: in these he impressed all observers as being modest, frank, and straightforward. He was

*See Charcot and Darboux, *Mémoires de l'Institut, Comptes Rendus*, 1892, vol. CXIV, pp. 275, 528, 578; and Binet, *Révue des deux Mondes*, 1892, vol. CXI, pp. 905–924.

then ignorant of reading and writing: these arts he subsequently acquired.

His earlier performances were not specially remarkable as compared with those of similar calculating prodigies, but with continual practice he improved. Thus at Lyons in 1873 he could multiply together almost instantaneously two numbers of three digits. In 1874 he was able to multiply a number of six digits by another number of six digits. Nine years later he could work rapidly with numbers of nine or ten digits. Still later, in Paris, asked by Darboux to cube 27, he gave the answer in 10 seconds. In 13 seconds he calculated how many seconds are contained in 18 years 7 months 21 days 3 hours; and he gave immediately the square root of one-sixth of the difference between the square of 4,801 and unity. He also calculated with ease the amount of wheat due according to the traditional story to Sessa, who, for inventing chess, was to receive 1 grain on the first cell of a chess-board, 2 on the second, 4 on the third, and so on in geometrical progression.

He could find the integral roots of equations and integral solutions of problems, but proceeded only by trial and error. His most remarkable feat was the expression of numbers less than 10^5 in the form of a sum of four squares, which he could usually do in a minute or two; this power was peculiar to him. Such problems were repeatedly solved at private performances, but the mental strain caused by them was considerable.

A performance before the general public rarely lasted more than 12 minutes, and was a much simpler affair. A normal programme included the subtraction of one number of twenty-one digits from another number of twenty-one digits: the addition of five numbers each of six digits; the multiplying of a number of four digits by another number of four digits; the extraction of the cube root of a number of nine digits, and of the fifth root of a number of twelve digits; the determination of the number of seconds in a period of time, and the day of the week on which a given date falls. Of course, the questions were put by members of the audience. To a professional calculator these problems are not particularly difficult. As each number was announced, Inaudi repeated it slowly to his assistant, who wrote it on a blackboard, and then slowly read it aloud to make sure that it was right. Inaudi then repeated the number once more. By this time he had generally solved the problem, but if he wanted longer time he made a few remarks of a general character, which he was able to do without interfering with his mental calculations. Throughout the exhibition he faced the audience: the fact that he never even glanced at the blackboard added to the effect.

It is probable that the majority of calculating prodigies rely on the speech-muscles as well as on the eye and the ear to help them to recollect the figures with which they are dealing. It was formerly believed that they all visualized the numbers proposed to them, and certainly some have done so. Inaudi, however, trusted mainly to the ear and to articulation. Bidder also relied partly on the ear, and when he visualized a number, it was not as a collection of digits, but as a concrete collection of units divisible, if the number was composite, into definite groups. Rückle relied mainly on visualizing the numbers. So it would seem that the memories of calculators work in different ways. Inaudi could reproduce mentally the sound of the repetition of the digits of the number in his own voice, and was confused, rather than helped, if the numbers were shown to him in writing. The articulation of the digits of the number also seemed necessary to enable him fully to exhibit his powers, and he was accustomed to repeat the numbers aloud before beginning to work on them – the sequence of sounds being important. A number of twenty-four digits being read to him, in 59 seconds he memorized the sound of it, so that he could give the sequence of digits forwards or backwards from selected points – a feat which Mondeux had taken 5 minutes to perform. Numbers of about a hundred digits were similarly memorized by Inaudi in 12 minutes, by Diamandi in 25 minutes, and by Rückle in under 5 minutes. This power is confined to numbers, and calculators cannot usually recollect a long sequence of letters. Numbers were ever before Inaudi: he thought of little else, he dreamed of them, and sometimes even solved problems in his sleep. His memory was excellent for numbers, but normal or subnormal for other things. At the end of a séance he could repeat the questions which had been put to him and his answers, involving hundreds of digits in all. Nor was his memory in such matters limited to a few hours. Once, eight days after he had been given a question on a number of twenty-two digits, he was unexpectedly asked about it, and at once repeated the number. He was repeatedly examined, and we know more of his work than of any of his predecessors, with the possible exception of Bidder.

Most of these calculating prodigies find it difficult or impossible to explain their methods. But we have a few analyses by competent observers of the processes used: notably one by Bidder on his own work; another by Colburn on his work; and others by Müller and Darboux on the work of Rückle and Inaudi respectively. That by Bidder is the most complete, and the others are on much the same general lines.

Bidder's account of the processes he had discovered and used is contained in a lecture* given by him in 1856 to the Institution of Civil Engineers. Before describing these processes there are two remarks of a general character which should, I think, be borne in mind when reading his statement. In the first place, he gives his methods in their perfected form, and not necessarily in that which he used in boyhood; moreover, it is probable that in practice he employed devices to shorten the work which he did not set out in his lecture. In the second place, it is certain, in spite of his belief to the contrary, that he, like most of these prodigies, had an exceptionally good memory, which was strengthened by incessant practice. One example will suffice. In 1816, at a performance, a number was read to him backwards: he at once gave it in its normal form. An hour later he was asked if he remembered it: he immediately repeated it correctly. The number was: 2,563,721,987,653,461,598,746,231,905,607,541,128,975,231.

*Institution of Civil Engineers, Proceedings, London, 1856, vol. xv, pp. 251–280. An early draft of the lecture is extant in MS.; the variations made in it are interesting, as showing the history of his mental development, but are not sufficiently important to need detailed notice here.

Of the four fundamental processes, addition and subtraction present no difficulty and are of little interest. The only point to which it seems worth calling attention is that Bidder, in adding three or more numbers together, always added them one at a time, as is illustrated in the examples given below. Rapid mental arithmetic depended, in his opinion, on the arrangement of the work, whenever possible, in such a way that only one fact had to be dealt with at a time. This is also noticeable in Inaudi's work.

The multiplication of one number by another was, naturally enough, the earliest problem Bidder came across, and by the time he was six years old he had taught himself the multiplication table up to 10 times 10. He soon had practice in harder sums, for, being a favourite of the village blacksmith, and constantly in the smithy, it became customary for the men sitting round the forge-fire to ask him multiplication sums. From products of numbers of two digits, which he would give without any appreciable pause for thought, he rose to numbers of three and then of four digits. Halfpence rewarded his efforts, and by the time he was eight years old, he could multiply together two numbers each of six digits. In one case he even multiplied together two numbers each of twelve digits, but, he says, "it required much time," and "was a great and distressing effort."

The method that he used is, in principle, the same as that explained in the usual text-books, except that he added his results as he went on. Thus to multiply 397 by 173 he proceeded as follows:

We have				$100 \times 397 = 39,700,$		
to this must be added				$70 \times 300 = 21,000$ making	60,700,
,,	,,	,,	,,	,,	$70 \times 90 = 6,300$,,	67,000,
,,	,,	,,	,,	,,	$70 \times 7 = 490$,,	67,490,
,,	,,	,,	,,	,,	$3 \times 300 = 900$,,	68,390,
,,	,,	,,	,,	,,	$3 \times 90 = 270$,,	68,660,
,,	,,	,,	,,	,,	$3 \times 7 = 21$,,	68,681.

We shall underrate his rapidity if we allow as much as a

second for each of these steps, but even if we take this low standard of his speed of working, he would have given the answer in 7 seconds. By this method he never had at one time more than two numbers to add together, and the factors are arranged so that each of them has only one significant digit: this is the common practice of mental calculators. It will also be observed that here, as always, Bidder worked from left to right: this, though not usually taught in our schools, is the natural and most convenient way. In effect he formed the product of $(100+70+3)$ and $(300+90+7)$, or $(a+b+c)$ and $(d+e+f)$ in the form $ad+ae\ldots+cf$.

The result of a multiplication like that given above was attained so rapidly as to seem instantaneous, and practically gave him the use of a multiplication table up to 1000 by 1000. On this basis, when dealing with much larger numbers, for instance, when multiplying 965,446,371 by 843,409,133, he worked by numbers forming groups of 3 digits, proceeding as if 965, 446, etc., were digits in a scale whose radix was 1000: in middle life he would solve a problem like this in about 6 minutes. Such difficulty as he experienced in these multiplications seems to have been rather in recalling the result of the previous step than in making the actual multiplications.

Inaudi also multiplied in this way, but he was content if one of the factors had only one significant digit: he also sometimes made use of negative quantities: for instance, he thought of 27×729 as $27(730-1)$; so, too, he thought of 25×841 in the form 84100/4; and in squaring numbers he was accustomed to think of the number in the form $a+b$, choosing a and b of convenient forms, and then to calculate the result in the form $a^2+2ab+b^2$.

In multiplying concrete data by a number Bidder worked on similar lines to those explained above in the multiplication of two numbers. Thus to multiply £14 15s. $6\frac{3}{4}d$. by 787 he proceeded thus:

We have   £(787)(14) = £11018 0s. 0d.
to which we add (787)(15) shillings = £590 5s. 0d. making £11608 5s. 0d.
to which we add (787)(27) farthings = £22 2s. $8\frac{1}{4}d$. making £11630 7s. $8\frac{1}{4}d$.

Division was performed by Bidder much as taught in school-books, except that his power of multiplying large numbers at sight enabled him to guess intelligently and so save unnecessary work. This also was Inaudi's method. A division sum with a remainder presents more difficulty. Bidder was better skilled in dealing with such questions than most of these prodigies, but even in his prime he never solved such problems with the same rapidity as those with no remainder. In public performances difficult questions on division are generally precluded by the rules of the game.

If, in a division sum, Bidder knew that there was no remainder, he often proceeded by a system of two-digit terminals. Thus, for example, in dividing (say) 25,696 by 176, he first argued that the answer must be a number of three digits, and obviously the left-hand digit must be 1. Next he had noticed that there are only 4 numbers of two digits (namely, 21, 46, 71, 96) which when multiplied by 76 give a number which ends in 96. Hence the answer must be 121, or 146, or 171, or 196; and experience enabled him to say without calculation that 121 was too small and 171 too large. Hence the answer must be 146. If he felt any hesitation, he mentally multiplied 146 by 176 (which he said he could do "instantaneously"), and thus checked the result. It is noticeable that when Bidder, Colburn, and some other calculating prodigies knew the last two digits of a product of two numbers, they also knew, perhaps subconsciously, that the last two digits of the separate numbers were necessarily of certain forms. The theory of these two-digit arrangements has been discussed by Mitchell.

Frequently also in division, Bidder used what I will call a digital process, which a priori would seem far more laborious than the normal method, though in his hands the method was extraordinarily rapid: this method was, I think, peculiar to him. I define the *digital* of a number as the digit obtained by finding the sum of the digits of the original number, the sum of the digits of this number, and so on, until the sum is less than 10. The digital of a number is the same as the digital of the product of the digitals of its factors. Let us apply this in

Bidder's way to see if 71 is an exact divisor of 23,141. The digital of 23,141 is 2. The digital of 71 is 8. Hence if 71 is a factor, the digital of the other factor must be 7, since 7 times 8 is the only multiple of 8 whose digital is 2. Now the only number which multiplied by 71 will give 41 as terminal digits is 71. And since the other factor must be one of three digits and its digital must be 7, this factor (if any) must be 871. But a cursory glance shows that 871 is too large. Hence 71 is not a factor of 23,141. Bidder found this process far more rapid than testing the matter by dividing by 71. As another example let us see if 73 is a factor of 23,141. The digital of 23,141 is 2; the digital of 73 is 1; hence the digital of the other factor (if any) must be 2. But since the last two digits of the number are 41, the last digits of this factor (if any) must be 17. And since this factor is a number of three digits and its digital is 2, such a factor, if it exists, must be 317. This on testing (by multiplying it by 73) is found to be a factor.

When he began to exhibit his powers in public, questions concerning weights and measures were, of course, constantly proposed to him. In solving these he knew by heart many facts which frequently entered into such problems, such as the number of seconds in a year, the number of ounces in a ton, the number of square inches in an acre, the number of pence in a hundred pounds, the elementary rules about the civil and ecclesiastical calendars, and so on. A collection of such data is part of the equipment of all calculating prodigies.

In his exhibitions Bidder was often asked questions concerning square roots and cube roots, and at a later period higher roots. That he could at once give the answer excited unqualified astonishment in an uncritical audience; if, however, the answer is integral, this is a mere sleight of art which anyone can acquire. Without setting out the rules at length, a few examples will illustrate his method.

He was asked to find the square root of 337,561. It is obvious that the root is a number of three digits. Since the given number lies between 500^2 or 250,000 and 600^2 or 360,000, the left-hand digit of the root must be a 5. Reflection had shown him that the only numbers of two digits whose squares end in 61 are 19, 31, 69, 81, and he was familiar with this fact. Hence the answer was 519, or 531, or 569, or 581. But he argued that as 581 was nearly in the same ratio to 500 and 600 as 337,561 was to 250,000 and 360,000, the answer must be 581, a result which he verified by direct multiplication in a couple of seconds. Similarly in extracting the square root of 442,225, he saw at once that the left-hand digit of the answer was 6, and since the number ended in 225, the last two digits of the answer were 15 or 35, or 65 or 85. The position of 442,225 between $(600)^2$ and $(700)^2$ indicates that 65 should be taken. Thus the answer is 665, which he verified before announcing it. Other calculators have worked out similar rules for the extraction of roots.

For exact cube roots the process is more rapid. For example, asked to extract the cube root of 188,132,517, Bidder saw at once that the answer was a number of three digits, and since $5^3 = 125$ and $6^3 = 216$, the left-hand digit was 5. The only number of two digits whose cube ends in 17 is 73. Hence the answer is 573. Similarly the cube root of 180,362,125 must be a number of three digits, of which the left-hand digit is a 5, and the two right-hand digits were either 65 or 85. To see which of these was required, he mentally cubed 560, and seeing it was near the given number, assumed that 565 was the required answer, which he verified by cubing it. In general a cube root that ends in a 5 is a trifle more difficult to detect at sight by this method than one that ends in some other digit, but since 5^3 must be a factor of such numbers, we can divide by that and apply the process to the resulting number. Thus the above number 180,362,125 is equal to $5^3 \times 1,442,897$, of which the cube root is at once found to be 5×113, that is, 565.

For still higher exact roots the process is even simpler, and for fifth roots it is almost absurdly easy, since the last digit of the number is always the same as the last digit of the root. Thus if the number proposed is less than 10^{10}, the answer consists of a number of two digits. Knowing the fifth powers to 10, 20, ..., 90, we have, in order to know the first digit of the answer, only to see between which of these powers the number proposed lies, and the last digit being obvious, we can give the answer instantly. If the number is higher, but less than 10^{15}, the answer is a number of three digits, of which the middle digit can be found almost as rapidly as the others. This is rather a trick than a matter of mental calculation.

In his later exhibitions, Bidder was sometimes asked to extract roots, correct to the nearest integer, the exact root involving a fraction. If he suspected this, he tested it by "casting out the nines," and if satisfied that the answer was not an integer, proceeded tentatively as best he could. Such a question, if the answer is a number of three or more digits, is a severe tax on the powers of a mental calculator, and is usually disallowed in public exhibitions.

Colburn's remarkable feats in factorizing numbers led to similar questions being put to Bidder, and gradually he evolved some rules, but in this branch of mental arithmetic I do not think he ever became proficient. Of course, a factor which is a power of 2 or of 5 can be obtained at once and powers of 3 can be obtained almost as rapidly. For factors near the square root of a number he always tried the usual method of expressing the number in the form $a^2 - b^2$, in which case the factors are obvious. For other factors he tried the digital method already described.

Bidder was successful in giving almost instantaneously the answers to questions about compound interest and annuities: this was peculiar to him, but his method is quite simple, and may be illustrated by his determination of the compound interest on £100 at 5 per cent for 14 years. He argued that the simple interest amounted to £(14)(5), i.e. to £70. At the end of the first year the capital was increased by £5, the annual interest on this was 5s. or one crown, and this ran for 13 years, at the end of the second year another £5 was due, and the 5s. interest on this ran for 12 years. Continuing this argument, he had to add to the £70 a sum of $(13 + 12 + \ldots + 1)$ crowns. i.e. (13/2)(14)(5) shillings, or £22 15s. 0d., which, added to the £70 before mentioned, made £92 15s. 0d. Next the 5s. due at the end of the second year (as interest on the £5 due at the end of the first year) produced in the same way an annual interest of 3d. All these three-pences amount to (12/3)(13/2) (14)(3) pence, i.e. £4 11s. 0d., which, added to the previous sum of £92 15s. 0d., made £97 6s. 0d. To this we have similarly to add (11/4)(12/3)(13/2)(14)(3/20) pence, i.e. 12s. 6d., which made a total of £97 18s. 6d. To this again we have to add (10/5)(11/4)(12/3)(13/2)(14)(3/400) pence, i.e. 1s. 3d., which made a total of £97 19s. 9d. To this again we have to add (9/6)(10/5)(11/4)(12/3)(13/2)(14)(3/8000) pence, i.e. 1d., which made a total of £97 19s. 10d. The remaining sum to be added cannot amount to a farthing, so he at once gave the answer as £97 19s. 10d. The work in this particular example did in fact occupy him less than one minute – a much shorter time than most mathematicians would take to work it by aid of a table of logarithms. It will be noticed that in the course of his analysis he summed various series.

In the ordinary notation, the sum at compound interest amounts to $£(1·05)^{14} \times 100$. If we denote £100 by P and ·05 by r, this is equal to $P(1+r)^{14}$ or $P(1 + 14r + 91r^2 + \dots)$, which, as r is small, is rapidly convergent. Bidder in effect arrived by reasoning at the successive terms of the series, and rejected the later terms as soon as they were sufficiently small.

In the course of this lecture Bidder remarked that if his ability to recollect results had been equal to his other intellectual powers, he could easily have calculated logarithms. A few weeks later he attacked this problem, and devised a mental method of obtaining the values of logarithms to seven or eight places of decimals.* He asked a friend to test his accuracy, and in answer to questions gave successively the logarithms of 71, 97, 659, 877, 1297, 8963, 9973, 115249, 175349, 290011, 350107, 229847, 369353, to eight places of decimals; taking from thirty seconds to four minutes to make the various calculations. All these numbers are primes. The greater part of the answers were correct, but in a few cases there was an error, though generally

*See W. Pole, *Institution of Civil Engineers, Proceedings*, London, 1890–91, vol. CIII, p. 250.

of only one digit: such mistakes were at once corrected on his being told that his result was wrong. This remarkable performance took place when Bidder was over 50.

Alexander Craig Aitken provides an appropriate climax for this chapter, because he was not only one of the greatest calculating prodigies, but a first-rate mathematician as well: the author of four books and about seventy papers. He was born in 1895 in Dunedin, New Zealand, and served in the First World War till he was seriously wounded in 1917. He left New Zealand for Edinburgh in 1923. His thesis for the PH.D. was so remarkable that he was awarded the higher degree of D.SC. instead, and he remained there for the rest of his life, succeeding Sir Edmund Whittaker as Professor of Mathematics in 1946. He was a most inspiring teacher, and took a keen personal interest in his students. His phenomenal power of mental calculation* was partly based on his unusual memory. He could repeat long passages from Virgil or Milton. He once remarked that he had to be careful what he read for entertainment, because of the difficulty of forgetting it afterwards. He gave occasional demonstrations before an audience, performing almost instantaneous multiplication, division, and extraction of square and cube roots, and rapidly writing on the blackboard (from memory) the 707 digits of Shanks's calculation of π (see page 357). When Ferguson showed in 1945 that Shanks had gone wrong at the 528th place, he easily memorized the correct value to 1000 places (or possibly, according to J.C.P. Miller, to 2000 places). He once remarked that this process of recall was "largely rhythmic." A more unusual virtuosity was the quick evaluation of special determinants.

He competed successfully with Wim Klein, a Dutch prodigy who had memorized the multiplication table up to 100×100 but lacked the mathematical knowledge to employ clever short cuts. Aitken often made subconscious calculations. He

*A.C. Aitken, "The Art of Mental Calculation; with Demonstrations," in the *Transactions of the Royal Society of Engineers*, London, 1954, vol. XLIV, pp. 295–309. See also the Obituary in the *Proceedings of the Edinburgh Mathematical Society*, 1968, vol. XVI (series II), part 2, pp. 151–176.

told of results that "came up from the murk," and would say of a particular number that it "feels prime," as indeed it was. He was one of the few to whom integers were personal friends. He noticed, for instance, an amusing property of 163: that $e^{\pi\sqrt{163}}$ differs from an integer by less than 10^{-12}. As he himself once put it, "Familiarity with numbers, acquired by innate faculty sharpened by assiduous practice, does give insight into the profounder theorems of algebra and analysis."

His friends and students warmly remember his great kindness and patience. In addition to his many other accomplishments, he was a gifted musician. He played the violin and viola, composed songs, piano pieces, and some orchestral works ("rigorously suppressed", as he said).

He retired from his Chair of Mathematics in 1965 because of poor health, and died two years later.

ADDENDUM

Note.

$$e^{\pi\sqrt{163}} = 262537\ 412640\ 768743 \cdot 999999\ 999999\ 25 \cdots .$$

There is no need to assume that Aitken carried out this stupendous computation: it all drops out from the theory of elliptic modular functions (see H.J.S. Smith, *Report on the Theory of Numbers,* Chelsea, New York, 1965).

CHAPTER XIV

CRYPTOGRAPHY AND CRYPTANALYSIS

The art of writing secret messages – intelligible to those who are in possession of the key and unintelligible to all others – has been studied for centuries. The usefulness of such messages, especially in time of war, is obvious; on the other hand, their solution may be a matter of great importance to those from whom the key is concealed. But the romance connected with the subject, the not uncommon desire to discover a secret, and the implied challenge to the ingenuity of all from whom it is hidden have attracted to the subject the attention of many to whom its utility is a matter of indifference.

Although it is possible to communicate information in many different ways, we shall be concerned here only with secret communications which are in writing or in some other permanent form. Thus, small muscular movements, such as breathing long and short in the Morse dot and dash system, or signalling with a fan or stick, will not concern us. The essential feature is that the information conveyed remain hidden from all those who may have been enabled to obtain copies of the messages transmitted but are not in possession of the key. Thus we would consider secret those messages which are outwardly intelligible, but really convey a different, hidden meaning.* On the other hand, a communication in a foreign language or in any recognized notation like shorthand is *not* a secret message.

The famous diary of Samuel Pepys is commonly said to have been written in cipher, but in reality it was written in

---

*The reader is referred, for two classic examples of this type, to *Cryptographie Pratique*, A. de Grandpré, Paris, 1905, p. 57.

CRYPTOGRAPHY AND CRYPTANALYSIS 389

shorthand according to a system invented by J. Shelton.* It is, however, somewhat difficult to read, for the vowels were usually omitted, and Pepys used some arbitrary signs for terminations, particles, and certain frequent words. Further, in certain places, where the matter was such that it could hardly be expressed with decency, he changed from English to a foreign language, or inserted non-significant symbols. Shelton's system had been forgotten when attention was first attracted to the diary. Accordingly, we may say that, to those who first tried to read it, it was written in cipher; but Pepys' contemporaries would have properly described the diary as being written in shorthand, although it involved a few modifications of his own invention.

The mere fact that a message is concealed or secretly conveyed does not make it a secret message. The majority of stories dealing with secret communication are concerned with the artfulness with which the message is concealed or conveyed, and have nothing to do with cryptography. Many of the ancient instances of secret communication are of this type. Herodotus tells of the practice of shaving a slave's head and inscribing the message on his scalp; then sending the slave to deliver the message after his hair had grown back again. More modern illustrations are to be found in messages conveyed by pigeons or written on the paper wrapping of a cigarette.

Cryptographic systems. Every method for converting a plain-text message into a secret message consists of two parts: (1) a basic, invariant method, called the *general system*, and (2) a variable, keying element usually consisting of a word, phrase, or sequence of numbers, called the *specific key*. It is ordinarily assumed that the *enemy* (any person who has obtained unauthorized possession of the messages and is attempting to solve them) has full knowledge of the general system. (This assumption is based on the fact that no large communication system can hope to keep its general procedure

Tachy-graphy, by J. Shelton, first edition 1620, sixth edition used by Pepys 1641. A somewhat similar system by W. Cartwright was issued by J. Rich under the title *Semographie*, London, 1644.

secret very long, nor can it attempt to vary its procedure at will, in view of the great difficulty of training personnel in new methods.) The relative security of any cryptographic system is then considered to be proportional to the length of time required to determine the specific key. In this connection it must be appreciated that it is ordinarily not practicable to change specific keys more than once a day. Consequently, if all the messages (or most of them, at any rate) are intercepted, the enemy would sometimes have several hundred messages available for the determination of one specific key. This fact is often overlooked, and many systems whose security is quite low are given exaggerated importance, in the minds of their inventors, because they may be very difficult or even impossible to solve when only one message is available.

According to the method of treating the original message, general systems are divided into two main classes. If the characters of the plain-text message are merely rearranged without suffering any change in identity – i.e. if some permutation is applied to the original characters – the system is called *transposition*. If, however, the characters themselves are replaced by equivalents in the form of letters, or figures, or arbitrary symbols, without introducing any change in their original sequence, the system is called *substitution*. Both of these systems may be combined in a single cryptogram, by applying one to the result obtained by the other.

In the short treatment of the subject which will be given in this chapter, an attempt will be made to present a brief description of the classic systems of enciphering messages. In each case, a few words will be added with reference to the method of solution. The reader should note in this connection the difference in the use of the words *deciphering* and *solving*. The former refers to the procedure followed by the bona-fide correspondent who is in possession of all the details of the system and who merely reverses the steps followed in enciphering. The latter refers to the method used by an enemy of obtaining an unauthorized translation by applying the principles of the science now called *Cryptanalysis*.

Transposition systems. Practically every transposition system involves a geometric figure in which the plain-text message is inscribed according to one route and then transscribed according to another route.

The following is an example of encipherment, in a system commonly called *route transposition*. Suppose the message to be transmitted is

I MUST HAVE ANOTHER HUNDRED DOLLARS

The general system, we will say, requires a completely filled rectangle of eight columns, so that it is necessary to add dummy letters if the number of letters in the message is not a multiple of eight. We therefore add two such letters, say XX, at the end of the message, in order to have 32 letters in all. We will assume next that the route for inscription is horizontal, starting at the upper left-hand corner and proceeding alternately to the right and to the left. The following arrangement will then be obtained:

I	M	U	S	T	H	A	V,
R	E	H	T	O	N	A	E
H	U	N	D	R	E	D	D
X	X	S	R	A	L	L	O

If the method of transcription is vertical, beginning at the upper right-hand corner and proceeding alternately down and up, the final cipher message, prepared for transmission, would be VEDOL DAAHN ELARO TSTDR SNHUM EUXXH RI. The reversal of the above steps, in deciphering, would involve no particular difficulty.

It is not amiss to observe at this point that, because of international telegraphic regulations governing the cost of transmitting messages, the final text of cryptograms is usually divided up into regular groups of five letters. This consideration is obviously of no consequence from a cryptanalytic standpoint.

392 MATHEMATICAL RECREATIONS AND ESSAYS

A system of the type just described possesses very little security and has the further disadvantage of not being readily adaptable to change. Even if the dimensions of the rectangle and the routes of inscription and transcription were to be changed regularly, an enemy would have little difficulty in reading intercepted messages. The procedure in solution is essentially one of trial, but the number of possibilities is so limited that with a little experience a solution can be obtained in a very short time.

A very widely used variation of route transposition is known as *columnar transposition*. In this system the geometric figure is again a rectangle but with the message inscribed in the *normal* manner of writing. The transcription is vertical, the columns being selected in the order determined by a numerical key, the width of the rectangle being equal to the length of the key. To illustrate, suppose the key is 3–2–7–1–4–6–5 and the message is THE PRISONERS HAVE SEIZED THE RAILWAY STATION.

The first step would be to write the message in a rectangle whose columns are headed by the key numbers:

3	2	7	1	4	6	5
T	H	E	P	R	I	S
O	N	E	R	S	H	A
V	E	S	E	I	Z	E
D	T	H	E	R	A	I
L	W	A	Y	S	T	A
T	I	O	N			

It is inadvisable to complete the last row of the rectangle with dummy letters, as that would give the enemy a clue to the length of the key. Then, transcribing the columns in order according to the numerical key and simultaneously preparing the message for transmission by dividing it up into five-letter groups, we have:

PREEY NHNET WITOV DLTRS IRSSA EIAIH ZATEE SHAO

The first step in deciphering is to determine how many long columns there are – that is, how many letters occur on the last row of the rectangle. This is accomplished by dividing the number of letters in the message by the length of the key; the remainder obtained from this division is the desired number. When it has been calculated, the cipher letters can be put into their proper positions, and the message will reappear.

In cases where the key is fairly long and it is inadvisable to reduce it to writing, it is found convenient to derive the numerical sequence from an easily remembered key word or phrase. This may be done in any one of several ways, the most common being to assign numbers to the letters in accordance with their normal alphabetical order. To illustrate, suppose the key word is CRYPTOGRAPHY. The letter A which appears in it is numbered 1. Then, since there is no B in the word, the letter C is numbered 2. The next letter, alphabetically, is G, and this is numbered 3, etc. This procedure is continued until a number has been assigned to every letter in the word. Whenever a letter appears more than once, the several appearances are numbered consecutively from left to right. The complete numerical key thus obtained is

C	R	Y	P	T	O	G	R	A	P	H	Y
2	8	11	6	10	5	3	9	1	7	4	12

The procedure in solving a columnar transposition is based on the fact that the letters of an entire column have been transcribed as a unit. Let us suppose that we are trying to solve the cipher message just obtained. The first step is to fit together two groups of consecutive letters which will form good plain-text combinations. The most frequent *digraph* (two-letter combination) in English is TH, and so we might begin by setting every T next to every H, in turn. In each such case, several letters just preceding and just following the T and the H would also be juxtaposed. For example, if we place the first T in the message against the last H, we have the following combinations:

```
Y  A
N  T
H  E
N  E
E  S
T  H
W  A
I  O
T
O
```

The number of letters to be placed in each column is as yet unknown, but certain limitations are sometimes imposed by the positions of the particular letters being used. In the above instance, the O at the foot of the first column may be discarded, since the long columns of a transposition rectangle exceed the short ones by one letter. The T just above the O will be correct only if it is the last letter of the message, for it would end the last long column. As another example of how limitations may be placed upon the length of the columns corresponding to a given assumption, suppose we had placed the first T of the message against the first H, which is only three letters removed. Then only three digraphs would have been obtained. Consequently with this assumption no column of the rectangle could contain more than four letters.

There would be one pair of columns such as those given above for every possible TH combination, and it is next necessary to choose that pair which is composed of the best selection of high-frequency combinations. The experienced cryptanalyst can do this at sight, but the same result can be obtained by a purely mathematical process. Extended studies of the relative frequencies of single letters, digraphs, trigraphs, and polygraphs have been made by various authors. These relative frequencies may be considered an invariant property of the corresponding plain-text combinations, since they are derived from studies of a very large amount of text. By assigning to each digraph a weight equal to its relative frequency, it then becomes a very simple matter to select that pair of columns which yields the greatest average weight per digraph.

Such a test will show that the arrangement already set up is an exceptionally good one. The only digraphs which are not high-frequency combinations are YA and WA, and the former may be beyond the limits of the column. If we assume that this arrangement is correct, we then try to add a third column, either on the right or on the left, so as to obtain good plain-text trigraphs.

Now, IO is almost always followed by N. In addition the TH suggests the word THE. This prompts us to seek an E followed at two intervals by an N, and we find just one such place in the cipher message. Note how the addition of the corresponding letters to the combinations already obtained delimits the columns.

$$
\begin{array}{lll}
\text{H} & \text{E} & \text{P} \\
\text{N} & \text{E} & \text{R} \\
\text{E} & \text{S} & \text{E} \\
T & H & E \\
\text{W} & \text{A} & \text{Y} \\
I & O & N \\
T & &
\end{array}
$$

Should there be several possibilities instead of only one, these may again be compared by summing the weights of the trigraphs in each case and choosing the arrangement which gives the greatest average. In the present case, however, it is obvious that we are on the right track, and the solution will follow very readily.

The procedure which has just been carried out is greatly facilitated by a knowledge of particular words which would be likely to appear in the message. Such a knowledge is quite usual in military cryptanalysis, for example. Indeed, many writers on cryptanalysis refer frequently to the *Probable word* or *Intuitive* method as a means of solution for many different systems.

What has been said thus far about the method of solving

columnar transpositions has implied that the cryptanalyst is in possession of only one message. It has already been pointed out that in actual practice there may be many messages in the same key available for study. In such cases several methods can be used to obtain a rapid solution.

Among these the following is of note, since it is applicable to any transposition system whatever. Suppose two or more messages of *identical lengths* are subjected to the same transposition. Then, no matter how complicated the system may be, it is obvious that letters which were in corresponding positions in the plain-text will be in corresponding positions in the cipher text. Suppose the cipher messages are superimposed so that the first letters of all the messages are in one column, the second letters of all the messages are in a second column, etc. Then if it is assumed that two particular letters of one message are sequent in the plain-text, the letters in the corresponding positions of each of the other messages will combine in the same order. They will thus afford a check on the correctness of the first assumption, in the same way that the combinations arising from two columns in columnar transposition served as a check on the juxtaposition of a particular pair of letters in those columns. To each of these digraphs a third letter may be added to form trigraphs, etc. The idea is essentially one of *anagramming* the entire columns of the superimposed messages, and the mathematical method of summing frequencies may be used to advantage.

A good start can often be obtained by choosing an invariant combination, such as QU, or a digraph composed of two letters which have a great affinity for one another, such as TH or RE. If one of the letters of the digraph is infrequent – as, for example, in VE – the number of possibilities is very considerably reduced. Given a minimum of four messages of identical lengths, one can safely be assured of obtaining a solution of those messages. However, it is not always true that this solution will permit the cryptanalyst to translate additional messages of different lengths. In order to accomplish this, he would have to obtain some information about the general system and the specific key from the messages already solved.

A second procedure which is very fruitful in solving columnar transpositions is applicable to two or more messages involving a long repetition. Such repetitions are often found in the messages of large communication systems where there is a tendency to use stereotyped phraseology. In order to appreciate the procedure more fully, let us encipher, with the numerical key 8-6-4-1-5-3-2-7 the following two messages containing the repetition THE FIRST AND SECOND DIVISIONS.

8	6	4	1	5	3	2	7
W	H	A	T	A	R	E	T
H	E	O	R	D	E	R	S
F	O	R	T	H	E	F	I
R	S	T	A	N	D	S	E
C	O	N	D	D	I	V	I
S	I	O	N	S			

8	6	4	1	5	3	2	7
T	H	E	F	I	R	S	T
A	N	D	S	E	C	O	N
D	D	I	V	I	S	I	O
N	S	W	I	L	L	L	E
A	V	E	A	T	O	N	C
E							

The two cryptograms available for study by the cryptanalyst are

```
        1       2       3        4       5
1. TRTAD NERFS VREED IAORT NOADH NDSHE
        6     7     8
   OSOIT SIEIW HFRCS
      1      2      3      4      5      6      7      8
2. FSVIA SOILN RCSLO EDIWE IEILT HNDSV TNOEC TADNA E
```

He would notice and underline the italicized repetitions.

Note how the letters which make up the repetition appear in the two messages. The number of portions into which it has been broken up is equal to the number of columns in the rectangle, and the figure over each repeated portion is the key number which heads the corresponding column. The letters *RCS* in cipher message 1 are at the foot of their column. They indicate that the repeated phrase is at the end of the corresponding plain-text message. On the other hand, the letters *FSV*, which are at the head of their column in cipher message 2, indicate that the repetition is at the beginning of the corresponding plain-text message. As a result of this information, the length of each column is at once ascertainable. In the first message, for example, column 1 contains the letters TR*TADN*, column 2 contains the letters ER*FSV*, etc. In the second message column 1 contains the letters *FSV*IA, column 2 contains the letters *SOI*LN, etc. We thus have determined which are the long columns and which are the short columns of each of the two messages.

In addition to the above information, which in itself is of considerable value, one can almost always get portions of the key, and sometimes the whole key, from these repetitions. In order to appreciate how this is done, let up assume for a moment that the original rectangles are known. Note then that the letter T, which is the first letter of the repetition, appears in the fourth column of message 1 and in the first column of message 2. As a result, any letter of the repetition will appear, in message 1, three columns to the right of the position of the corresponding letter in message 2.

The letters *FSV* which begin the second cipher message must come from the column headed by number 1 of the key. The letters *FSV* of message 1 come from the column headed by number 2. But, if these letters in the first message are three places in advance of the corresponding letters of the second message, it follows that the numerical key contains the sequence 1–?–?–2 (where the question marks indicate undetermined numbers). The portion of the repetition which appears in the second message in the column headed by the number 2 is *SOI*, and corresponds in message 1 to the column headed by the number 6. Hence, by the same reasoning as before, the key must contain the sequence 1–?–?–2–?–?–6. Column 6 in the second message corresponds to column 5 in the first. Remembering that the key is of length 8, we may now say that it contains the sequence 1–5–?–2–?–?–6–?.

Continuing in this way, we build up the entire key 1–5–3 2–7–8–6–4, which is a cyclic permutation of the correct key. If it is applied to the first cipher message, the result will be:

1	5	3	2	7	8	6	4
T	A	R	E	T	W	H	A
R	D	E	R	S	H	E	O
T	H	E	F	I	F	O	R
A	N	D	S	E	R	S	T
D	D	I	V	I	C	O	N
N	S				S	I	O

and the proper starting-point is at once determined by inspection. Another method of determining the beginning of the cycle is to note how the cipher messages break up into long and short columns. Since the columns which have one extra letter are all at the left in the original rectangle, the cycle begins with the first long column that follows a short column.

The reader will appreciate from what has preceded that we were enabled to derive the entire key because of the fact that the relative displacement of corresponding letters in the two repetitions was prime to the length of the key. If these two numbers had had a common factor, the key would have broken up into as many partial cycles as the greatest common divisor of the two numbers. It would then have become necessary to join the partial cycles into one complete cycle. This would have introduced several possible solutions, but a moment's trial would have determined the correct one.

Now, in obtaining the solution just described, we assumed that the relative displacement was known. This is sometimes the case. For example, consider two messages whose endings are identical. After the length of the key has been determined by the number of sections into which the repetition breaks up, the number of long columns in each of the rectangles is at once obtained. The difference between these two numbers is the sought displacement. In the two messages just studied it is also possible to determine the relative displacement between the two repetitions, because of the fact that one repetition is at the beginning and the other at the end of their respective messages. Since the repetition is 26 letters long, its last letter must be in the second column of the second message. Moreover, this same letter is in the last long column – i.e. the fifth column – of the first message. The displacement is therefore three columns.

If the relative positions of the repetitions are the same, so that there is no displacement, as would happen if they were both at the beginnings of their messages, no information whatever about the actual numerical key would be obtainable by this method.

In those cases where the relative displacement is indeterminate, the assumption of each possibility in turn would still require a relatively small number of tests before the correct answer would be obtained.

The security of columnar transposition is very greatly increased if the resulting cipher is put through a second columnar transposition. This second step may use the same or a different numerical key. In either case, the superimposition of one columnar transposition upon a second yields a result obtainable from the original plain-text by a single transposition, but this latter process is much more complicated than columnar transposition. Such a system is usually called *double transposition*, and a single message so enciphered has a very high degree of security. However, two or more messages of identical lengths can be solved without difficulty in the manner already explained. The numerical keys involved in a double transposition can be obtained if the cryptanalyst is in possession of a single cipher message with its translation.

As a last example of transposition, a few words will be said about a classic system known as the *grille*, an example of which is shown below. It is a perforated card, usually square, which when put over a sheet of paper permits only certain portions of the sheet to be visible. In encipherment, the letters of the message are entered through these perforations. In decipherment, the message is written out in a diagram of the proper dimensions, the grille is superimposed, and only the desired letters of the plain-text will show through.

Grilles may be used in two different ways. With the first of these methods, the final cipher text will involve only the letters of the original plain-text message. One way of accomplishing this is to arrange the grille in such a way that if it is used successively in different positions, every cell of the underlying sheet of paper will be occupied. In the second method, on the other hand, it would be arranged that only certain cells will be filled, and the encipherer then has the further task of surrounding these significant letters with *false text*. This is generally a very difficult procedure, and has the further disadvantage of making the cipher text considerably longer than the plain-text. The inscription and transcription in grille systems may, of course, follow any prearranged routes.

The grille shown above is an example of a *revolving grille* of 36 cells. If it is successively rotated through 90° after the cells exposed by each position have been filled, every cell will be found to have been filled when the grille has returned to its starting position. The numbers which have been entered in the various cells indicate the method of construction. From each concentric band of cells, one cell must be cut to correspond to each number. If the number of cells on a side is odd, the central square of the grille must remain uncut.

Grille are not very practicable when there is any considerable amount of communication. Besides, their security is quite low. For example, all messages enciphered with a given revolving grille would break up into portions of identical lengths, each of which has been treated identically. The method of anagramming columns is therefore applicable.

In all of the transposition systems which have been described thus far the unit of cryptographic treatment has been a single letter. There is no reason why it could not be changed to a regular group of letters, or to syllabic groupings, or even to whole words. Such systems are by no means uncommon. An outstanding historical instance was the use of a route transposition on words by the Federal Army in the American Civil War, 1861–1865, equivalents being substituted for proper names. Also grilles of a rectangular shape, with rather long perforations, are sometimes used, so that a syllable or even a whole word can be entered at a time.

Substitution systems. The simplest type of substitution system is that in which the same plain-text letter is always represented by the same equivalent. It is probably the best-known type of cipher, and seems to be the very first idea that comes to the mind of a beginner in cryptography. Just what form the equivalent takes is obviously of no consequence, and yet it is surprising how many people get the notion that the use of complicated arbitrary symbols yields a more secure system than would be obtained with letters or figures.

The simplest way to employ such a system is to set down the equivalents in the form of a *substitution alphabet*, which consists of a *plain-sequence* and a *cipher sequence*, superimposed one above the other. The plain-text letter is found in the plain sequence, and replaced by the corresponding character in the cipher sequence.

For example, if we employ the substitution alphabet

Plain   A B C DE F GH I J.K L MN OP Q R S TU V W X Y Z
Cipher A D G J M P S V Y B E H K N QT W Z C F I L O R U X

the sentence COME HOME ALL IS FORGIVEN would be
enciphered GQKMV QKMAH HYCPQ ZSYLM N.

In the usual method of writing a substitution alphabet, the
plain sequence is the *normal* alphabet. In such cases, the assignment of the cipher sequence alone is sufficient to define
the entire substitution alphabet. Depending upon the manner
in which the cipher sequence is constructed, substitution alphabets are divided into three different classes.

1. *Standard alphabets.* Here the cipher sequence is a cyclic
permutation of either the normal alphabet or the normal alphabet *reversed.* This is the oldest type of substitution alphabet
known. Some writers call cipher systems using a standard
alphabet *the system of Julius Caesar.* However, Caesar used
but one of the possible standard alphabets; his cipher sequence
was always the normal alphabet beginning with the letter D.*

2. *Systematically mixed alphabets.* The difficulty arising from
the use of standard alphabets is obvious. One or two identifications are sufficient to determine the whole alphabet. To
circumvent this difficulty and still avoid the necessity of
reducing the cipher sequence to writing, some systematic
procedure must be employed for disarranging the normal
alphabet. Cryptographic literature is full of devices for accomplishing this end, but only one will be described here,
merely for the sake of example. Let a key word be chosen
which has no repeated letters in it – say FISHER. Write all
the remaining letters of the alphabet in order in a rectangular
array under the key word:

*Of some of Caesar's correspondence, Suetonius says (cap. 56) *si quis in-
vestigare et persequi velit, quartam elementorum literam, id est, d, pro a, et
perinde reliqua commutet.* And of Augustus he says (cap. 88) *quoties autem per
notas scribit, b pro a, c pro b, ac deinceps eadem ratione, sequentes literas ponit
pro x autem duplex a.* ("If anyone wishes to investigate and follow it up, let him
replace the fourth letter of the alphabet, that is, D, by A and in like manner
with the others." "Whenever he writes in cipher, he puts B for A, C for B and
continues in the same way; for X, however, he puts AA.")

2	4	6	3	1	5
F	I	S	H	E	R
A	B	C	D	G	J
K	L	M	N	O	P
Q	T	U	V	W	X
Y	Z				

Then transpose the letters of this figure by columns according to the numerical key based on the word FISHER. The resulting cipher sequence is

E G O W F A K Q Y H D N V I B L T Z R J P X S C M U

3. *Random alphabets.* Here the letters of the cipher sequence are chosen at random, and in such a sequence no aid in determining unknown letters can be obtained from known identifications. The objection to this type of sequence is, of course, that it must be reduced to writing.

The method of solution of a system which uses only one substitution alphabet or, as we shall call it, a *monoalphabetic* system, is fairly well known. It is based on the relative frequencies of the individual letters of the alphabet, together with the relative frequencies of their combinations with one another.

In English, the relative frequencies of occurrence of the individual letters, most frequent digraphs, and trigraphs are as follows:*

E ·131	R ·067	F ·028	G ·014	Q ·001
T ·090	S ·065	U ·028	B ·013	Z ·001
O ·082	H ·059	M ·026	V ·010	
A ·078	D ·044	P ·022	K ·004	
N ·073	L ·036	Y ·015	X ·003	
I ·068	C ·029	W ·015	J ·001	

*P. Valerio, *De la Cryptographie*, Paris, 1893, pp. 202, 204. See also A. Conan Doyle, *The Return of Sherlock Holmes*, Frome, 1924, pp. 58–83.

TH ·034	ER ·019	IN ·014	AT ·013	HA ·012
HE ·026	ON ·019	ED ·014	OF ·013	EN ·011
AN ·019	RE ·017	ND ·014	OR ·012	NT ·011

THE ·015	HAT ·003	FOR ·003	NDE ·003
AND ·005	EDT ·003	ION ·003	HAS ·002
THA ·004	ENT ·003	TIO ·003	MEN ·002

Further aids are: (1) the determination of the vowels as being those high-frequency letters which rarely combine with one another, and hence show definite interval relationships; (2) the selection of unusual combinations of letters which suggest certain definite plain-text words or phrases. Examples: THAT, WHICH, AS SOON AS, BEGINNING, etc. These are detected in the cipher text by the intervals between the repeated letters, their appearance being referred to as *word patterns*; (3) the search for probable words, which are always very helpful to a cryptanalyst.

Once a few identifications have been made by using any one of the above ideas, a complete solution follows very readily.

The relatively low security possessed by a monoalphabetic system is the result of the fact that each plain-text letter has only one equivalent. If we insist that this equivalent may not represent any other letter, and wish at the same time to provide additional equivalents for the individual letters, it becomes necessary to have more than 26 cipher characters. If, for example, we use every two-digit number as a cipher unit, we have 100 possible equivalents: if we use a two-letter combination as a cipher unit, we have 676 possible equivalents. It then becomes possible with the introduction of additional cipher equivalents to represent each plain-text letter by any one of several *variant* values. If the number of equivalents assigned to each letter is proportional to its relative frequency of occurrence in plain-text, the resulting system has much more security than an ordinary monoalphabet.

However, it may be solved without too much trouble. The procedure is to reduce the cipher text to monoalphabetic terms by determining which cipher characters are equivalent to one another. This is done in two ways: (1) it will be found by frequency studies that certain characters combine in the same sort of way with all the others. A detailed study of all their appearances will establish the equivalence of those characters which represent the same plain-text letter; (2) a careful study of repetitions will disclose places where the same word has been differently enciphered. Thus, one might find such occurrences as

$$
\begin{array}{ccccccc}
11 & 22 & 27 & 75 & 89 & 16 & 31 \\
11 & 22 & 27 & 61 & 89 & 16 & 31 \\
11 & 22 & 45 & 75 & 82 & 16 & 31
\end{array}
$$

The obvious conclusion is that 27 and 45, 61 and 75, 82 and 89 are equivalent pairs. An extended study along these lines will yield considerable information, and before many equivalents have been established, the cryptanalyst will be able to take advantage of frequency considerations and word patterns to obtain plain-text identifications.

A cipher system in which one plain-text letter is replaced by two or more cipher characters is called *polyliteral*. Sometimes polyliteral systems are used without taking advantage of variant values. Francis Bacon's cipher, for example, was penta-literal, each plain-text letter being replaced by a five-letter group consisting of A's and B's only. Since that allowed 32 possibilities (2^5), there were 6 symbols to spare. These were not used by Bacon. The Morse code is an example of a poly-literal system in which the cipher equivalents are not all of the same length.*

A more effective way to suppress frequency than by the use of variants is to make use of more than one substitution alphabet. Such systems are often called *polyalphabetic*, and take two very distinct forms, according as the alphabets are used periodically or aperiodically.

*The Morse alphabet is listed here for those who are not familiar with it.

A ·–	G ––·	M ––	S ···	Y –·––
B –···	H ····	N –·	T –	Z ––··
C –·–·	I ··	O –––	U ··–	
D –··	J ·–––	P ·––·.	V ···–	
E ·	K –·–	Q ––·–	W ·––	
F ··–·	L ·–··	R ·–·	X –··–	

As a simple example of the first type, suppose the correspondents have constructed five different random alphabets:

Plain	A	B	C	D	E	F	G	H	I	J	K	L	M	N	O	P	Q	R	S	T	U	V	W	X	Y	Z
1	R	K	S	L	D	P	A	V	E	C	O	G	J	U	B	N	I	F	X	Z	W	H	Y	M	Q	I
2	J	Y	B	I	W	Q	H	X	D	P	T	F	A	Z	G	O	C	L	S	U	E	K	N	V	M	R
3	F	K	U	D	N	H	T	A	Z	L	G	Q	Y	R	B	I	J	V	M	O	C	W	E	X	S	P
4	N	T	Q	I	C	G	S	E	J	O	P	A	W	B	V	U	X	D	L	F	R	H	K	Z	M	Y
5	K	E	M	A	J	B	T	L	V	D	N	U	F	S	C	Q	G	O	X	I	P	Y	W	Z	R	H

(first column label: Cipher)

The plain-text message is written out in five columns, and the first alphabet is used to encipher the letters of the first column, the second alphabet is used to encipher the letters of the second column, etc. The effect of this procedure is that the ath alphabet enciphers every letter whose position in the plain-text is a number of the form $5k + a$. With the alphabets given above, the message THREE SHIPS SAILED TODAY would be enciphered ZXVCJ XXZUX XJZAJ LUBIK Q.

Instead of having a fixed number of alphabets, the correspondents may provide arrangements for varying their alphabets by regular changes of a key word. One means of attaining this end will now be explained. It is described by most of the older authors, and is called by some of them the *double key* system. One key word – for example, COPYRIGHT – is used to construct the following diagram, known as a Vigenère Square (see below). If we consider each row of the square a cipher sequence, and the normal alphabet above the square the plain sequence, the diagram gives us 26 substitution alphabets. Let each of these alphabets be designated by the letter in the first column of the square. Then a second key word – say AUGUST – is employed to select the particular alphabets to be used and the order in which to use them. The first substitution alphabet would be the A alphabet, and would be used to encipher each letter whose position in the plain-text is given by a number of the form $6k + 1$; the second substitution alphabet would be the U alphabet, and would be used to encipher each letter whose position in the plain text is given by

A	B	C	D	E	F	G	H	I	J	K	L	M	N	O	P	Q	R	S	T	U	V	W	X	Y	Z
C	O	P	Y	R	I	G	H	T	A	B	D	E	F	J	K	L	M	N	Q	S	U	V	W	X	Z
Z	C	O	P	Y	R	I	G	H	T	A	B	D	E	F	J	K	L	M	N	Q	S	U	V	W	X
X	Z	C	O	P	Y	R	I	G	H	T	A	B	D	E	F	J	K	L	M	N	Q	S	U	V	W
W	X	Z	C	O	P	Y	R	I	G	H	T	A	B	D	E	F	J	K	L	M	N	Q	S	U	V
V	W	X	Z	C	O	P	Y	R	I	G	H	T	A	B	D	E	F	J	K	L	M	N	Q	S	U
U	V	W	X	Z	C	O	P	Y	R	I	G	H	T	A	B	D	E	F	J	K	L	M	N	Q	S
S	U	V	W	X	Z	C	O	P	Y	R	I	G	H	T	A	B	D	E	F	J	K	L	M	N	Q
Q	S	U	V	W	X	Z	C	O	P	Y	R	I	G	H	T	A	B	D	E	F	J	K	L	M	N
N	Q	S	U	V	W	X	Z	C	O	P	Y	R	I	G	H	T	A	B	D	E	F	J	K	L	M
M	N	Q	S	U	V	W	X	Z	C	O	P	Y	R	I	G	H	T	A	B	D	E	F	J	K	L
L	M	N	Q	S	U	V	W	X	Z	C	O	P	Y	R	I	G	H	T	A	B	D	E	F	J	K
K	L	M	N	Q	S	U	V	W	X	Z	C	O	P	Y	R	I	G	H	T	A	B	D	E	F	J
J	K	L	M	N	Q	S	U	V	W	X	Z	C	O	P	Y	R	I	G	H	T	A	B	D	E	F
F	J	K	L	M	N	Q	S	U	V	W	X	Z	C	O	P	Y	R	I	G	H	T	A	B	D	E
E	F	J	K	L	M	N	Q	S	U	V	W	X	Z	C	O	P	Y	R	I	G	H	T	A	B	D
D	E	F	J	K	L	M	N	Q	S	U	V	W	X	Z	C	O	P	Y	R	I	G	H	T	A	B
B	D	E	F	J	K	L	M	N	Q	S	U	V	W	X	Z	C	O	P	Y	R	I	G	H	T	A
A	B	D	E	F	J	K	L	M	N	Q	S	U	V	W	X	Z	C	O	P	Y	R	I	G	H	T
T	A	B	D	E	F	J	K	L	M	N	Q	S	U	V	W	X	Z	C	O	P	Y	R	I	G	H
H	T	A	B	D	E	F	J	K	L	M	N	Q	S	U	V	W	X	Z	C	O	P	Y	R	I	G
G	H	T	A	B	D	E	F	J	K	L	M	N	Q	S	U	V	W	X	Z	C	O	P	Y	R	I
I	G	H	T	A	B	D	E	F	J	K	L	M	N	Q	S	U	V	W	X	Z	C	O	P	Y	R
R	I	G	H	T	A	B	D	E	F	J	K	L	M	N	Q	S	U	V	W	X	Z	C	O	P	Y
Y	R	I	G	H	T	A	B	D	E	F	J	K	L	M	N	Q	S	U	V	W	X	Z	C	O	P
P	Y	R	I	G	H	T	A	B	D	E	F	J	K	L	M	N	Q	S	U	V	W	X	Z	C	O
O	P	Y	R	I	G	H	T	A	B	D	E	F	J	K	L	M	N	Q	S	U	V	W	X	Z	C

a number of the form $6k + 2$; etc. The resulting cipher would be a polyalphabet of six alphabets, of which only five are distinct. With these two keywords the message NEED REINFORCEMENTS AT ONCE would be enciphered VZBXD EMTDA DBFHB TFCAJ STVE.

The first step in solving periodic polyalphabetic systems is to determine the number of alphabets. In order to understand how this is accomplished, suppose a word is repeated several times in a message. Any two appearances for which the relative positions of this word are the same with regard to the key will yield identical cipher text. Two appearances placed differently with respect to the key will not yield cipher repetitions. Hence, the intervals separating cipher repetitions are multiples of the length of the key. The exact length of the key is the greatest common divisor of all the intervals separating such repetitions. Suppose this number has been determined. Then if the message is written out in as many columns as the length of the key, all the letters in any one column will have been enciphered in a single substitution alphabet.

The second step is the analysis of these different mono-alphabets. It involves essentially the same procedure as that described in connection with monoalphabetic systems. If some of the alphabets are the same, this can be determined by statistical methods, and the corresponding frequencies combined. In addition, if the different alphabets are related as in a Vigenère square, certain properties of symmetry can be used to advantage, so that identifications in one alphabet will simultaneously yield identifications in others.

Quite recently there has been considerable research carried on in an attempt to invent cipher machines for the automatic encipherment and decipherment of messages. Most of them employ periodic polyalphabetic systems. The most recent machines are electrical in operation, and in many cases the period is a tremendously large number; in addition, the number of independent alphabets is often of the same order of magnitude as the period. These machine systems are more rapid and accurate than hand methods. They can even be combined with printing and transmitting apparatus so that, in enciphering, a record of the cipher message is kept and the message transmitted; in deciphering, the secret message is received and translated, all automatically. So far as present cryptanalytic methods are concerned, the cipher systems derived from some of these machines are very close to practical insolvability.

The study of aperiodic polyalphabetic systems would lead us into very involved cryptanalysis, and therefore only two such systems will be mentioned.

1. Suppose a Vigenère square has been constructed, giving 26 different alphabets, each designated by a single letter. Then the letters of the plain-text message itself may serve as the key. For example, suppose the first alphabet is, by prearrangement,

known to be the A alphabet. Then the first letter is enciphered in this alphabet. Thereafter, each succeeding letter is enciphered in the alphabet designated by the plain-text letter just preceding. Such a system is called *auto-key*. With the Vigenère square based on the word COPYRIGHT, and A the first substitution alphabet, the message BRIDGE DYNAMITED would be enciphered BOETM BKALN UZXEK.

2. The system known as *running key* is similar to the above except that the key is quite distinct from the message. In this system the key may be the text of a book or periodical beginning on some prearranged page and line. Or, it may be a random collection of letters of which each correspondent has a copy.

Aperiodic systems such as the first of the two just described have a very serious disadvantage from the standpoint of practicability. If for any reason an error should be introduced into a message, causing a single letter to be incorrectly received, every subsequent letter will be affected. Since the average number of errors involved in just the transmission of a message may be as high as 5%, this makes it difficult, and sometimes impossible, to read a cipher message.

The increased security gained in polyalphabetic systems is basically due to the fact that they suppress monoalphabetic frequencies. But these will reappear if the cryptanalyst can find ways of breaking up the message into its component monoalphabets. This possibility will always exist if the unit of cryptographic treatment is a single letter, as it has been in all the substitution systems described thus far. We are thus led to the idea of *polygraphic substitution* – the replacement of a plain-text polygraph by a cipher group of the same number of letters.

As a first example we give the classic *digraphic* system known as the *Playfair Cipher*. In this system, 25 cells arranged in the form of a square are filled in some prearranged manner with the letters of the alphabet (one letter, say Q, is omitted). The square given below contains a systematically mixed sequence based on the word MANCHESTER.

M	A	N	C	H
E	S	T	R	B
D	F	G	I	J
K	L	O	P	U
V	W	X	Y	Z

The plain-text message is divided into pairs of letters, and in order to prevent any pair consisting of the same two letters, a dummy letter like Z is introduced whenever necessary. Thus, if the first word of a message is BATTALION, it would be broken up as follows BA TZ TA LI ON. In encipherment, if both letters of a digraph appear in the same horizontal (or vertical) line of the square, each of them is replaced by the letter in the square immediately to its right (or below it), the letters in each row and column being treated as in cyclical order. If the letters of a pair do not appear in the same row or column, they must necessarily be at opposite corners of some rectangle, and they are then replaced by those at the other corners of the rectangle, each by that which is on the same horizontal line. Thus the message WILL MEET YOU AT NOON would first be written WI LZ LM EZ ET YO UA TN OZ ON. With the square shown above, the message would then be enciphered YFUWK ABVSR XPLHG TUXXT.

The security of this system, or, for that matter, of any digraphic system, is quite low. There exists sufficient variation in the relative frequencies of plain-text digraphs to permit a cryptanalyst to obtain a great deal of information from considerations of frequency alone. The correct identification of only a few digraphs is sufficient to ensure solution. In addition, the cryptanalyst can make considerable use of probable words and of word patterns. Of course, he is limited in the latter to patterns based on digraphs. Examples: *RE* FE *RE* NC E, P *RE* PA *RE*, *DE* CI *DE*.

In the Playfair Cipher there are further aids. If it has been determined that TH is replaced by BN, then HT is replaced by NB. In general if $A_1A_2 = B_1B_2$, then $A_2A_1 = B_2B_1$. More-over, if these four letters A_1, A_2, B_1, B_2 are known to be at the corners of a rectangle, the following additional identifications will result at once: $B_1B_2 = A_1A_2$; $B_2B_1 = A_2A_1$. These facts are quite helpful in special frequency studies – for example, in distinguishing the equivalent of a high-frequency digraph such as TH whose reversal (HT) is infrequent from one such as ER, whose reversal (RE) is frequent. It may be pointed out that in Playfair ciphers the complete square can be recon-structed from just a few identifications.

The security of polygraphic substitution increases very rapidly as the unit of crytographic treatment increases in size. When the unit becomes as large as five or six letters, the security is very high. Unfortunately, a serious practical diffi-culty enters at this point. An error in one letter of a cipher group will give an incorrect translation for the entire unit of plain-text letters, so that, in effect, it garbles five or six letters at once. Since it is not possible to introduce means for correct-ing errors in such a system, four or five errors in the transmis-sion of a message may render the entire message unintelligible. Such a difficulty is prohibitive. From a theoretical standpoint, however, such a system is quite interesting, and a very general mathematical treatment of it has been given by Professor Lester S. Hill.*

In every cryptographic system that has been considered thus far, the cipher text has been at least as long as the plain-text. The chief concern was secrecy, and considerations of economy did not enter. But such considerations are often of prime importance, and have led to the development of very highly specialized substitution systems called *code*. In these systems each correspondent is given a copy of a *code book* containing a long list of words, phrases, and sentences, each of which has an arbitrary equivalent set alongside it. These equivalents are most often five-letter groups, although regular groupings of a smaller number of letters and, more rarely, groups of figures are sometimes used. Provision is made for

*American Mathematical Monthly, 1931, vol. XXXVIII, pp. 135–154.

spelling out words which may not appear in the code *vocabulary*, by assigning equivalents to the individual letters of the alphabet and to the most frequent syllables.

Generally, a code book is constructed for a particular industry or for a group of people having special interests, and, as a consequence, long phrases, and even whole sentences, apt to be often repeated, are included in the vocabulary, and can be replaced by one code group. It results quite frequently that a code message is only one-fourth or one-fifth as long as the corresponding plain-text message.

By way of illustration we reproduce below a portion, under the *caption* ARRIVAL, from the Western Union Traveller's Cable Code.

ADAUX	Am awaiting arrival of . . .
ADAVY	Arrived all right
ADAXA	Arrived all right, address letters to care of . . .
ADBAE	Arrived all right, telegraph (cable) me in care of . . .
ADBEI	Arrived all right, pleasant passage, advise friends
ADBIM	Arrived all right, pleasant passage, am writing

etc.

This power of condensation brings with it the difficulty that an error in a single code group may cause the loss of a considerable portion of plain-text. In order to avoid the necessity of having the message repeated, provision is made for the correction of errors. What is done is to select the code groups from a special type of code construction table,* which assures that every code group will differ from every other by at least two letters. It results therefrom that a code group is completely determined by any four of its letters. If, then, one letter is garbled in transmission, the resulting group will not be found in the book, and the decoder is thus informed of the presence of an error. By assuming each letter in turn to be the incorrect one, there will be at most five possible translations

*An example of such a table, together with some interesting notes about code, will be found in a paper by W.F. Friedman and C.J. Mendelsohn, *American Mathematical Monthly*, 1932, vol. xxxix, pp. 394–409. See also p. 275 above.

for the garbled word, and the context will indicate the correct translation. This procedure is based on the assumption that only one letter of the code group which has been incorrectly received is wrong, and, fortunately, this is almost always true.

A further improvement in code-word construction tables, which assists in the correction of errors, assures that no two code words in the book will differ by only an interchange of two letters. The necessity for such a feature arises in the tendency of some clerks, particularly typists, to introduce such an error, without realizing it, when working rapidly.

Secrecy in code communication is largely dependent upon keeping the code book out of unauthorized hands. But code books are often compromised without passing out of the hands of their rightful owners; for example, they may be copied or photographed. If there is any possibility whatever of such compromise, recourse may be had to *superencipherment*. This means applying some one or more of the possible cipher systems to the code text, treating the code message as though it were plain-text. These systems and the specific keys applying to them may be changed as often as desired.

Determination of cryptographic system. It has been remarked that in most cryptanalytic studies it is assumed that the enemy is in possession of the general system. Naturally this is not always true, and the cryptanalyst sometimes has the problem of determining what general system applies to a set of messages under consideration. This is one of the most difficult problems of the entire field, and in order to prevent the present discussion from becoming too involved, a few words will be said about the methods for distinguishing general systems from one another *when only one process is involved*. Given a set of messages for study, the first step is to make a systematic search for repetitions, and to study the external appearance of the messages. If we limit ourselves to repetitions at least five or six letters in length, we can feel assured that we are omitting any repetitions which may be accidental. Suppose it is found that the lengths of all the repetitions which appear and the intervals separating them are all multiples of the same number n. Then the system is either polyliteral, polygraphic, or code.

If we tabulate all the n-letter groups which occur and make a frequency distribution of their appearances, we can then distinguish between these three systems as follows. A polyliteral system which does not use variants will involve at most 26 different groups, whose frequencies will correspond with the normal frequencies of the individual letters of plain-text. If variants are used, this can be determined by establishing equivalents in the manner already described. Code systems differ from polygraphic systems, in that the n-letter groups of code are regular in form and structure, possess a two letter difference, and may be made to fit a code-word construction table, whose size limits the total number of groups. In a polygraphic system, on the other hand, *every n-letter combination is a bona-fide cipher group*. The number of these groups actually obtained in the set of messages at hand is of the same order as the number of distinct plain-text polygraphs of n letters to be expected in such an amount of text.

If the cipher unit is not a group of two or more letters, the system is monographic, and must be either transposition or substitution. If it is transposition, it is merely a permutation of the original letters of the message, and the relative frequency of each individual letter will be the same as in plain-text. Hence, the proportion of vowels must be the same as in plain-text – i.e. about 40%. Very few, if any, long repetitions will be found, as the transposition process has a tendency to break up words and scatter the individual letters throughout the message.

If the system is not transposition, it must be substitution. A monoalphabetic substitution can be distinguished by the fact that the frequencies of the individual letters are of the same relative magnitudes and show the same degree of variation as plain-text frequencies. The noticeable difference in the cipher lies in the fact that the high frequencies may be exhibited by letters which are infrequent in plain-text, and, similarly, letters which are very frequent in a plain-text may be infrequent in the cipher-text. The proportion of vowels generally is much less than in plain-text, since the vowels are all high-frequency letters in plain-text. Then, too, a mono-alphabet will contain a considerable number of repetitions, since every plain-text repetition yields a cipher repetition; and these repetitions will contain recognizable word patterns. The intervals between them will, of course, have no common factor.

If the general system is one of polyalphabetic substitution, the frequency distribution will be quite *flat* – that is, the cipher letters will all have approximately equal frequencies. This is the result of the use of more than one alphabet, causing a particular cipher equivalent sometimes to represent a frequent plain-text letter, and at other times an infrequent one. It remains to determine whether the alphabets are being used periodically or aperiodically. If they are used periodically, the intervals between repetitions will all be multiples of the period. When the messages are broken up into sections equal to the length of the period, and these sections are superimposed, the letters in a single column will constitute a mono-alphabet. If, however, the intervals between repetitions show no common factor, then the system is aperiodic.

Further study of any one of these systems with a view towards solution depends upon the particular system encountered, and upon any other elements of pertinent information which may be available to the cryptanalyst as a result of the particular situation.

A few final remarks. In most cases, the information contained in a set of cryptograms is valuable only for a very short period. In fact, a system is generally considered sufficiently secure if the delay caused the enemy in its attempts at solution is long enough to make the information valueless. However, it sometimes happens that the application of the principles of cryptanalysis will result in obtaining material of definite historic importance. Consider, for example, as a first type of material, the ancient systems of writing. In the real sense of the word, these were not intended to be *secret* writing. Nevertheless they had to be treated as such when attempts were made to read them. On the other hand, historians are sometimes vitally interested in reading secret messages written many years ago. We might cite, in connection with this type of material, the diplomatic code messages transmitted during the period of the American Revolution.

Under each of the two heads mentioned above there is considerable material available for study, even to-day. In connection with ancient systems of writing there is the outstanding example of the Mayan *glyphs*. To be sure, some progress in their identification has been made. The numbers, the calendric symbols, the glyphs representing the various Mayan gods and a few others have been identified. But a considerable amount of work still remains to be done.

In conclusion, a word might be said about instances of attempts to solve ciphers whose very existence has been a matter of question. We might cite, in particular, all the work that has been done in connection with the Bacon-Shakespeare controversy. Very little knowledge of cryptanalysis is necessary to appreciate that the solutions which have been obtained are entirely subjective and that no two independent investigators using the methods proposed would arrive at identical results. It is quite possible for an investigator to become so obsessed with an idea that he actually forces it to work out by selecting a system sufficiently flexible to permit any desired result. A tragic instance of this phenomenon is provided by the case of Prof. W.R. Newbold,* who imagined that he had succeeded in reading the Voynich Manuscript. The latter is a beautifully written and profusely illustrated medieval treatise of over two hundred pages, discovered in Italy by Wilfrid M. Voynich, about 1912. It remains an outstanding challenge to linguists and cryptanalysts.

*The Cipher of Roger Bacon, Philadelphia, 1928. There are photostatic copies of the Voynich Manuscript in the British Museum and the New York Public Library. (It is no longer attributed to Roger Bacon.)

ADDENDUM

REFERENCES FOR FURTHER STUDY

BAZERIES, ÉTIENNE
Les chiffres secrets dévoilés, Paris, 1901
Encyclopaedia Britannica
"Cryptology"
FIGL, ANDREAS
Système des chiffrierens, Graz, 1926
GIVIERGE, COL. MARCEL
Cours de Cryptographie, Paris, 1925
GRANDPRÉ, A. DE
La cryptographie pratique, Paris, 1905
LANGIE, ANDRÉ, and SOUDART, E.A.
Traité de Cryptographie, Paris, 1925
LANGIE, ANDRÉ
De la cryptographie, Paris, 1918 (English translation by J.C. H. Macbeth. *Cryptography,* New York, 1922)
RIVEST, R.L.
Advances in Cryptography, Santa Barbara, CA, 1982
SINKOV, ABRAHAM
Elementary Cryptanalysis, New York, 1968
SLOANE, N.J.A.
Error-correcting Codes and Cryptography, in *The Mathematical Gardner,* ed. D.A. Klarner, Belmont, CA, 1981
VALERIO, P.L.E.
De la cryptographie: Part I, Paris, 1893; Part II, Paris, 1896
YARDLEY, HERBERT O.
The American Black Chamber, Indianapolis, 1931

INDEX

Abel, N.H., 62
Aces paradox, 44
Agrippa, Cornelius, 194
Ahrens, W., 194
Aitken, A.C., x, 46, 386–387
Aix, labyrinth at, 258
Alcuin, 3, 118
Alexander, J.W., 232
Alkborough, labyrinth at, 259
Alphabet, Morse, 406
Ammann, R., 161
Anallagmatic pavements, 107–109
Analysis situs, 222–270
Anaxagoras, 350
Andreas, J.M., vii, 128, 154, 171, 215, 239
Andreini, A., 147, 160
Andrews, W.S., 194, 212
Anning, N., 171
Antipho, 350
Antiprism, 130
Apollonius, 343, 351
Arago, F.J.D., 372
Archibald, R.C., 65
Archimedean solids, 136–140, 159
Archimedes of Syracuse, 340–343, 347, 350
Archytas, on Delian problem, 341
Arithmetical fallacies, 41–45
Arithmetical prodigies, 360–387
Arithmetical puzzles, 4–50
Arithmetical recreations, 3–75, 312–337
Arithmetical restorations, 20–26
Arnoux, G., 206, 210
Arya-Bhata, 88, 351
Asenby, labyrinth at, 259
Augustus, 403
Auto-key system, 410

Axis of symmetry, 131, 135

Bachet's *Problèmes,* 3, 5, 6, 8, 11, 12, 17, 18, 28, 30, 33, 50, 118, 326, 328
Bacon, Francis, 406, 417
Bacon, Roger, 417
Baker, H.F., 323
Ball-piling, 149–152
Barbette, E., 212
Barrau, J.A., 108
Baudhayana, on π, 351
Bazeries, E., 418
Belevitch, V., 308
Bell, E.T., 57, 61
Bennett, G.T., 81
Bergholt, E.G.B., 49, 211
Berlekamp, E.R., 303
Bernoulli, John, 41, 349
Bernoulli numbers, 71
Bertrand, J.L.F., 41
Bertrand, L. (of Geneva), 177
Berwick, W.E.H., 23
Besicovitch, A.S., 100
Bewley, E.D., 49
Bhaskara, on π, 352
Bidder, G.P., 367–372, 378–386
Bidder family, 371–372
Biering, C.H., 339
Bilinski, Stanko, 143, 300
Bilguer, von, on chess pieces, 165
Binet, A., 375
Birthdays problem, 45
Bishop's re-entrant path, 187
Block design, 276
Bordered magic squares, 200
Bose, R.C., 292, 302
Boughton Green, labyrinth at, 259
Bouniakowski, V., on shuffling, 323

Bourlet, C.E.E., 31
Bouton, C.L., 37
Bouwkamp, C.J., 113, 116
Brahmagupta, on π, 351
Breton, on mosaics, 258
Brillhart, J.D., 62, 66–69
Bromton, 256
Brooks, R.L., 116
Brouncker, Viscount, on π, 355
Brown, B.H., 27
Bruck, R.H., 311
Brückner, M., 147, 154
Brun, Viggo, 62
Bryso, 350
Busschop, P., 89
Bussey, W., 282
Buxton, Jedediah, 361–364

C-matrix, 308
Caesar, Julius, 403
Calculating prodigies, 360–387
Calendar problems, 26
Callet, J.F., on π, 357
Cantor, M., on π, 349
Cardan, G., 3, 318, 320
Cards, problems with, 18–20, 47, 49, 191, 322–337
Cartwright, W., 389
Cauchy, A.L., 145, 372
Cayley, A., 146, 223, 238, 261, 336, 355
Cells of a chess-board, 162
Ceulen, van, on π, 353
Characteric, Euler-Poincaré, 233
Charcot, J.M., 375
Chartres, labyrinth at, 259
Chartres, R., 41, 83, 359
Chasles, Michel, on trisection of angle, 346
Cheney, W.F., 94
Chernick, J., 211
Chess: maximum pieces problem, 172; minimum pieces problem, 172; number of initial moves, 163
Chess-board: games on, 124–127, 162–192; knight's move on, 175–186; notation of, 162
Chess pieces, value of, 163–165
Chifu-Chemulpo Puzzle, 117, 129

Chilcombe, labyrinth at, 259
Chilton, B.L., 143
Chinese, 196, 352
Chinese rings, 318–322
Chrystal, G., 41, 57
Ciccolini, T., on chess, 181
Ciphers, 388–418
Circle, quadrature of, 347–359
Cissoid, 341, 343
Clairaut, on trisection of angle, 346
Claus, 316
Clausen, on π, 357
Clerk Maxwell, J., 238
Close-packing: hexagonal, 151; spherical, 149
Cnossos, coins of, 257
Cocoz, M., 78, 212
Code, Morse, 406
Code-book ciphers, 404–406, 412–414
Colburn, Z., 365–370, 378, 381, 384
Cole, F.N., 65, 281
Cole, R.A.L., 335
Collini, on knight's path, 181
Colour-cube problem, 113
Colouring maps, 222–242
Comberton, labyrinth at, 259
Combinatorial designs, 271–311
Compass problems, 96
Compound polyhedra, 134, 136, 146
Conchoid, 341, 345
Conics, 346
Continued fractions, 55, 58, 86
Contour-lines, 238
Convergents to a continued fraction, 56, 58, 87
Conway, J.H., 308
Counters, games with, 103–105, 121–127
Coxeter, H.S.M., 39, 57, 99, 143, 152, 159, 237, 280, 296, 304
Creation, 66
Cretan labyrinth, 258
Cryptography and cryptanalysis, 388–418
Cube, duplication of, 93, 339–344
Cubes: coloured, 113–115; magic, 217–221; skeleton, 48, 295; tessellation of, 147, 152

Cuboctahedron, 137, 147
Cudworth, W., on Sharp, 356
Cunningham, Frederic, 101
Cusa, on π, 353
Cutting cards, problems on, 19
Cyclotomy, 94–96

D'Alembert, J., 43
Darboux, G., 376, 378
Dase, J.M.Z., 357, 372–374
Davenport, H., 45, 108
Davis, Chandler, 267
Decimals, 53–55
Decimation, 32–36
De Coatpont, 89
De Fonteney, on ferry problem, 119
De Fouquières, Becq, 104
De Grandpré, A., 388
De Haan, B., on π, 349, 354
Dehn, M., 93
De Lagny, on π, 356
De la Hire, P., 202, 203
De la Loubère, S., 195, 204–205
Delannoy, 119, 123
De Lavernède, J.E.T., 182
Delens, P., 21
Delian problem, 93, 339–344
Deltoid, 100
De Lury, D.B., 44
Dembowski, Peter, 311
De Moivre, A., 175
De Montmort, 1, 46, 175
De Morgan, A., 223, 347, 349, 359
Denary scale of notation, 10–17, 20–26, 53
De Parville, on Tower of Hanoï, 317
De Polignac, A., 64, 184
Derangements, 46, 337
Derrington, on queens problem, 171
De St. Laurent, T., 323
Desargues configuration, 305
Descartes, René, 344, 346, 356
Diabolic magic squares, 203, 206, 208–212
Diamandi, P., 375, 377
Dice, cubic and octahedral, 214
Dickson, L.E., 67, 73, 279, 332
Digital process, 381
Digits, missing, 20–26

Diocles, on Delian problem, 343
Diodorus, on Lake Moeris, 256
Dirichlet, Lejeune, 71
Dissection problems, 87–94
Dodecahedron: great, 145, 234; pentagonal, 132, 231, 235, 262–263; rhombic, 137, 150; stellated, 144
Dodgson, C.L., on parallels, 77
Dominoes, 30, 214, 251–254
Donchian, P.S., 141
Double transposition, 400
Doubly-magic squares, 212
Dragon designs, 266–270
Drayton, 256
Dual maps, 234
Dubner, H., 64, 66
Dudeney, H.E., 35, 39, 49, 59, 79, 88, 92, 96, 118, 127, 187, 189, 211, 214
Duijvestijn, A.J.W., 116
Duplication of cube, 93, 339–344
Dürer, A., 194, 202

e, 47, 56, 87, 348
Eddington, A.S., 69
Eden, M., 112
Eichler, M., 75
Eigenvalue, 302
Eight queens problem, 166–171
Eisenlohr, A., on Ahmes, 350
Elements, the four, 133
Enantiomorphism, 130, 135, 139, 147, 159, 215, 238, 241
Enciphering and deciphering, 390
Encke, J.F., 373
Eneström, G., on π, 349
Equiangular lines, 299
Eratosthenes, 340
Erdös, Paul, 63
Error-correcting code, 274, 297
Escott, E.B., 214, 358
Estermann, T., 64
Etten, H. van, 12
Euclid, 60, 66, 132, 350
Euclidean construction, 95–97, 139, 338
Euler, L., 58, 61, 66, 68, 70, 175–179, 203, 349, 356
Eulerian cubes, 192, 219
Eulerian squares, 190–192, 212, 290

Euler's formula, 131, 146, 229, 238; proof of, 232–233
Euler's officers problem, 192
Euler's unicursal problem, 243–247
Euzet, M., 89
Exploration problems, 32

Factorization, 61, 62
Fallacies: arithmetical, 41–44; geometrical, 76–84
Fano, Gino, 282
Fedorov, E.S., 141–143
Feller, W., 324
Ferguson, D.F., on π, 358
Fermat numbers, 67, 94
Fermat's criterion for primes, 61
Fermat's Last Theorem, 69–73
Ferry-boat problems, 118–120
Fibonacci numbers, 57, 86, 161
Fifteen puzzle, 312–315
Finite affine plane, 283
Finite projective space, 284
Fisher, R.A., 192
Five-disc problem, 97
Fonteney, on ferry problem, 119
Fouquières, Becq de, on games, 104
Four-colour conjecture, 222–231
Four "4's" problem, 16
Four digits problem, 15
Fourrey, E., 27
Fox, Captain, on π, 359
Franklin, P., 237
Frènicle (de Bessy), B., 200
Friedman, W.F., 413
Frost, A.H., 203
Fuller, T., 354

Galois fields, 73–75, 107, 191, 219, 282
Galton, F., 41
Games: dynamical, 116–129; statical, 103–116
Gardner, Martin, 110
Gauss, C.F., 68, 75, 94, 365, 373
Genus, 234
Geodesic problems, 120–121
Geography, physical, 238
Geometrical fallacies, 76–84

Geometrical problems, three classical, 338–359
Geometrical recreations, 76–161, 312–337
Gergonne's problem, 328–333
Gerwien, P., 89
Ghersi, I., 186
Gillies, D.B., 65
Glaisher, J.W.L., 167, 349
Golay, M., 303
Goldbach's theorem, 64
Goldberg, Michael, 90, 96, 155
Golden section, 57, 132
Golomb, S.W., 109
Gosper, R.W., 55
Gosset, Thorold, 171
Gradshtein, I.S., 67
Grandpré, A. de, 388
Great dodecahedron, 145, 234
Great icosahedron, 145
Great Northern puzzle, 116, 129
Greenwich, labyrinth at, 259
Gregory, Jas., 348, 356
Gregory of St. Vincent, 343
Gregory's series, 356
Grienberger, on π, 355
Grilles, 401
Grinbergs, E. Ya., 264–265
Gros, L., on Chinese rings, 320–322
Group, icosahedral, 158
Guarini's problem, 189
Guitel, E., 89
Günther, S., 166, 194
Guthrie, on colouring maps, 223
Guy, R.K., 39

Haan, B. de, on π, 349, 354
Hadamard, Jacques, 62, 274, 296
Halberstam, H., 62
Hall, Marshall, 278, 283
Halley, on π, 356
Hamiltonian game, 262–266
Hampton Court, maze at, 259
Hanoï, Tower of, 316–317
Hardy, G.H., 68
Heath, R.V., 13, 200, 201, 212, 216, 221
Heath, T., 132, 134

Heawood, P.J., on maps, 224, 227–232, 236–238
Hegesippus, on decimation, 32
Hein, Piet, 113
Henry, Ch., on Euler's problem, 243
Hero of Alexandria, 341, 351
Herodotus, on Lake Moeris, 256
Herschel, Sir John, 368
Herschel, Sir William, 369
Hess, E., 155
Hexagons, interlocked, 215–217
Hilbert, D., 88, 237
Hill, L.S., 412
Hills and dales, 238
Hippias, 350
Hippocrates of Chios, 341, 350
Hoppenot, F., 58
Houdin, J.E.R., 374
Hudson, C.T., 332
Hudson, W.H.H., 324
Hurwitz, Adolf, 40, 65
Hutton, C., 4, 356
Huygens, C., 343, 347, 355
Hyper-cubes, 141
Hypocychoid, three-cusped, 100

Icosahedron: regular, 132, 238–242, 300; rhombic, 143; stellated, 146–147
Icosidodecahedron, 137
Inaudi, J., 375–378
Incidence matrix, 272
Infeld, Leopold, 73
Ingham, A.E., 63, 235
Interlocked hexagons, 215–217
Intermediate convergents, 87
Inwards on the Cretan maze, 257

Jaenisch, C.F. de, 173, 185
Jephson, T., 368
Johnson, W., on fifteen puzzle, 312
Jones, W., on π, 349, 356
Josephus problem, 32–36
Julian's Bowers, 259
Julius Caesar, 403

Kakeya, S., 99
Kaleidoscope, 155–160

Kamke, E., 44
Kayles, 39
Kempe, A.B., on colouring maps, 224
Kepler, J., 133
Kepler-Poinsot polyhedra, 144–146, 234
Khajuraho, 203
King's re-entrant path, 185
Kirkman, T.P., 288
Kirkman's problem, 287–289
Klarner, D.A., 112
Klein, F.C., 69, 86, 237, 338
Knight's re-entrant path, 174–185
Knight of the Round Table, 49
Knuth, Donald, 267, 352
Knyghton, 256
König, Dénes, 266
Königsberg problem, 243
Kraïtchik, M., 13, 28, 65, 73, 106, 186, 200, 202, 207
Kummer, E.E., 71

Labyrinths, 254–260
Ladder graph, 277
Lagny, de, on π, 356
La Hire, P., 202, 203
La Loubère, S., 195, 204–205
Lamé, G., 71
Landau, E., 324
Landry, F., 68
Latin cubes, 192
Latin squares, 189, 289–293
Lattice graph, 305
Lattice points, 86
La Vallée Poussin, C.J. de, 62–63
Lavernède, J.E.T. de, 182
Leake, 12, 17, 28
Lebesgue, H., 71, 99
Leech, John, 308
Legendre, A.M., 62, 71, 177, 185
Lehmer, D.H., 61, 72
Lehmer, D.N., 58
Lehmer's machine, 61
Leibniz, G.W., 2, 61
Leonardo of Pisa, on π, 352
Leslie, J., 341, 345
Leurechon, J., 3, 12
Linde, A. van der, 175
Lindemann, on π, 348

Lines of slope, 238
Lint, J.H. van, 307
Listing, J.B., 128, 244
Littlewood, J.E., 63
London and Wise, 259
Longuet-Higgins, M.S., 146
Lo-shu, 196
Loubère, S. De la, 195, 204–205
Loyd, S., 29, 39
Lucas, E., 50, 59, 65, 119, 124, 125, 145
Lucas, labyrinth at, 258

Macaulay, W.H., 93
McClintock, E., 203
Machin's series for π, 356
MacLane, Saunders, 73
MacMahon, P.A., 113
Magic cubes, 217–221
Magic squares, 193–221
Mallison, H.V., 313
Mangiamele, V., 372
Mann, H.B., 297
Map colour theorems, 222–231, 235–237
Margossian, 208, 221
Mars, 297
Mascheroni, L., 96
Maxwell, J. Clerk, 6, 238
Mayer, J., 230
Mazes, 254–260
Medema, P., 116
Medieval problems, 27–36
Menaechmus, 342
Ménage problems, 50
Mendelsohn, C.J., 413
Menon, P.K., 293
Mersenne's numbers, 65
Mesolabum, 341
Metius, A., on π, 353
Méziriac, see Bachet
Mikami, Y., 201
Miller, G.A., 279
Miller, J.C.P., 53, 138, 146, 153
Minos, 254, 340
Minotaur, 257
Mirrors, 155–161
Missing digits, 20–26
Miyazaki, K., 161

Möbius, A.F., 159
Möbius strip, 128, 237
Models, geometrical, 130, 134, 153, 158
Mohammed's sign-manual, 248
Mohr, Georg, 96
Moivre, A. de, 175
Mondeux, H., 372, 377
Monge, on shuffling cards, 323
Monoalphabetic systems, 404
Montmort, de, 2, 46, 176
Montucla, J.E., 4, 89, 347, 355
Moore, E.H., 38, 278
Mordell, L.J., 69
Morehead, J.C., 68
Morgan, A. de, 223, 347, 349, 359
Moroń, Z., 115
Morrison, M.A., 68
Morse code, 406
Mosaic pavements, 107, 266–269
Moschopulus, 194
Mousetrap, game of, 336
Mulder, P., on Kirkman's problem, 288
Müller (Regiomontanus), 353
Müller, G.E., 360–378
Muncey, J.N., 211
Mydorge, 3

Nasik magic squares, 203, 206, 208–212
Nauck, F., 166
Neville, E.H., 98
Newbold, W.R., 417
Newton, Isaac, 343, 344
New York Times, 294
Nicomedes, 341
Nim, 36–38
Nine digits problem, 15, 40
Non-orientable surface, 232
Normal piling, 149, 151
Noughts and crosses, 104
Numbers: perfect, 66–67; puzzles with, 4–50; theory of, 57–75, 231

O'Beirne, T.H., 39, 359
Octahedron, 132, 147, 215
Ogawa, T., 161
Orientability of surfaces, 232

Oughtred's *Recreations,* 12, 17, 28
Ovid, 256
Ozanam, A.F., on labyrinths, 258
Ozanam's *Récréations,* 3, 4, 12, 28, 85, 118, 176, 318

π, 347–359
Pacioli di Burgo, 3
Pairs-of-cards trick, 326–328
Pál, Julius, 99
Paley, R.E.A.C., 108, 309
Pandiagonal magic squares, 203, 206, 208–212
Pan-Kai, 111
Pappus, 134, 341, 345
Paradromic rings, 127, 232
Parker, E.T., 292
Parkin, T.R., 58
Pars, L.A., 227–230
Parville, de, 317
Pawns, games with, 124–126
Penrose, R., 161
Pentagram, 144, 248
Pepys, S., 389
Perelman, 40
Perfect code, 276
Perfect numbers, 66–67
Perigal, H., 88
Permutation problems, 48–50, 312–326
Pervusin, I.M., 65
Petersburg paradox, 44
Peterson, A.C., 373
Peterson, J., on graphs, 226, 306
Petrie, J.F., 147, 153, 157
Petrie polygon, 135, 145
Philo, 341
Philoponus, on Delian problem, 339
Photo-electric number-sieve, 61
Physical geography, 238
Pile problems, 328–333
Pirie, G., on π, 355
Pits, peaks, and passes, 238
Planck, C., 203, 210
Plane of symmetry, 131, 136
Planets, 194
Platonic solids, 130–135, 238
Plato, on Delian problem, 339–342
Playfair cipher, 410

Pliny, 256
Poitiers, labyrinth at, 258
Pole, W., 385
Polignac, A. de, 64, 184
Pollock, B.W., 71
Pólya, G., 106, 161, 261
Polyalphabetic systems, 406–410
Polygraphic substitution, 410–412
Polyhedra, 130–161
Polyliteral systems, 406
Polyominoes, 109–113
Portier, B., on magic squares, 212
Posner, E.C., 297
Poussin, De la Vallée, 62–63
Powers, R.E., 65
Pratt, on knight's path, 181
Prime pairs, 64
Primes, 60–66, 211
Probabilities, 44–48, 84, 337, 359
Projective plane, 234–236, 271, 310
Ptolemy, 351
Purbach, on π, 352
Puzzles, arithmetical, 4–52; geometrical, 77–128
Pyramidal numbers, 59, 149
Pythagoras' theorem, 57, 88

Quadratic residues, 108
Quadrature of circle, 347–359
Queens problem, eight, 166–171
Queens, problems with, 189

Raghavarao, D., 309
Railway puzzles (shunting), 116–118
Ramification, 260–262
Ravenna, labyrinth at, 258
Ray-Chandhuri, D.K., 288
Reciprocal polyhedra, 132–136, 152, 234
Reciprocal tessellations, 106
Re-entrant chess paths, 175–188
Reflexible figures, 130
Regiomontanus, on π, 253
Regular polygons, 131
Regular polyhedra, 131
Regular tessellations, 105, 134, 152, 266
Reimer, N.T., 339
Reiss, M., 127, 254

Renton, W., 84
Residues, 60, 108
Restorations of digits, 20–26
Reynolds, Osborne, 151
Rhind papyrus, 350
Rhombicuboctahedron and rhombicosidodecahedron, 137
Rich, J., 389
Richmond, H.W., 95
Richter, on π, 358
Riemann, G.F.B., 63
Riesel, H., 65
Right-angled triangles, 57, 88
Rilly, A., 185, 212
Ringel, Gerhard, 238
Robert-Houdin, J.E., 374
Robinson, G. de B., 159
Robinson, R.M., 64
Rockliff Marshes, labyrinth at, 259
Rodet, L., on Arya-Bhata, 351
Roget, P.M., 179–185
Romanus, on π, 353
Rome, labyrinth at, 258
Rook's re-entrant path, 187
Rosamund's Bower, 256
Rosen, F., on Arab values of π, 342
Rosser, J.B., 72, 203, 204
Rotating rings of tetrahedra, 153, 215, 234
Roth, K.F., 62
Round table, Knights of the, 49
Routes on chess-board, 175–188
Rückle, C., 375–378
Rudis, F., on π, 347
Running-key system, 410
Rutherford, on π, 357
Ryser, H.J., 311

Safford, T.H., 374
Saffron Walden, labyrinth at, 259
St. Laurent, T. de, on cards, 323
St. Omer, labyrinth at, 259
St. Petersburg paradox, 44
St. Vincent, Gregory of, 343
Sauveur, J., 203
Sayles, H.A., 211
Scale of notation: denary, 10–17; transforming to another, 55, 352
Schläfli, L., 141, 146

Schoenberg, I.J., 101
Schots, M.H., 212
Schubert, H., on π, 347
Schuh, H., 22
Schumacher, H.C., 373
Scripture, E.W., 360
Secret communications, 388–418
Seidel, J.J., viii, 307
Selander, K.E.I., on π, 349
Selberg, Atle, 62–63
Selfridge, J.L., 58, 62, 71
Seven-colour map theorem, 237
Shanks, W., on π, 357
Sharp, A., 130, 143, 356
Shelton, J., 389
Sherwin's tables, 356
Shrikhande, S.S., 292
Shuffling cards, 323–325
Shuldham, C.D., 211
Shunting problems, 116–118
Singer, James, 286
Sinkov, A., vii
Sixty-five puzzle, 85
Skeleton cubes, 48, 295
Sloane, N.J.A., 418
Smith, C.A.B., 39, 265–266
Smith, D.E., 196, 201, 203
Smith, H.J.S., 387
Snell, W., on π, 354
Snub solids, 139
Solitaire, 127
Soma, 112
Sommerville, D.M.Y., 141
Southwark, labyrinth at, 259
Space-filling curve, 269
Sponges, regular, 152–153
Sporus, on Delian problem, 343
Sprague, R., 115
Sprague-Grundy number, 36–37
Squaring the circle, 347–359
Squaring the square, 115
Stalker, R.M., 154
Standard alphabets, 403
Stark, H.M., 60
Staudt, K.G.C. von, 282
Steen, on the Mouse Trap, 336
Steiner, Jakob, 278, 288
Steiner triple systems, 278–289
Steinhaus, H., 232, 238

Steinitz, E., 136
Stella octangula, 135, 136, 146
Stellated polyhedra, 144–146
Storey, on the Fifteen Puzzle, 312
Stott, Mrs. A. Boole, 139
Strabo, on Lake Moeris, 256
Strasznicky, 373
Sturm, A., 339
Sub-factorial, 47
Substitution, 402–414
Suetonius, 403
Super-dominoes, 109–113
Surfaces, unbounded, 232
Swastika, 258
Swinden, B.A., 36
Sylvester, J.J., 67, 104
Symmetrical magic squares, 202
Symmetry, 130, 135, 151

Tait, P.G., 34, 123, 224–227, 230–231, 244
Tanner, L., on shuffling cards, 323
Tarry, G., 120, 212, 249–253
Tartaglia, 3, 28, 33, 56, 118
Taylor, B., 176
Taylor, Ch., 346
Taylor, H.M., 92, 163
Ten digits problem, 15, 40
Tessellation: plane, 105–116, 155, 266; solid, 112, 146–152, 159
Tetrahedral numbers, 59
Tetrahedron, 132, 148
Texeira, F.G., 338
Theaetetus, 132
Theory of numbers, 57–75, 231
Thibaut, G., on Baudhayana, 351
Three-in-a-row, 104
Three-pile problem, 328
Todd, John, 358
Torus, 216, 222, 237
Tower of Hanoï, 316–317
Trastevere, labyrinth at, 258
Travers, J., 93, 201
Trebly-magic squares, 213
Trees, 260–262
Treize, game of, 337
Triacontahedron, 137
Triangular graph, 304

Triangular number, 59
Tricks with numbers, 4–50
Trisection of angle, 344–347
Troitsky, 14
Trollope, E., on mazes, 257
Troy-towns, 259
Truncated solids, 138, 147, 153
Tsu Ch'ung-chih, 352
Tuckerman, Bryant, 65, 66, 67
Tuckerman, L.B., 242
Turton, W.H., 81, 174
Tutte, W.T., 115, 264
Two-digit process, 351

Uniform polyhedra, 136, 159

Valerio, P., 404, 418
Vallée Poussin, Ch. J. de la, 62–63
Van Ceulen, on π, 353
Vandermonde, A.T., 175, 180
Vandiver, H.S., 71
Van Etten, H., 12
Vase problem, 28
Veblen, O., 207, 282
Vega, on π, 357
Vieta, F., 343, 353
Vigenère Square, 407
Vinogradoff, L.A., 64
Virgil, 256
Volpicelli, P., on knight's path, 175
Von Bilguer, on chess pieces, 165
Voting, 49, 106
Voynich, W.M., 417

Walker, G.T., 41
Walker, R.J., 203, 204
Wallis, J., 318, 355, 361
Wantzel, P.L., 338
Warnsdorff, on knight's path, 181
Watch problem, 17
Watersheds and watercourses, 238
Watson, G.N., 59
Weights problem, 50
Weil, André, 75
Western, A.E., on binary powers, 68
Whateley, R., 365
Wheeler, A.H., 93, 94, 147
Wiedemann, A., on Lake Moeris, 256
Wieferich, A., 71

Williams, H.C., 66
Wilson, R.M., 288
Wilson's Theorem, 61
Window reader, 333–336
Wing, labyrinth at, 259
Winter, F., 113
Wright, E.M., 68
Wythoff, W.A., 39, 159

Yardley, H.O., 418
Yates, F., 192
Youngs, J.W.T., 238

Zach, F.X. von, on π, 357
Zamebone, U., 375
Zeta function, 63
Zonohedra, 141